Application of Neural Networks and Other Learning Technologies in Process Engineering

Application of Neural Networks and Other Learning Technologies in Process Engineering

Editors

I. M. Mujtaba

University of Bradford, United Kingdom

M. A. Hussain

University of Malaya, Malaysia

ICP

Imperial College Press

Published by

Imperial College Press
57 Shelton Street
Covent Garden
London WC2H 9HE

Distributed by

World Scientific Publishing Co. Pte. Ltd.

P O Box 128, Farrer Road, Singapore 912805

USA office: Suite 1B, 1060 Main Street, River Edge, NJ 07661

UK office: 57 Shelton Street, Covent Garden, London WC2H 9HE

British Library Cataloguing-in-Publication Data
A catalogue record for this book is available from the British Library.

APPLICATION OF NEURAL NETWORKS AND OTHER LEARNING TECHNOLOGIES IN PROCESS ENGINEERING

ISBN 1-86094-263-6

Printed in Singapore.

To my parents: Professor M. Ishaque and R. Akhter
My wife: Nasreen
And my children: Sumayya, Maria, Hamza and Usama

\- I.M. Mujtaba

To my parents: Hussain Mohamed and Khairun Haider
My wife: Fakhriani Hj. Yusof
And my children: Nor Daleela, Ahmad Nasruddin, Ahmad Zubair, Nor Sakeenah and Nor Ameenah

\- M.A. Hussain

Foreword

This book is a follow-up of the IChemE CAPESG workshop on "The Application of Neural Networks and Other Learning Technologies in Process Engineering" held on the 12th May 1999 at Imperial College, London. The interest showed by the participants especially those from the industries in these emerging technologies has inspired us to come up with this book. This is not only the proceedings of the workshop but an expanded and revised versions of the talks presented at the workshop as well as invited papers from other well known international researchers in this area. Hence in short, this book contains contributions in the field of neural networks and learning technologies from experts in different parts of the globe.

In summary the papers are arranged in this book in parts based on certain topic-related sequences. Part I (Papers 1 to 5) relates to the use of neural networks for identification and modelling purposes as well as some aspects of neural network training. Part II (Papers 6 to 8) discusses the utilisation of neural networks in hybrid schemes for modelling and control purposes. Part III (Papers 9 to 11) relates to the use of this technology for estimation and control of various chemical processes. Part IV (Papers 12 and 13) involves their usage in new and learning technologies strategies in chemical process systems while Part V (Papers 14 and 15) discusses the use of this technology in experimental and industrial applications.

Part I: Modelling and Identification

The first paper by Aldrich and Slater starts with the discussion on the use of neural networks for modelling of liquid-liquid extraction column and prediction of equilibrium data and kinetic coefficients. In the paper they show examples of the use of neural networks for dispersed phase holdup and drop size prediction in extraction columns and rotating disc contactors. They also demonstrated the modelling of extraction in a vortex ring batch cell as well as the performance monitoring of extraction in an industrial column using neural network methodology.

The next paper by Bomberger et al. is about utilising radial basis function (RBF) networks for the identification of a multivariable coplymerisation reaction in a continuous stirred tank reactor. The k-means clustering and stepwise regression analysis methods are used for the process of RBF modelling. The minimum model order is determined using the method of false nearest neighbours. The simulation is also performed utilising conditions similar to the actual plant to assess its practical approach.

The third paper by Eikens et al. demonstrates the use of unsupervised neural networks in the form of self-organising maps for process identification in a yeast fermentation system. The network was found to predict accurately the different physiological states in the fermentation process.

The forth paper by Kershenbaum and Magni is about the use of nonlinear techniques to determine the proper centre locations in radial basis function networks. The training approach is performed through the Bayesian method and done for a simulated continuous stirred tank reactor system and a kin robot arm utilising Gaussian and thin plate spline networks. This approach is found to improve the performance of the networks over that of the traditional unsupervised methods.

The fifth paper by Scheffer and Maciel Filho involves the use of a recurrent neural network for nonlinear identification of a fed-batch penicillin process. In this work the neural network is trained by a multiple stream extended Kalman filter methodology. This approach allows the processs to be identified in real time which is a useful tool for calculation of the optimal feeding strategy in real time.

Part II: Hybrid Schemes

Paper 6 by Eikens et al. utilises first principles parametric models with neural networks in a hybrid strategy to identify a fed-batch fermentation process. Different types of neural networks were integrated into the hybrid model structure in the simulation work for multi-step ahead predictions and these results were compared with utilising the traditional neural network approach.

Paper 7 by Greaves et al. discusses the use of neural networks in hybrid strategies for optimal control purposes. In this paper a hybrid model for an actual pilot batch distillation column is developed where the neural network is used to predict the plant-model mismatch of the system. With this hybrid model, a general optimisation framework is developed to find optimal reflux ratio policies which then minimises the batch time for a given separation task.

Paper 8 by Meleiro et al. discusses the use of the hierarchical neural fuzzy models in the simulation of an industrial plant. The models here consist of a set of radial basis function networks formulated as simplified fuzzy systems connected in cascade. This hybrid model approach is then applied for the modelling of a multi-input multi-output complex biotechnological process for ethyl alcohol production with long range prediction capabilities.

Part III: Estimation and Control

Paper 9 by Hussain involves the control of a continuous fermentation process, using the internal-model control strategy, wherein the neural network inverse model act as the controller in the closed loop system. The simulation for the control of the biomass concentration was performed for both set point tracking and disturbance rejection cases. The offsets obtained in these cases were eliminated by the use of an adaptive online control scheme, wherein the adaptation of the forward and inverse models was carried out.

Paper 10 by Aziz et al. demonstrates the use of neural networks for estimating the heat released in an exothermic batch reactor system. This estimation was then used in a generic model control scheme for controlling the reactor temperature by manipulating the jacket temperature. The set point tracking of the reactor temperature followed an optimum profile generated by the formulation of the reactor's optimal operation in the offline mode. Comparisons with the conventional dual mode strategy were also shown in this work.

The next paper (paper 11) by Zhang and Morris utilises the bootstrap aggregated stacked neural networks approach to nonlinear empirical modelling. This method is effective in building models from a limited data set. In their study, the robust neural network was utilised for inferential estimation of polymer in a batch polymerisation reactor. The estimation of the amount of reactor fouling during the early stage of the batch process was also done as well as the optimal control of the batch polymerisation process. Neural network models are used to provide inferential estimation of polymer quality as well as to predict the trajectory of polymer quality variables from the batch recipe and control profile, which provide appropriate control actions for the polymerisation process.

Part IV: New Learning Technologies

Paper 12 by Wilson and Martinez utilises the reinforcement learning method for optimisation and control of a semi-batch reactor process. They utilise the notion of the performance of the value function to achieve the target. For batch-to-batch learning and control, the value function is represented by wire fitting methods incorporating neural networks methodology.

The next paper (paper 13) by Wang demonstrates the use of the emerging data mining and knowledge discovery technology in analysing large volumes of data in a meaningful way. One case study involves utilising data from a refinery separation process to help operators in analysing the operational states of the process. The second case study involves utilising wavelet analysis for identifying feature extraction and operational states in a fluid catalytic cracking process while another study on a methyl tertiary butyl ether plant illustrates the clustering approach in identifying the operational states of the process.

Part V: Experimental and Industrial Applications

The paper 14 by Cabassud and Le Lann involves neural networks in three experimental applications. The first one involves utilising neural networks in an inverse model method to control a semi-batch chemical reactor pilot plant with time varying operating conditions. Various neural network designs were investigated in this study. The second study involves using neural networks in a mutivariable controller for controlling a liquid-liquid extraction column. The control strategy was done based on the inverse modelling approach. The results obtained showed improvement with regard to previous studies of using the conventional adaptive control method. The third study involves using neural networks to measure and control a low-pressure chemical vapour deposition reactor. A hybrid neural network model was developed to compute the deposition rate profile along the reactor. A mutivariable controller using inverse dynamic methodology was also developed to compute the local set points of the PID controllers.

The last paper 15 by Puigjaner discusses the use of neural networks in evolutionary optimization of a nonlinear, time-dependent process in combination with genetic algorithms. Neural network is used off-line to update real plant representation and for multilevel decision making online as well as in real time optimisation process. Results from various real industrial applications are reported and discussed in the paper.

Acknowledgements

Alhamdulillah- All praise to almighty Allah who made it possible for us to complete this book.

We thank IChemE CAPE subject group to give I. Mujtaba the opportunity to organize the symposium on "The Application of Neural Networks and Other Learning Technologies in Process Engineering" on 12 May 1999. The main inspiration for compiling such a book came from this symposium. Special thanks go to all the speakers of the symposium who accepted our invitation to contribute in this book.

This book includes contributions from Europe, North America, South America, Africa and Asia. We are sincerely grateful to all the contributors who had sacrificed their valuable time to prepare the manuscripts.

We would like to thank the reveiwers who made relentless efforts to review each manuscript carefully and to make useful comments.

We gratefully acknowledge the UK Royal Society financial support to: (i) M.A. Hussain in 1999 for his visit to Bradford University when the initial planning to compile such a book was made; (ii) I. Mujtaba to cover the expenses in Malaysia during the final editing stage of this book.

Finally, we thank to the publisher for publishing this book and sincerely acknowledge their support and help.

Contents

Part I Modelling and Identification

1. SIMULATION OF LIQUID-LIQUID EXTRACTION DATA WITH ARTIFICIAL NEURAL NETWORKS

C. ALDRICH

Department of Chemical Engineering, University of Stellenbosch, Stellenbosch, South Africa

M. J. SLATER

Department of Chemical Engineering, University of Bradford, Bradford, BD7 1DP, United Kingdom

Liquid-liquid extraction is not understood well enough to allow acceptably accurate design calculations to be made. Modelling and simulation can be difficult and time-consuming and is usually heavily dependent on empirical correlations of restricted range of applicability. The use of artificial neural networks to achieve more precise simulation has therefore been examined. Application to multicomponent equilibrium and diffusion coefficient data, extraction column hydrodynamic data (drop sizes and hold-up), mass transfer stage efficiency and performance prediction of an industrial extraction column has been carried out with widely varying degrees of success. The lack of data for building a neural network is the largest problem faced.

1. Introduction

Liquid-liquid extraction has long been an important mass transfer operation in chemical engineering (Thornton, 1992; Godfrey and Slater, 1994). It is sometimes superior to rectification, especially where azeotropic, temperature-sensitive or other refractory systems are concerned. Although extraction is widely applied in the food, metallurgical, petrochemical and nuclear industries, the design and control of extraction columns are still far from optimal. The design of columns is often hampered by both hydrodynamic and mass transfer constraints (Rückl and Marr, 1985), such as limited throughput and the type and geometry of the column. This can be attributed to the fact that the effects of column geometry and the rheological properties of multiphase extraction systems are at present not understood sufficiently well to permit exact column design and operation. Modelling and simulation of extraction columns therefore often involves costly and time-consuming procedures, which are not necessarily guaranteed to approximate the behaviour of process equipment with adequate accuracy. By making use of artificial neural networks the behaviour of liquid-liquid extraction systems can be simulated accurately and cost- effectively, as will be demonstrated in this paper. Better design and equipment control can thereby be achieved.

2. Artificial Neural Networks

Artificial neural networks are inspired by the architecture of biological nervous systems which consist of a large number of relatively simple nerve cells or neurons that function in parallel to facilitate rapid decisions. Likewise neurocomputers or artificial neural networks consist of a large number of primitive computational elements which are arranged in a massively parallel structure. These elements are connected by means of artificial synapses which are characterised by a matrix of weights or numeric values, which can typically be adjusted by a learning process. A major advantage is that neurocomputing devices do not have to be programmed, but instead they can learn to form distributed representations of complex relationships from examples. Artificial neural networks, connectionist systems, or neuromorphic computers as they are also known have proved to be highly successful in applications such as process control, modelling, simulation and system identification (Bhat and McAvoy, 1990; Bhat et al., 1990; Psichogios and Ungar, 1991; Hunt et al., 1992; Morris et al., 1994).

The field of neural networks had its inception in the 1940s when the paper of McCulloch and Pitts on the modelling of neurons, and Hebb's book *The Organization of Behaviour* first appeared in the 1940s. The interest sparked by these publications was further buoyed when Rosenblatt presented his Mark I Perceptron in 1958 and Widrow the ADALINE in 1960, but came to a dramatic end in 1969 when Minsky and Papert showed that the capabilities of the linear networks studied at the time were severely limited (Eberhart and Dobbins, 1990). These revelations caused a virtually total cessation in the availability of research funding and many talented researchers left the field permanently. The initial interest in neural networks was only revived again in the early 1980s, as a result of a breakthrough concerning the training of multilayer neural networks, and since then the field of neural networks has seen phenomenal growth, passing from a research curiosity to commercial fruition in less than a decade.

Neural networks are presently being investigated by researchers from as wide a range of disciplines as any field in the recent history of technology, i.e. mathematicians, scientists, engineers, physicists, psychologists, cognitive scientists and even a few philosophers and social scientists. To date these systems have been used in process engineering to generate non-linear models for the design of fixed or adaptive model-predictive control systems, the diagnosis of process faults and the identification of the root causes of these faults (Fan et al., 1993; Hoskins et al., 1991; Venkatasubramanian et al., 1990), the detection of errors in plant data (Aldrich and Van Deventer, 1993, 1994a) and data reconciliation (Aldrich and Van Deventer, 1994b), as well as the monitoring and interpretation of process trends (Karim and Riviera, 1992) and the evaluation of the performance of batch and continuous processes (Reuter et al., 1992, 1993; Su and McAvoy, 1992).

Despite the promise artificial neural networks appears to hold for the chemical and metallurgical processing industries, the first commercial applications for neural networks only saw the light in the early 1990s, with the implementation of a hybrid neural network-fuzzy control system from Pavilion Technologies in Eastman Kodak's refinery in Texas. Other commercial applications include hybrid control systems sold by Neural Applications Corporation, consisting of neural networks as well as expert systems, used in arc furnaces. These systems are used to optimise the positions of the electrodes of the arc furnaces used for the smelting of scrap metal in steel plants, and are estimated to save approximately $US 2 000 000 annually on the operating costs of each furnace.

In the UK process industry the control of a nuclear fusion reactor at AEA Technology's Culham Laboratory in Oxfordshire has recently been reported (Geak, 1993). The optimal conditions for fusion in the Compass tokamak reactor occur where the turbulence in the plasma is minimal, and cannot be calculated sufficiently fast by conventional computers, which can take hours or even days to compute the set-up of the magnetic fields needed to produce suitable plasma shapes in the reaction chamber. The problem is solved by making use of a neural network that can do the necessary calculations in approximately ten microseconds (significantly faster than the fluctuations in the plasma that typically last for a few hundred milliseconds). The Compass network obtains data from 16 magnetic field sensors inside the chamber and has four output nodes linked to the magnet controls of the system. An added advantage is the flexibility of the network, which can be retrained (by sets of approximately 2000 exemplars at a time) when the implementation of different control strategies are warranted. Conventional controllers in contrast, can only cope with narrow ranges of process conditions.

The popularity of neural networks for solving many different types of engineering problems can mainly be ascribed to the richness of the presentations they can capture (Boolean, qualitative, semi-quantitative, analytic, etc.), their high degree of parallelism bestowing on them supercomputing capabilities, as well as their relatively simple and flexible structures. Commercial software is available for carrying out neural network studies: the learning requirements and the ease of use are comparable to software for spreadsheet calculations for example.

2.1. Neurodynamics

Artificial neural networks have been described exhaustively in the literature, e.g. Lippmann (1987), Zurada (1992) and Haykin (1999), and the fundamentals are only considered briefly in this paper.

In essence neural networks consist of networks of primitive process elements (alternatively referred to as process or computational nodes or elements), as shown in Fig. 1. The nodes receive inputs (x) from other nodes in the network or from the outside, which are subsequently weighted and summed. These weighted sums ($w^T x$, also referred to as the potentials of the nodes) are then operated on by so-called node transfer functions $g(w^T x)$, which map or squash the potentials to smaller domains before passing the output to other nodes or the outside environment of the network. The structure of a basic feed-forward network is shown in Fig. 2. The network has

Figure 1. Model of a neural network node, with an input vector $x = [x_1, x_2, x_3, \ldots x_M]^T$, and a weight vector $w = [w_1, w_2, w_3, \ldots w_M]^T$.

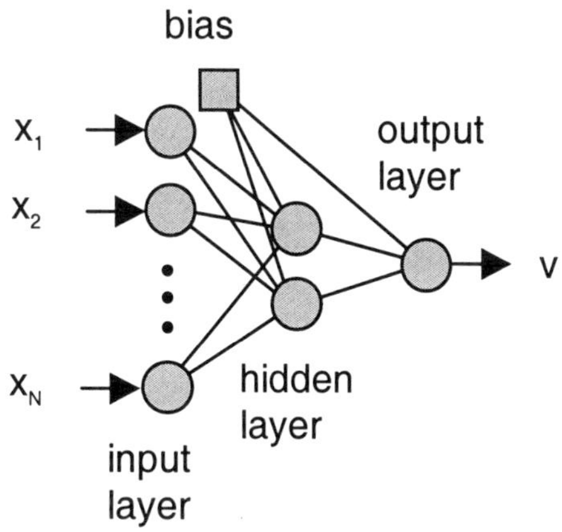

Figure 2. Generic structure of a simple feedforward neural network with a single hidden layer.

at least an input and an output layer, and possibly one or more hidden layers. Nodes in these layers are connected by means of artificial synapses, each of which is associated with a numerical value or weight. The network is trained (i.e. the weights are adapted) based on examples of the process.

More formally, computation in neural networks (such as the one shown in Fig. 2) is feed-forward and synchronous, i.e. the states of the computational elements in the layers nearest to the input layer of the network are updated before units in successive layers further down in the network. The activation rules or neurodynamics of the network determine the way in which the process units are updated and are typically of the form

$$v_i(t+1) = g[u_i(t)] \tag{1}$$

where $u_i(t)$ designates the potential of a process unit at time t, i.e. the difference between the weighted sum of all the inputs to the unit and the unit bias

$$u_i(t) = \Sigma_j w_{i,j} \cdot v_j(t) - \Theta_j \tag{2}$$

The form of the transfer function g may vary, but could be a linear, step or sigmoidal transfer function, among others, with a domain typically much smaller than that of the potential of the process unit, such as [0;1] or [-1;1], for example.

2.2. *Training*

The training of commonly-used back-propagation neural networks is an iterative process involving the changing of the weights of the network, typically by means of a gradient descent method, in order to minimise an error criterion, that is

$$w_{i,j}(t+1) = w_{i,j}(t) + \Delta w_{i,j}, \tag{3}$$

where

$$\Delta w_{i,j} = -\tau . \partial \xi / \partial w_{i,j} \tag{4}$$

and where τ is the learning rate and ξ the error criterion, i.e.

$$\xi = ½ \Sigma_j (T_{o,j} - v_{o,j})^2 \tag{5}$$

based on the difference between the desired ($T_{o,j}$) and the actual outputs ($v_{o,j}$) of the unit. Since the error ξ is propagated back through the network, these types of networks are widely known as back-propagation neural networks. Once the network is trained, its ability to generalise is validated against a test set of data not used in the training process. Provided that the training data are sufficiently representative of the process being modelled, the network will be able to predict underlying process trends with a high degree of accuracy.

3. General Approach To Process Plant Modelling With Neural Networks

The generalised plant modelling problem consists of two parts, namely the decomposition of the plant into sets of acyclic process circuits if necessary, followed by modelling of these irreducible subsystems. The decomposition of large or complex plants can be accomplished by various means which can among others be incorporated in connectionist structures (Aldrich et al., 1994) in order to take advantage of parallel processing strategies. Assuming the process system to be modelled to be acyclic, the problem concerned with the construction of a process unit or plant model can be expressed as follows:

$$\mathbf{Y} = \begin{matrix} y_{1,1} & y_{1,2} & .. & y_{1,q} \\ y_{2,1} & y_{2,2} & .. & y_{2,q} \\ .. & .. & .. & .. \\ y_{n,1} & y_{n,2} & .. & y_{n,q} \end{matrix} \qquad (6)$$

$$\mathbf{X} = \begin{matrix} 1, x_{1,1} & x_{1,2} & .. & x_{1,m} \\ 1, x_{2,1} & x_{2,2} & .. & x_{2,m} \\ .. & .. & .. & .. \\ 1, x_{n,1} & x_{n,2} & .. & x_{n,m} \end{matrix} \qquad (7)$$

where $y_{i,k}$ (i = 1,2, .. q) represent q variables dependent on m causal or independent variables $x_{j,k}$ (j = 1,2, .. m), based on n observations (k = 1,2, .. n). The variables $y_{i,k}$ are usually parameters which provide a measure of the performance of the plant, while the $x_{j,k}$ variables are the plant parameters on which these performance variables are thought to depend.

The problem is then to relate the matrix **Y** to some function of matrix **X**, in order to predict **Y** from **X**. The simplest approach, and a method often used on plants, is to assume a linear relationship between **X** and **Y**, i.e. $\mathbf{Y} = \mathbf{X}.\mathbf{b}$ and to find the coefficient vector **b** by ordinary least squares methods, that is $\mathbf{b} = (\mathbf{X}^T\mathbf{X})^{-1}.\mathbf{X}^T\mathbf{Y}$, provided that the elements of the columns $\mathbf{X}_j$ of matrix **X** are not correlated and that

the number of observations is larger than the number of coefficients that has to be estimated (i.e. n > m). If not, other techniques, such as partial least square methods (Qin and McAvoy, 1992) can be used to obviate the problem. Should the assumption of multi-linear relationships between the variables prove to be inadequate, they can be extended by the addition of suitable non-linear terms, the incorporation of spline methods, or replaced by non-linear regression methods.

The main advantage of modelling techniques based on the use of neural networks is that *a priori* assumptions with regard to the functional relationship between **x** and **y** are not required. The network learns this relationship instead, on the basis of examples of related **x-y** vector pairs or exemplars.

Wherever possible, fundamental knowledge of the process should always be included in the network. This can be done by making use of hybrid neural network systems (Aldrich and Van Deventer, 1994c) in which neural networks are explicitly combined with phenomenological process models, or by structuring the inputs to the network in such a way that previous knowledge is incorporated in the network via the training process.

4. Modelling Of Liquid-Liquid Extraction Equipment With Neural Networks

Liquid-liquid extraction columns are typically ill-defined systems in that the physical phenomena underlying the extraction process are complex and generally difficult to model on a first-principles basis. As a result it is not an easy task to identify the essential features of the processes involved in extraction plant operations, and hence to simulate and control the plant effectively. The development of process models based on plant data (often at small scale) is usually not cost-effective and the data are usually analysed by means of multiple linear or non-linear regression techniques. Since these techniques require explicit process models, they are not always suitable for modelling of the complex behaviour that industrial plants so often exhibit. In contrast, neural networks do not suffer from this drawback and (provided they are presented with sufficient representative data) constitute an efficient means for the construction of implicit models of ill-defined processes. In spite of these well-known attributes (Venkatasubramanian and McAvoy, 1992), little has been published in the chemical engineering literature with regard to the use of neural networks as far as extraction equipment is concerned except for the work of Boger and Ben-Haim (1993) who have described an application to a mixer-settler plant, and Woinaroschy (1998) who has investigated the use of a neural network for the dynamic simulation of multistage countercurrent extraction with immiscible solvents.

The possible use of neural networks for the simulation of liquid-liquid extraction systems is subsequently described using some very simple examples serving as illustrations. The modelling of an extraction column requires information on equilibria, kinetic coefficients, the hydrodynamics of the column, and mass transfer processes in the column. Each of these is amenable to neural network simulation.

5. Equilibrium Data

Multicomponent system equilibrium data can be modelled well using UNIQUAC if the binary interaction parameters are obtained using appropriate experimental data. However, even for quaternary systems this can be an expensive exercise. For systems such as lube oil refining or aliphatics/aromatics separation such an approach is impracticable and other modelling tools might prove useful. Many complex processes have been developed on the basis of pilot plant and full-scale plant experience but at high cost. Complex metal separation processes can rarely be modelled without detailed knowledge of chemical mechanisms of extraction. Even the apparently simple process of zinc extraction/stripping with D2EHPA/H_2SO_4 has proved difficult to model (Sainz-Diaz et al., 1996; Corsi et al., 1999). The separation of rare earths poses a more difficult problem; the equilibrium data have been simulated successfully using neural network techniques with advantage over other possible approaches (Giles et al., 1996).

6. Kinetic coefficients

The prediction of mass transfer coefficients depends on knowledge of molecular diffusion coefficients. In multicomponent systems the Stefan-Maxwell (rather than Fickian) diffusion coefficients are required in rate-based calculations (Taylor and Krishna, 1993). Fickian diffusion coefficients vary markedly with composition; the dependency can be estimated using thermodynamic principles and can be simulated using neural networks. The limited work done so far only serves to demonstrate the difficulty of the problem due to shortage of data (von Reden, 1998).

7. Column hydrodynamics

The hydrodynamics of an extraction column are important in that they determine the column diameter, but also directly influence the mass transfer characteristics of the system (i.e. the height of the column). At present, columns are often overdesigned to compensate for a lack of knowledge regarding the process variables and dynamics, which inevitably results in the specification of less than optimally sized process equipment.

In the following simple examples the use of neural networks to illustrate the simulation of dispersed phase hold-up and drop size in extraction columns is demonstrated. The expected benefit may lie in improving control systems rather than in design.

7.1. Example 1. Systems with no mass transfer: modelling of hold-up and drop size

Two systems were considered, namely a cumene/isobutyric acid/water system, as well as a butanol/succinic acid/water system, as used by Bailes et al. (1986) for rotating disc contactor studies, the column being 152 mm diameter with 23 compartments. The physical properties of these systems (with and without mass transfer) are summarised in Table 1. These systems differ considerably as far as their behaviour in extraction columns is concerned, mainly owing to their different interfacial tensions.

Table 1. Physical properties

γ [mN/m]	μ_c [mPa.s]	μ_d [mPa.s]	ρ_c [kg/m^3]	ρ_d [kg/m^3]
cumene/isobutyric acid/water system (no mass transfer)				
18	1.05	0.81	1000	868
butanol/succinic acid/water system (no mass transfer)				
0.75	1.61	3.93	1000	876
cumene/isobutyric acid/water system (with mass transfer)				
16-20	1.05	0.81	1000	868
butanol/succinic acid/water system (with mass transfer)				
0.75-1.5	1.55	3.65	991	865

The experimental data comprised examples of the process behaviour of the form {inputs γ, N, F_d ; outputs d_{32}, h}, where γ is the interfacial tension, N is the speed of the rotor [s^{-1}], F_d the flow rate [cm^3/s] of the dispersed solvent phase at a fixed flow ratio ($F_d = F_c/3$ for cumene and $F_d = F_c$ for butanol), d_{32} the Sauter mean diameter of the drops in the column and h the fractional hold-up of the dispersed phase in the column.

The back-propagation network consisted of an input layer, with three nodes (associated with the three inputs to the network, i.e. γ, N, F_d), a hidden layer with two nodes and an output layer with two process nodes (associated with the two outputs of the network, i.e. d_{32} and h). All the nodes in the hidden and output layers had sigmoidal transfer functions of the form $g(u) = 1/(1+e^{-u})$. This structure was set up using NeuralWorks Professional II software; other software is commercially available.

Since only 38 exemplars were available, the network was trained by means of a leave-one-out basis, i.e. the network was trained on all the data, except one which was held out for testing or validating the performance of the network. This procedure (also known as jack-knifing or hold-out) was repeated until the network had been validated against all the data. The alternative procedure of using say half the data for simulation and half for validation provides insufficient information for close simulation with three input variables at several different levels of each.

Despite the limited availability of data, the network was capable of predicting holdup (h) and the mean drop size (d_{32}) with average absolute errors of 13% and 14% respectively, as shown in Figs. 3 and 4. These errors are comparable to the experimental errors in the output data, which were about 20% for both d_{32} and h.

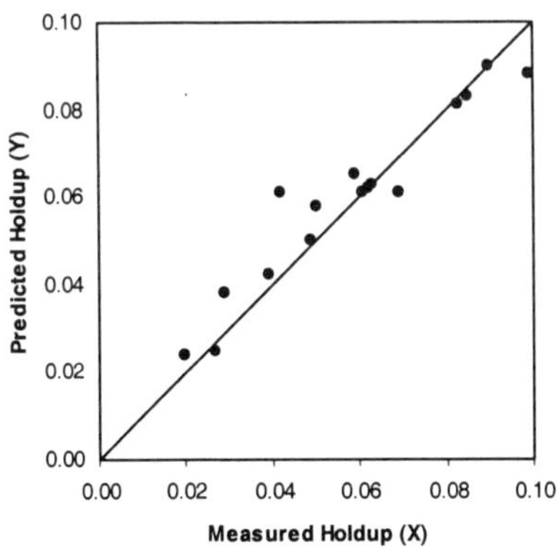

Figure 3. The ability of a back 3:2:2 propagation neural network to predict fractional hold-up

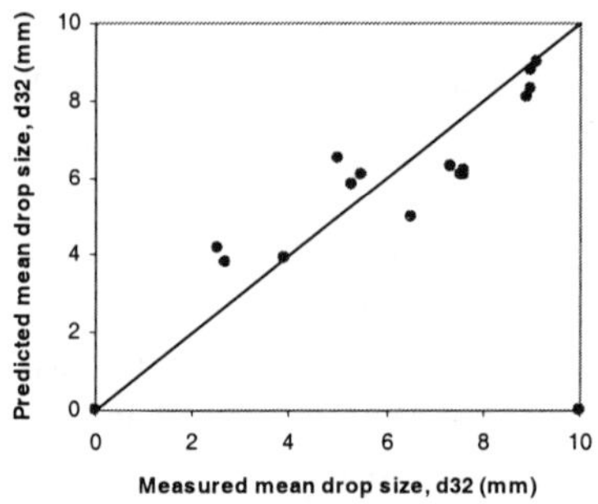

Figure 4. The ability of a back 3:2:2 propagation network to predict the Sauter mean diameter of drops.

7.2. Example 2. Modelling Of Average Drop Sizes In Rotating Disc Contactors In General

In this example the Sauter mean diameter (d_{32}) of solvent drops was modelled in terms of the geometry of the column, as well as the interfacial tensions of the systems. Published experimental data (Chang-Kakoti et al., 1985) from six systems were considered (dispersed/continuous), viz. n-butanol/water, iso-butanol/water, cumene/water, toluene/water, kerosine/water and Clairsol 350/water. A range of column sizes is involved. This allowed the construction of a training (74 exemplars) and a test data set (15 exemplars) against which the performance of the network could be measured. The data sets had the form $\{\gamma, D_T, D_S, D_R, Z_C, N \mid d_{32}\}$, i.e. the drop size was modelled in terms of the interfacial tension of the system, as well as the geometry of the column. The differences in densities and viscosities were ignored for simplicity.

The network consisted of an input layer with 6 input nodes (for the six input variables), a sigmoidal hidden layer with 4 hidden nodes, and a single-node sigmoidal output layer (for the output variable, d_{32}). The network converged rapidly (after approximately 10 000 iterations or less than a minute of training on a 486 DX personal computer or better), and could then be used to predict d_{32}-values based on data not used during training of the network. The results are shown in Fig. 5.

As can be seen from this figure, the network was capable of significantly more accurate prediction (average absolute error approximately 25%) of the mean drop sizes than could be obtained with empirical equations (average absolute error approximately 40%) proposed for example by Chang-Kakoti et al. (1985).

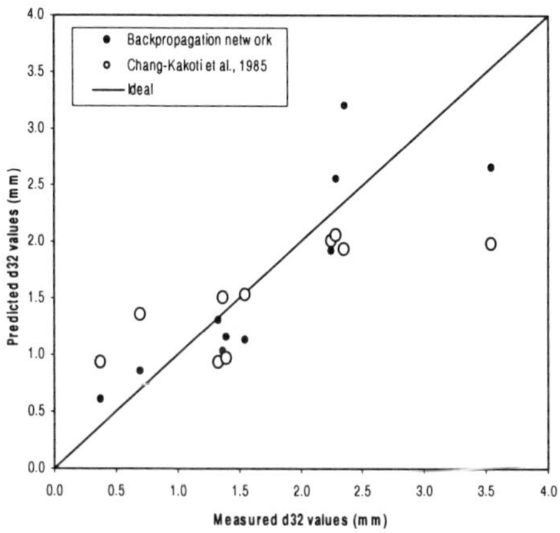

Figure 5. Neural network (6:4:1) modelling of the drop size in a rotating disc contactor (example 2).

These and other results suggest that a simple back-propagation neural network can be employed for the construction of useful models of the hydrodynamics of an extraction column. Although no physical meaning can be attached to the weights in the network, their collective effect enables the network to accurately predict drop sizes and hold-up in an existing column. Trial and error procedure is required to set up the simplest network possible (to minimize the number of weighting parameters), but this is not a long process. An alternative procedure using a mechanistic model such as that discussed by Cauwenberg et al. (1997) requires much more computational effort and has no potential for learning the consequences of inevitable changes with time often experienced in industrial plants.

8. Mass Transfer

Neural network models of the hydrodynamics of the column can be incorporated into conventional simulators to enhance the simulation of mass transfer. Alternatively, as will be demonstrated in this example, neural networks can also be used to model mass transfer directly.

8.1. Example 3. Modelling of extraction in a vortex ring batch cell

Baird et al. (1992) investigated the unsteady state mass transfer rates of benzoic acid from kerosene to water in a batch extraction cell. Vortex rings were formed in the cell by means of a horizontal plate with a single orifice, which was periodically subjected to a compressed air pulse as shown in Fig. 6. The plate was located in the aqueous phase and mass transfer was induced by discrete rings of water rising through the organic phase.

The fractional extent of extraction E depended on the pulse frequency of the column, as well as a dimensionless rate constant k' which in turn depended on the oscillation stroke and the velocity of the impulse. Such equipment is by no means easily modelled in a fundamental sense.

By making use of a simple back-propagation neural network, the extraction process can be related to the pulse frequency of the column, i.e. $E = f_{NN}(t,p)$. Fort three (43) exemplars of the form $\{t,p|E\}$ were used to train the network on a leave-one-out basis. The network consisted of a two-node input layer, a hidden layer with three sigmoidal nodes and an output layer with a single sigmoidal node. The experimental data were used in training. The ability of the network to represent the extraction process is shown in Fig. 7. The network was able to predict the degree of extraction with an average absolute error of less than 3% (5.86).

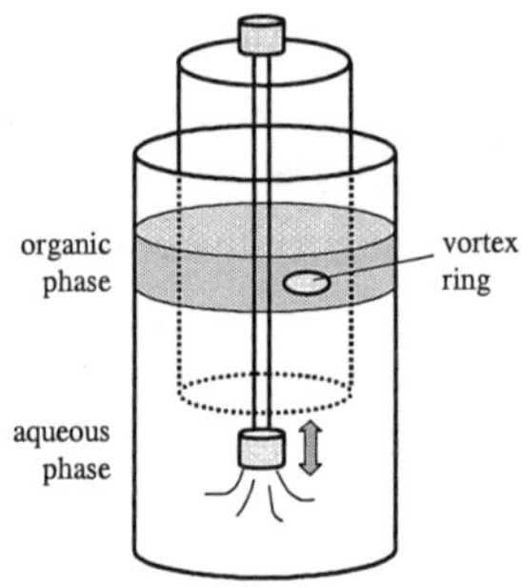

Figure 6. Diagrammatic representation of a vortex ring batch cell (example 3).

9. Monitoring of The Performance of An Industrial Column

By using a neural network model, historic data representing the operational behaviour of a large industrial liquid-liquid extraction column could be used to prototype the system. Daily values of stream flow rates, temperatures, feed and product compositions, the level of the interface, pressure, as well as the degree of impurities in the solvent were collected over a period of five years. The temperature gradients and flow rates in the column were controlled in order to maintain a product of constant quality and composition, despite process disturbances associated with changes in the compositions of the feed streams.

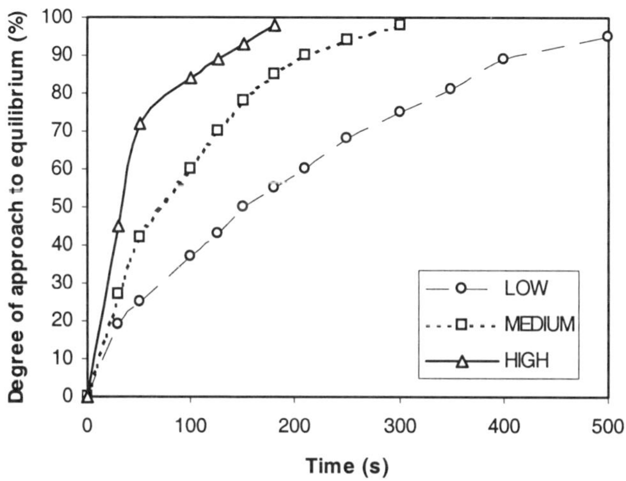

Figure 7. Neural network model (2:3:1) of the extraction efficiency of a vortex ring batch cell (example 3).

Although a linear model could explain most of the variation in the flow of the treated raffinate stream, it could only explain approximately 53% of the variation in the composition of the product stream. The reason for this can be seen in Fig. 8, which shows the clustered structure of the data. These clusters are not handled well by standard regression methods. However, by training a radial basis function neural network model on a subset of the available data, approximately 72% of the variation in the composition of the product stream (portrayed in Fig. 9) could be explained.

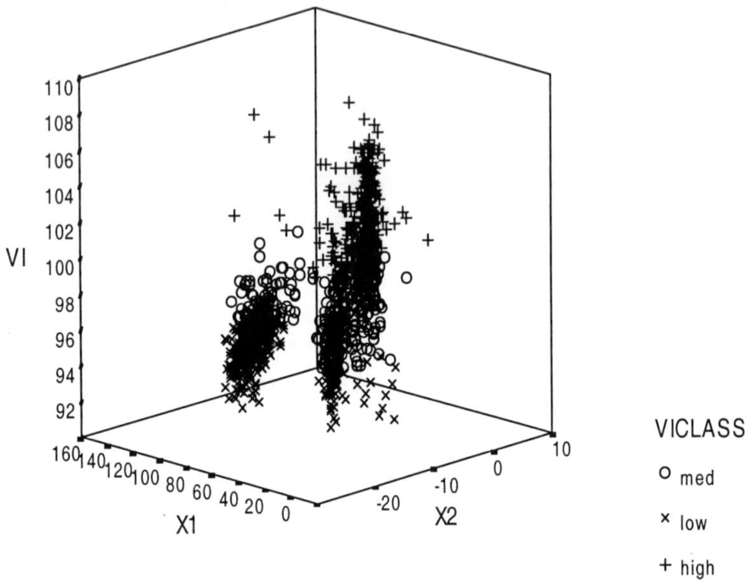

Figure 8. Distribution of product composition variable (VI) as a function of two explanatory variables x_1 and x_2.

10. Summary

As demonstrated by these deliberately simple examples, neural networks are able to represent certain liquid-liquid extraction data with sufficient accuracy for predictive and control purposes. Neural networks are essentially data driven, and their performance depends on the quantity and quality of available data. Provided that these data are sufficiently representative of the process, neural networks can be used to model the phenomena.

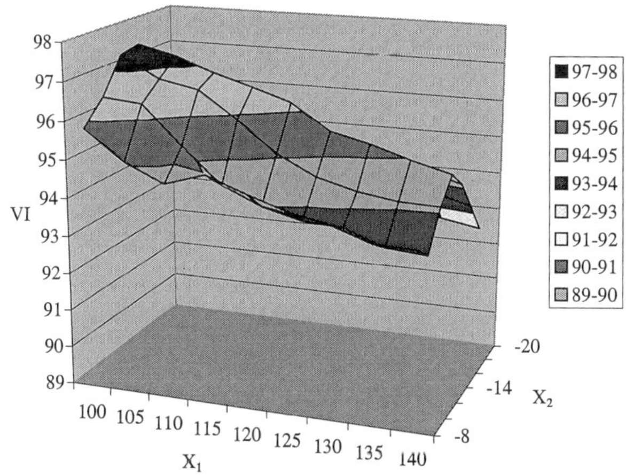

Figure 9. Modelling of the product composition (VI) with a radial basis function neural network with 15 hidden nodes and a linear output node.

References

Aldrich, C. and Van Deventer, J.S.J., *International Journal of Mineral Processing*, **39** (1993), 173-197.

Aldrich, C. and Van Deventer, J.S.J., *Chem. Eng. Sci.* **49** (1994a), 1357-1968.

Aldrich, C. and Van Deventer, J.S.J., *The Chemical Engineering Journal.* **54** (1994b),125-135.

Aldrich, C. and Van Deventer, J.S.J., *Thermochimica Acta.* **257** (1994c) 127-137.

Aldrich, C. et al., *Minerals Engineering.* **7** (1994), 793-809.

Bailes, P.J. et al., *Chem. Eng. Res. Des.* **64** (1986)., 43-55.

Baird, M.H.I. et al., *Chem. Eng. Res. Des.* **70** (1992), 323-332.

Bhat, N.V. and McAvoy, T.J., *Comput. Chem. Eng.*, **14** (1990), 573-583.

Bhat, N.V. et al., *IEEE Control Systems Magazine.* **10** (1990), 24-30.

Boger, Z. and Ben-Haim, M., *in ISEC '93, Solvent Extraction in the Process Industries*, eds. D.H. Logsdail and M.J. Slater, (Elsevier Applied Science, 1993), 1198-1205.

Cauwenberg, V. et al., *Can. J. Chem. Eng.* **75** (1997), 1046-1055, 1056-1066.

Chang-Kakoti, D.K. et al., *Journal of Separation Process Technology*. **6** (1985), 40-48.
Corsi, C. et al., *Hydrometallurgy*. **50** (1999), 125-141.
Eberhart, R.C. and Dobbins, R.W., *IEEE Engineering in Medicine and Biology*. **9** (1990), 15-18.
Fan, J.Y. et al., *AIChE Journal*. **39** (1993), 82-88.
Geak, E. *New Scientist*. **138** (1993), 5.
Giles, A.E. et al., *Hydrometallurgy*. **43** (1996), 241-255.
Godfrey, J.C. and Slater, M.J., *Liquid-liquid extraction equipment* (John Wiley and Sons, Chichester, 1994).
Haykin, S., *Neural Networks* (MacMillan College Publishing Company, Englewood Cliffs, NJ, USA, 1999).
Hoskins, J.C. et al., *AIChE Journal*. **37** (1991), 137-141.
Hunt, K.J. et al., *Automatica*. **28** (1992), 1083-1112.
Karim, M.N. and Riviera, S.L., *in European Symposium on Computer Aided Process Engineering - 1*, 24-28 May 1992, Elsinore, Denmark, 1992), S369-S377.
Lippmann, R.P., *IEEE ASSP Magazine*. **35**4-21.
Morris, A.J. et al., *Trans. IChemE*. **72** (1994), 3-19.
Psichogios, D.C. and Ungar, L.H., *IEC Res*. **30**, (1991), 2564-2573.
Qin, S.J. and McAvoy, T.J., *Comput. Chem. Eng*. **16**, (1992), 379-391.
Reuter, M.A. et al., *Metallurgical Transactions B*. **23B** (1992) 643-650.
Reuter, M.A. et al., *Chem. Eng. Sci*. **48** (1993), 1281-1297.
Rückl, W. and Marr, R., *German Chemical Engineering*. **8** (1985), 27-31.
Sainz-Diaz, C.I. et al., *Hydrometallurgy*. **42** (1996), 1-12.
Su, H.-T. and McAvoy, T.J. *IEC Res*. **31** (1992), 1338-1352.
Taylor, R., Krishna, R., *Multicomponent mass transfer*. (Wiley-Interscience, New York, 1993).
Thornton, J.D., *Science and Practice of Liquid-liquid Extraction* (Oxford Science Publications, Oxford, UK, 1992).
von Reden, C., Ph.D. thesis no. D290 (University of Dortmund, Shaker Verlag, Aachen, 1998).
Venkatasubramanian, V. and McAvoy, T.J., *Comput. Chem. Eng*., **16** (1992), v-vi.
Venkatasubramanian, V. et al., *Comput. Chem. Eng*. **14** (1990), 699-712.
Woinaroschy, A., *Hung. J. Ind. Chem*. **26** (1998), 121-123.
Zurada, J.M., *Introduction to Artificial Neural Systems*. (PWS Publishing Co., 20 Park Plaza, Boston, MA, U.S.A, 1992).

Nomenclature

d_{32}	Sauter mean diameter of drops
D_T	diameter of column in rotating disc column
D_S	diameter of stator in rotating disc column
D_R	diameter of rotor in rotating disc column
E	extent of extraction as a percentage of equilibrium extraction value
$f_{NN}(\ )$	relationship represented by a neural network
F_c	flow rate of continuous phase
F_d	flow rate of dispersed phase
g	transfer function
h	dispersed phase hold-up in column
k'	rate constant
k_1,k_2	constant parameters
m	causal variables
n	number of observations
N	speed of rotor
p	pulse frequency of vortex ring contactor
q	number of variables
t	time
$T_{o,j}$	target output value associated with the j'th output node
$u_i(t)$	potential of the i'th process node at time t in an artificial neural network
$v_i(t)$	the output of process node i at time t in an artificial neural network
$w_{i,j}$	the weight associated with the connection from the i'th process node to the j'th process node in an artificial neural network
$x_{i,k}$	the k'th observation of the i'th independent variable x
X	a vector of independent real variables in general $[x_1,x_2,..x_N]$
$y_{i,k}$	the k'th observation of the i'th dependent variable y
Y	a vector of dependent real variables in general $[y_1,y_2,..y_N]$
Z_C	height of compartment in rotating disc column

Greek symbols

ξ	error criterion
γ	interfacial tension
μ_c	dynamic viscosity of continuous phase
μ_d	dynamic viscosity of dispersed phase
ρ_c	density of continuous phase
ρ_d	density of dispersed phase

Θ_I the bias associated with process node i

τ learning coeficient controlling the rate at which weights in the neural network are adjusted

Acknowledgements

U. Pleschiutschnig carried out work on the industrial column simulation as part of his Diplomarbeit carried out at Bradford. Dr. Aldrich was on leave from Stellenbosch University for the period of the work done at Bradford.

2. RBFN IDENTIFICATION OF AN INDUSTRIAL POLYMERIZATION REACTOR MODEL

J. D. BOMBERGER
E. I. DuPont de Nemours & Co., Inc., Experimental Station
Wilmington, DE 19880-0101, U.S.A.

D. E. SEBORG
Department of Chemical Engineering, University of California
Santa Barbara, CA 93101, U.S.A.

B. A. OGUNNAIKE
E. I. DuPont de Nemours & Co., Inc., Experimental Station
Wilmington, DE 19880-0101, U.S.A.

Methods developed for radial basis function network (RBFN) identification are applied to a complex multiple-input, multiple-output (MIMO) simulation of a solution copolymerization reactor. For RBFN identification, k-means clustering and stepwise regression analysis are used. The practicality of applying these methods to large industrial identification problems is discussed, considering the restrictions of industrially practical input sequence design. The RBFN model has three inputs and two outputs, and the dimensionality of the identification problem poses some difficulties for nonlinear empirical model identification; specifically, the large amount of data required is a problem for plant testing and may cause computational difficulties for identification algorithms as well.

1. Introduction

Often, radial basis function network identification is applied only to single-input, single-output (SISO) example processes or simulations. This is also true for applications to real processes (Keulers, 1993; Eikens and Karim, 1994; Bomberger *et al.*, 1996). Application to SISO problems may be illuminating, but also can be somewhat misleading regarding the ease of use and data requirements of a particular identification method. In this chapter, RBFN identification methods are applied to a multivariable simulation of a solution copolymerization in a continuous stirred tank reactor (CSTR) developed by Congalidis *et al.* (1989). The simulation has seven input variables, four output variables, and a single disturbance variable. The simulation is treated as if it is an actual chemical process, and limitations on the amount of time available for plant identification tests and the type and duration of imposed changes in the plant operating conditions are considered.

To model the multivariable simulation, multiple-input, single-output (MISO) RBFNs are used. An RBFN takes the form of Fig. 1, and is a simple type of network model with single hidden layer of nodes and a linear output layer, with unity weighting between the input and hidden layer (Moody and Darken, 1989). Alternatively, an RBFN can be expressed as a weighted sum of radial basis functions

$$\hat{y}(i) = b_0 + \sum_{j=1}^{M} b_j\, \phi\left\{\left[\mathbf{x}(i) - \mathbf{c}_j\right], \beta_j\right\} \tag{1}$$

where b_0 is the bias term, b_j are the connection weights between the j-th hidden layer node and the scalar output as shown in Fig. 1, $\mathbf{c}_j$ is the center of the j-th radial basis function (RBF), and β_j is a parameter controlling the width of the j-th RBF. The radial basis function used in this work is the Gaussian RBF:

$$\phi\left\{\left[\mathbf{x}(i) - \mathbf{c}_j\right], \beta_j\right\} = \exp\left\{-\frac{\left[\mathbf{x}(i) - \mathbf{c}_j\right]^{\mathrm{T}}\left[\mathbf{x}(i) - \mathbf{c}_j\right]}{\beta_j}\right\} \tag{2}$$

The RBF center is analogous to the Gaussian mean, and the width parameter to the variance.

Nonlinear autoregressive models with exogenous inputs (NARX models) are used to build a dynamic, discrete time model of the simulation (Sjöberg *et al.*, 1995). For the NARX model, the RBFN inputs at sample i, $\mathbf{x}(i)$, are a vector of past values of the process inputs, $\mathbf{u}(i)$, and outputs, $\mathbf{y}(i)$. The relevant inputs and outputs are included in the network input vector,

$$\begin{aligned}\mathbf{x}(i) = \big[y_1(i-1) \;\cdots\; y_1(i-m_1) \;\cdots\; y_{n_y}(i-1) \;\cdots\; y_{n_y}(i-m_{n_y}) \;\; u_1(i-\theta_1-1) \\ \cdots\; u_1(i-\theta_1-n_1) \;\cdots\; u_{n_u}(i-\theta_{n_u}-1) \;\cdots\; u_{n_u}(i-\theta_{n_u}-n_{n_u})\big]^{\mathrm{T}}\end{aligned} \tag{3}$$

Parameters m_j and n_k are the number of past values of the j-th output or k-th input that are included in the NARX model; the number of process inputs and outputs are n_u and n_y, respectively. Model orders of $m_j = 0$ or $n_k = 0$ indicate that the corresponding j-th output or k-th input is not relevant to the process output being modeled. A scalar network output is used so that one MISO RBFN model has to be identified for each output of the process, and each MISO model may have a different input vector. The dead time for each input is θ_k, and may also be different for each input and each model. For RBFN modeling, the input and output data are usually normalized between zero and one.

The next section reviews the methods of RBFN identification used in this chapter. In Section 3 the multivariable simulation used is briefly described. The results of RBFN model identification using the data from the simulation are discussed in Section 4.

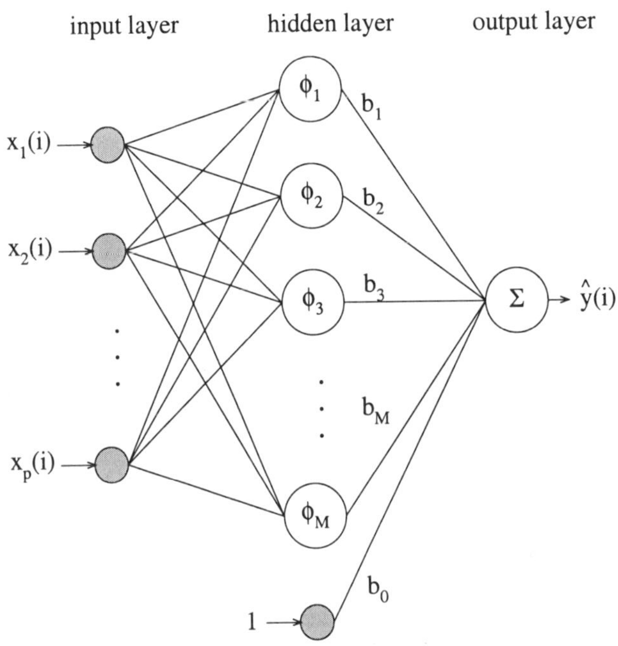

Figure 1. Radial basis function network.

2. RBFN Identification Methods

The chief feature of many RBFN identification methods is the placement of the RBF centers. Once the centers have been chosen and the other model parameters (m, n, θ, and β) have been specified in some manner, it is possible to use linear estimation methods, such as linear least squares, to determine the connection weights of the network. Two center selection techniques that are based on this approach are k-means clustering and linear regression methods.

The k-means clustering algorithm (Moody and Darken, 1989; Leonard and Kramer, 1991; Chen, *et al.*, 1992) partitions the data set used for identification into subsets of clusters, and finds a set of cluster centers. The cluster centers are used as the centers of the RBF nodes in the hidden layer of the network. The number of RBF centers (*i.e.*, the number of hidden nodes in the network), M, must be specified. The data are partitioned such that each data point in the network input space defined by $\mathbf{x}(i)$ in Eq. 3 is assigned to the cluster with the nearest center. There are both "off-line" (Leonard and Kramer, 1991) and "on-line" (Moody and Darken, 1989; Chen, *et al.*, 1992) versions of the clustering algorithm, but the two versions both minimize the total squared distance between the data points assigned to each cluster and the cluster centers. In the off-line version, the center of each cluster is initialized to a randomly chosen point in the input space. Next, each data point is assigned to the cluster with the nearest center. When all the data points have been assigned, each cluster center is moved to the mean of all the data points belonging to the cluster. The data points are then re-assigned to the cluster with the nearest center, and the procedure repeats until it converges. In general, the final center locations are not members of the identification data set, but are located within the spacial boundaries of the set.

Linear regression is another common method for determining the RBF centers. In this method, the RBF centers used in the network are chosen iteratively from a set of possible centers, which is typically the data set used for identification with each data point expressed in the manner of Eq. 3, and the connection weights are simultaneously determined using a version of least squares. Each RBF center describes the associated RBF, provided that the model orders, time delay, and width parameter are already specified. Two common examples of linear regression used for RBFN identification are forward regression and stepwise regression. In forward regression (Chen *et al.*, 1991; Zhu and Billings, 1996) the next RBF to be added to the network is chosen to be the one that explains the most of the variance of the prediction error. The procedure is iterated until an error tolerance is satisfied or additional RBFs do not improve the model. Stepwise regression (Pottmann and Seborg, 1992) operates similarly, but at each iteration all the RBFs in the network are statistically analyzed and those which no longer contribute substantially to prediction error reduction are removed. Both linear regression methods automatically determine the number of hidden layer nodes, M.

In this paper, a k-means clustering algorithm and a stepwise regression algorithm with orthogonal least squares are both used for RBFN identification. For the k-means clustering algorithm, the initial cluster centers are chosen from the data set used for identification, expressed as the network input vector $\mathbf{x}(i)$. Furthermore, data clusters with no members (other than the originally chosen random data point) are removed and not included in the final RBFN model. The p-nearest-neighbor heuristic is used to determine the value of the width parameter β_j (Moody and Darken, 1989); this heuristic accounts for the desired amount of overlap between RBF nodes, and can be used after the RBF center positions are fixed. The stepwise algorithm used is similar to the method of Pottmann and Seborg (1992), but with fewer statistical tests; only overall and partial F-tests (Draper and Smith, 1981) and the overall and partial Akaike Information Criteria (Akaike, 1972) are used. The number and placement of the RBFs in the network are chosen from a set of possible RBFs, the data set $\mathbf{x}(i)$, for $i = 1, ..., N$; the connection weights and bias term are then estimated using orthogonal least squares. If long identification data sets were to be used, however, stepwise regression with each sample in the identification set as a possible RBF center would be impractical. Two remedies for this problem are the use of a subset of randomly chosen samples from the identification data set as possible centers in the model, or the use of clustering algorithms to reduce the large data set to a number of clusters, the means of which can act as possible centers. These modifications reduce the computational requirements of the stepwise regression algorithm without necessarily reducing the model accuracy (Bomberger, 1997). The value of the width parameter is the same for all the RBFs in the network, $\beta_j = \beta$, and is specified before stepwise regression begins. The value for β may be estimated by trial and error, from various heuristics, or from approximate gradient norms (Bomberger and Seborg, 1997). In this paper, however, different values of β are tried for each model identified using stepwise regression.

For both identification methods, the number of past inputs and outputs used in the network input vector $\mathbf{x}(i)$ in Eq. 3 for dynamic NARX models can be determined using the method of false nearest neighbors (Kennel *et al.*, 1992; Bomberger and Seborg, 1998). The difficulty with this method is that large amounts of data are typically required to produce meaningful model orders (Rhodes and Morari, 1997; Bomberger *et al.*, 1998). In this chapter, model orders are chosen to ensure that each model input has the expected relationship with each model output.

3. Solution Copolymerization Simulation

The simulation is based on a dynamic model of the solution copolymerization of methyl methacrylate (MMA) and vinyl acetate (VAc) in a benzene solvent with azobisisobutyronitrile (AIBN) as initiator and acetaldehyde as the chain transfer agent. The polymerization takes place in a jacketed, perfectly mixed CSTR into which the monomers, initiator, solvent, and chain transfer agent are continuously added. A coolant flows through the jacket to remove the heat of polymerization. Inhibitors, such as *m*-dinitrobenzene (*m*-DNB) may be present in the monomer streams as an impurity. A schematic is shown in Fig. 2 (Congalidis *et al.*, 1989). The dynamic, first principles model has 12 state variables, and is based on a free radical mechanism reaction mechanism with 27 separate reactions. Simplifying assumptions include quasi-steady state, long chain hypothesis, and perfect mixing and constant volume and density in the reactor vessel.

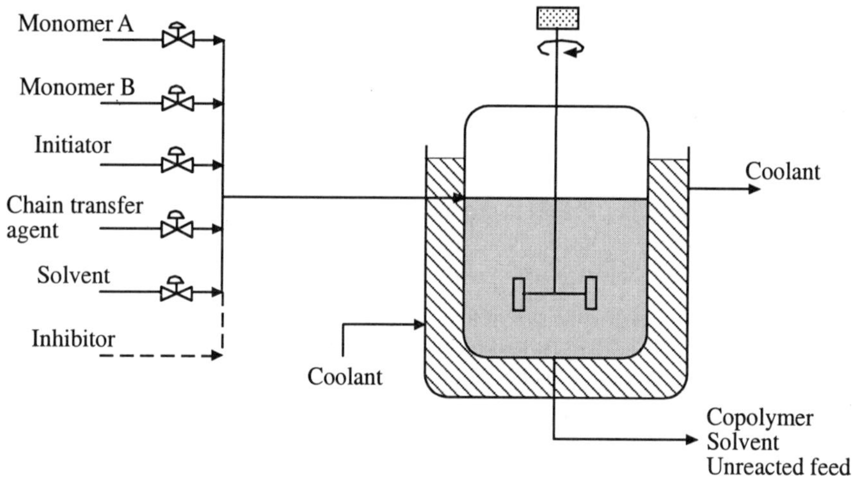

Figure 2. Stream diagram of the copolymerization reactor.

It is assumed that a properly tuned PI controller is used to manipulate the coolant flowrate in order to obtain any desired value of jacket temperature; therefore, the jacket temperature, T_j, and not the coolant flowrate is specified as one of the inputs to the simulation. The other inputs are the monomer A (MMA) feed rate G_a , the monomer B (VAc) feed rate G_b, the initiator feed rate G_i, the chain transfer agent feed rate G_t, the solvent feed rate G_s, the inhibitor flowrate G_z (which is used to represent the inhibitor impurity in the monomer streams), and the reactor feed temperature T_{rf}. The simulation outputs are the important reactor output variables for product quality control: the polymer production rate G_{pi}, the mole fraction of monomer A in the copolymer Y_{ap}, the weight average molecular weight

M_{pw}, and the reactor temperature T_r. These variables and their units are given in Table 1.

Table 1. Process variables for the copolymerization simulation.

type	name	description	units
output	G_{pi}	polymer production rate	kg / h
output	y_{ap}	mole fraction A in copolymer	-
output	M_{pw}	weight average mol. weight	-
output	T_r	reactor temperature	K
input	G_a	monomer A (MMA) feed rate	kg / h
input	G_b	monomer B (VAc) feed rate	kg / h
input	G_i	initiator (AIBN) feed rate	kg / h
input	G_t	chain transfer agent (Ac) feed rate	kg / h
input	G_s	solvent (Benzene) feed rate	kg / h
input	G_z	inhibitor flowrate (*m*-DNB)	kg / h
input	T_j	reactor jacket temperature	K
input	T_{rf}	reactor feed temperature	K

In Banerjee *et al.* (1997) the solvent feed rate G_s and the reactor feed temperature T_{rf} are treated as measured disturbance variables, and are used along with the other simulation inputs for state estimation. The inhibitor flowrate is treated as an unmeasured disturbance, and is not used for estimation purposes. The same tactic can be pursued for RBFN identification: G_s and T_{rf} are considered as possible inputs to the RBFN models, but G_z is not. Therefore, $n_u = 11$ and $n_y = 4$, giving a total of 11 possible inputs to the network and four MISO RBFN models. However, identification results for such an RBFN model using limited data (of the type available from plant tests or long-term historical data) have proven to be less than satisfactory (Bomberger *et al.*, 1998). In this chapter, a subset of possible model inputs and outputs will be used: G_a, G_b, and G_i will be used as simulation inputs, with the other flowrates and temperatures set to constant values, and y_{ap} and M_{pw} will be used as outputs for model identification.

For simulation, a MATLAB® Simulink® model of the process is used. A sampling period of 1 h is used to approximate the frequency of sampling necessitated by offline laboratory analysis of polymer properties. Gaussian additive measurement noise is added to all the simulated input and output variables, with a standard deviation of 1% of the nominal value of the variable. The nominal values for the input and output variables are shown in Table 2. (Congalidis *et al.*, 1989; Banerjee *et al.*, 1997).

Table 2. Nominal operating point

G_{pi}	23.9 kg / h
y_{ap}	0.551
M_{pw}	33783
T_r	353.59 K
G_a	18.0 kg / h
G_b	90 kg / h
G_i	0.18 kg / h
G_t	2.7 kg / h
G_s	36.0 kg / h
G_z	0 kg / h
T_j	336.15 K
T_{rf}	353.15 K

4. Identification of Copolymerization Simulation

As mentioned previously, an effort is made to treat the solution copolymerization simulation as an actual chemical plant for identification purposes; self-imposed constraints on the input sequence design and limitations of test length are respected, the sampling time is long, and noise is added to the simulated measurements. The identification is performed for open-loop conditions. In an actual polymer plant without an advanced control system, “recipe control” would probably be used, where the process inputs are specified to ensure a particular product quality, but feedback or feedforward control to correct for disturbances is not attempted. Instead, infrequent adjustments to the process would be made based on laboratory analysis of the product (Congalidis *et al.*, 1989).

4.1. Identification and Validation Data

Two types of input sequence designs are used: RUS and RDS. The RUS is a random, uniformly distributed sequence of step changes. A clock is used to time the steps, and at each clock period there is a specified probability for a step change in the input. The end value of the step change is uniformly distributed between lower and upper limits on the input. The RDS is a random, discretely distributed sequence of step changes. Again, a clock is used to time the steps, but a step must be made at the end of every clock period. The end value of the step change is one of a finite number of specified values set between lower and upper limits. The lower and upper limits on the variable inputs (G_a, G_b, and G_i) used in all three data sets are shown in Table 3.

Two sets of data are used for RBFN identification. Set 1 in Fig. 3 uses RUS inputs and is 360 samples long. Set 2 in Fig. 4 uses RDS inputs and is 360 samples long. A validation data set, Set 3, is also used; it has RUS inputs and is 360 samples long (Fig. 5).

Table 3. Lower and upper limits on variable flowrates (kg /h)

name	lower	upper
G_a	16	20
G_b	60	120
G_i	0.12	0.24

4.2. Results

Stepwise regression with orthogonal least squares and k-means clustering are used to identify RBFN models from these data sets. For each of the models, the model orders $\mathbf{m} = [\ 2\ 2\ ]^T$ and $\mathbf{n} = [\ 1\ 1\ 1\ ]^T$ are used, where

$$\mathbf{m} = \begin{bmatrix} m_1 & \cdots & m_{n_y} \end{bmatrix}^T \tag{4}$$

$$\mathbf{n} = \begin{bmatrix} n_1 & \cdots & n_{n_u} \end{bmatrix}^T \tag{5}$$

These vectors contain the number of past values to use for each output and input. The input vector is defined to be $\mathbf{u} = [\ G_a\ G_b\ G_i\ ]^T$ and the output vector is $\mathbf{y} = [\ y_{ap}\ M_{pw}\ ]^T$. In all cases, the dead time is $\theta_k = 0$, for $k = 1, \ldots, n_u$.

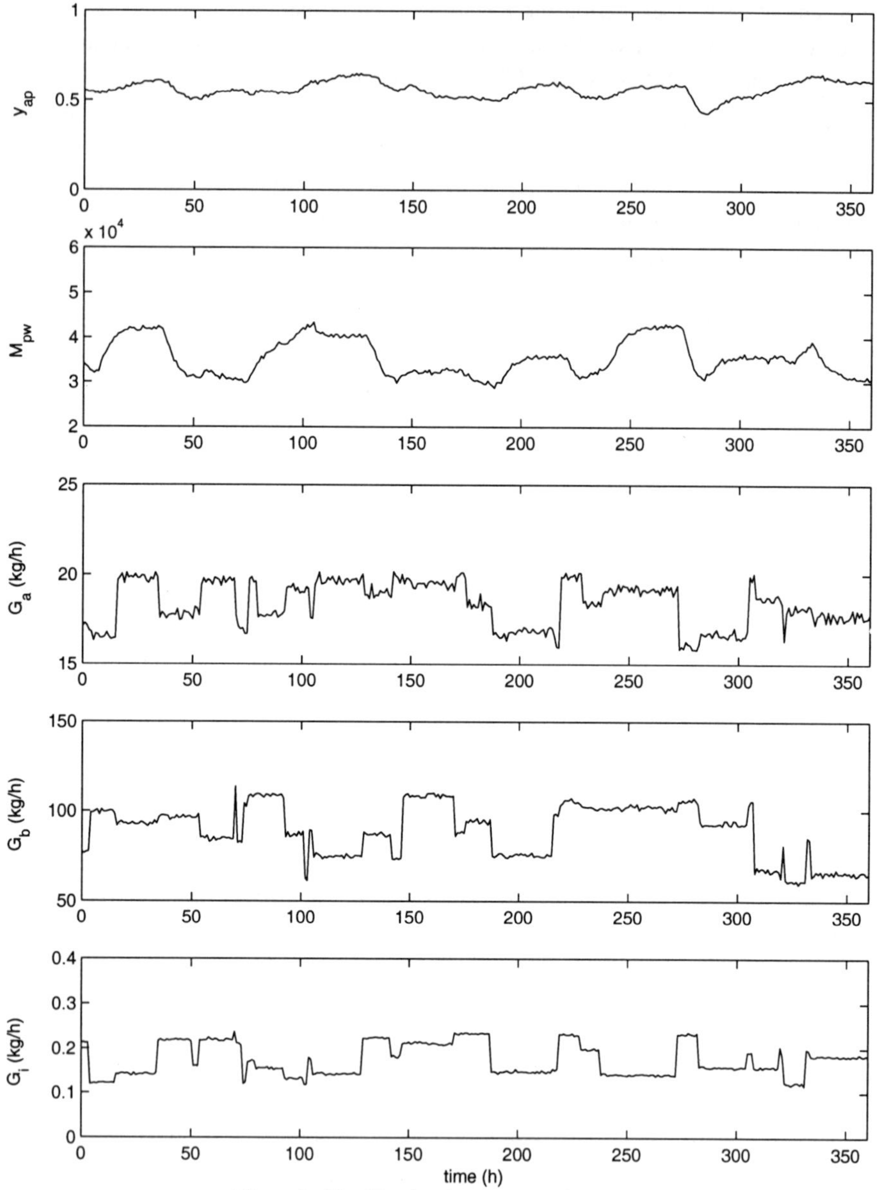

Figure 3. Identification data Set 1: RUS inputs.

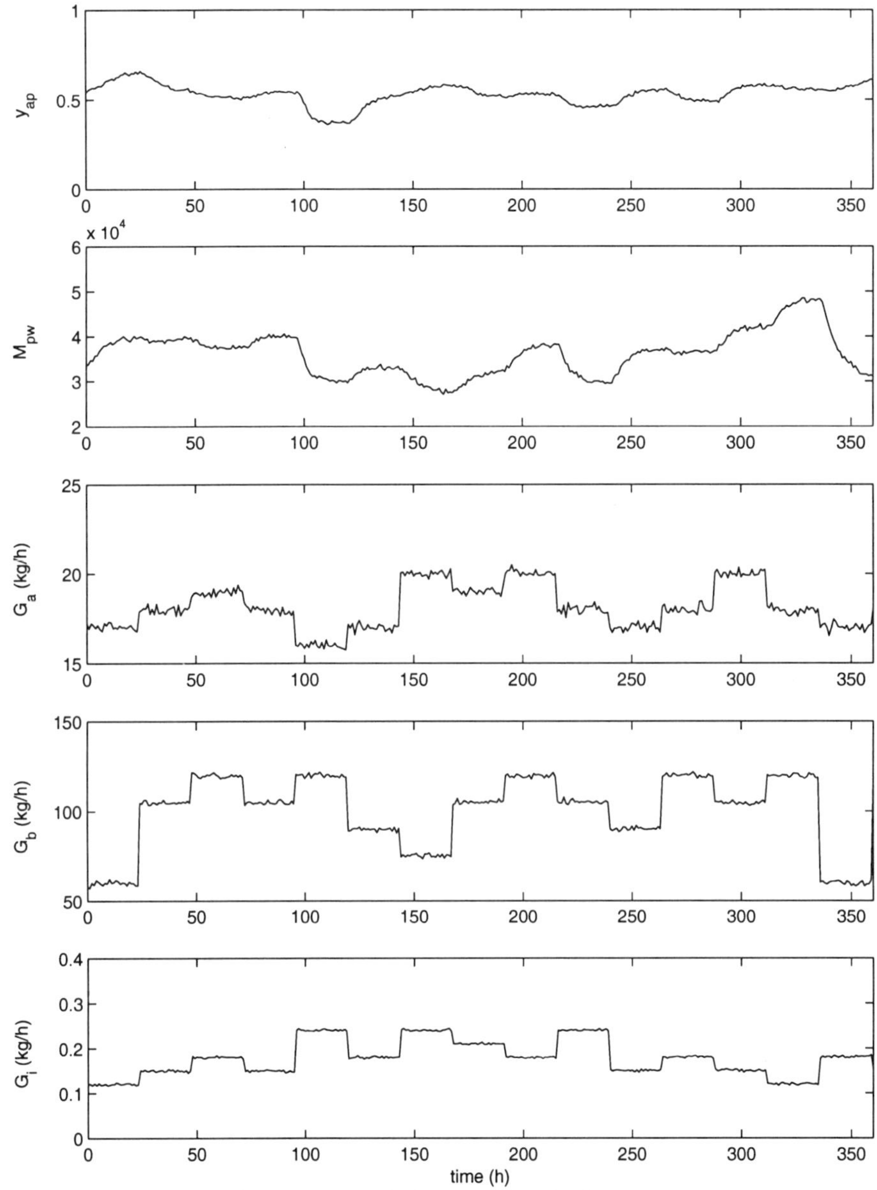

Figure 4. Identification data Set 2: RDS inputs.

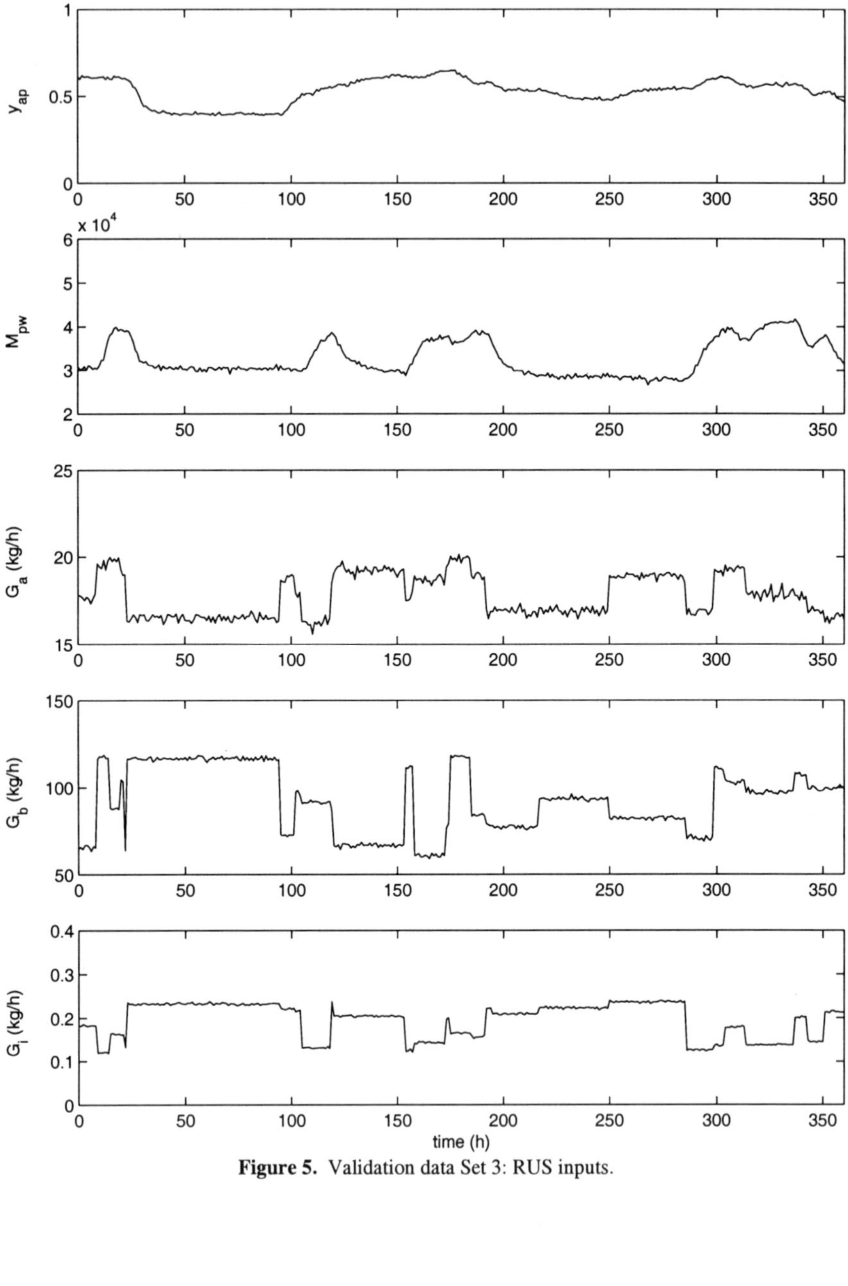

Figure 5. Validation data Set 3: RUS inputs.

Table 4. RBFN identification results using stepwise regression

output	β	M	iterations	I_1 (%)
identification of Set 1				
y_{ap}	0.1	40	80	2.45
	0.5	19	26	2.87
	1.0	16	24	3.03
	5.0	4	4	3.79
M_{pw}	0.1	47	94	1.26
	0.5	22	39	1.56
	1.0	17	26	1.53
	5.0	4	5	1.86
identification of Set 2				
y_{ap}	0.1	34	48	1.13
	0.5	22	42	1.35
	1.0	17	27	1.60
	5.0	9	12	1.81
M_{pw}	0.1	34	95	0.71
	0.5	24	37	0.68
	1.0	12	18	1.08
	5.0	5	5	1.07

The identification results for data in Sets 1 and 2 using stepwise regression with varying values for the width parameter β are shown in Table 4. Shown in the table are: the output for each MISO model; the width parameter β; the number of RBFs in the model, M; the number of iterations required for training; and the error index I_1 for one-step-ahead prediction error over the identification data set. The error index for each output is the variance of the prediction error divided by the variance of the modeled output:

$$I_{type} = \frac{\sum_{i=1}^{N}\left[\hat{y}(i) - y(i)\right]^2}{\sum_{i=1}^{N}\left[y(i) - \bar{y}\right]^2} \times 100\% \qquad (6)$$

where N is the number of data points in the data set. For one-step-ahead predictions, *type* = 1, and for multiple-step-ahead predictions *type* = m.

Table 5. RBFN identification results using k-means

output	M_0	M	$\overline{\beta}$	iterations	I_1 (%)
identification of Set 1					
y_{ap}	10	10	0.5864	8	10.43
	20	20	0.4919	9	7.72
	30	29	0.4210	13	4.34
	40	38	0.4040	7	2.78
M_{pw}	10	10	0.5959	9	5.22
	20	20	0.4911	9	3.36
	30	30	0.4084	7	2.39
	40	40	0.3486	10	2.02
identification of Set 2					
y_{ap}	10	9	0.5692	9	9.66
	20	20	0.4530	11	4.35
	30	29	0.3255	8	2.78
	40	40	0.2738	9	1.96
M_{pw}	10	10	0.5908	9	6.57
	20	20	0.4442	5	4.00
	30	30	0.3464	8	1.47
	40	40	0.3011	7	1.01

In Table 5 the results for RBFN model identification using k-means clustering on data Sets 1 and 2 are shown. In addition to the values shown in Table 4, the number of initial data clusters, M_0, is shown. The number of RBFs in the model may not match the number of initial clusters, because clusters that do not contain data points are pruned during the identification procedure. For each RBF in the network, the width parameter β_j is determined using the p-nearest neighbor heuristic with p = 2; the mean value of the width parameters for all the RBFs in the network, $\overline{\beta}$, is shown in Table 5 as well.

The RBFN models were validated on data Set 3, which was not used for training, as an indication of model accuracy. The validation tests used one-step-ahead prediction. Results for the models identified using stepwise regression and k-means clustering are shown in Tables 6 and 7, respectively. It is also possible to validate the models using multiple-step-ahead prediction. It is a prediction of the model output given only initial conditions and the system inputs as a function of time, $u_k(i)$. To make multiple-step-ahead predictions, the network input vector $\mathbf{x}(i)$ in Eq. 3 is rewritten as a function of predicted system outputs and inputs:

Table 6. Error index for one-step-ahead prediction on validation data Set 3

	I_1 (%)	
β	y_{ap}	M_{pw}
identified on Set 1 using stepwise regression		
0.1	81.68	55.61
0.5	22.43	5.38
1.0	9.67	3.74
5.0	3.28	2.88
identified on Set 2 using stepwise regression		
0.1	22.97	31.92
0.5	6.27	3.43
1.0	3.02	3.91
5.0	1.35	1.90

Table 7. Error index for one-step-ahead prediction on validation data Set 3

	I_1 (%)	
M_0	y_{ap}	M_{pw}
identified on Set 1 using k-means		
10	37.07	27.89
20	32.85	18.16
30	40.71	33.84
40	34.01	15.75
identified on Set 2 using k-means		
10	13.25	12.56
20	7.59	13.90
30	6.52	7.50
40	8.45	6.38

$$\mathbf{x}(i) = \big[\hat{y}_1(i-1) \;\cdots\; \hat{y}_1(i-m_1) \;\cdots\; \hat{y}_{n_y}(i-1) \;\cdots\; \hat{y}_{n_y}(i-m_{n_y}) \;\; u_1(i-\theta_1-1) \\ \cdots\; u_1(i-\theta_1-n_1) \;\cdots\; u_{n_u}(i-\theta_{n_u}-1) \;\cdots\; u_{n_u}(i-\theta_{n_u}-n_{n_u})\big]^{\mathrm{T}} \tag{7}$$

Multiple-step-ahead predictions are a much tougher test of model accuracy than one-step-ahead prediction tests. Additionally, in the case of a multivariable system, one MISO model is needed for each system output when making multiple-step-ahead predictions. This permits many possible combinations of models when more than one MISO model is available for each output. In this case, each MISO model in Tables 4 and 5 was paired with the all models for the other output in each data set and for each identification method. The results were analyzed to pick the pairings that resulted in the most accurate RBFN models on both the identification and validation data sets. The most accurate models are shown in Table 8. The error indices $I_{m,val}$ and $I_{m,val}$ are the averages of the error indices for y_{ap} and M_{pw} for the validation and identification data sets, respectively; the overall error index, $I_{m,o}$, is the average of $I_{m,val}$ and $I_{m,val}$. Comparing to Table 8 to Tables 4-7, it is worthwhile to note that good one-step-ahead prediction accuracy is not necessarily indicative of good multiple-step-ahead prediction accuracy.

Table 8. Error indices for multiple-step-ahead prediction

β for y_{ap}	β for M_{pw}	$I_{m,id}$ (%)	$I_{m,val}$ (%)	$I_{m,o}$ (%)
stepwise regression models identified on Set 1				
0.5	5.0	6.29	18.03	12.16
5.0	5.0	9.60	16.37	12.99
stepwise regression models identified on Set 2				
5.0	5.0	1.65	23.01	12.33
0.5	0.5	9.44	16.37	12.91

M_0 for y_{ap}	M_0 for M_{pw}	$I_{m,id}$ (%)	$I_{m,val}$ (%)	$I_{m,o}$ (%)
k-means clustering models identified on Set 1				
40	40	16.40	21.28	18.84
30	40	18.14	20.89	19.52
k-means clustering models identified on Set 2				
30	30	8.78	20.42	14.60
30	40	11.84	24.08	17.96

4.3. Discussion

It is apparent that the RBFN model identification does not work equally well in all the cases presented. The accuracy of the model depends on the model parameters (especially the width parameter β and the number of RBFs *M*), the identification data, and the identification method.

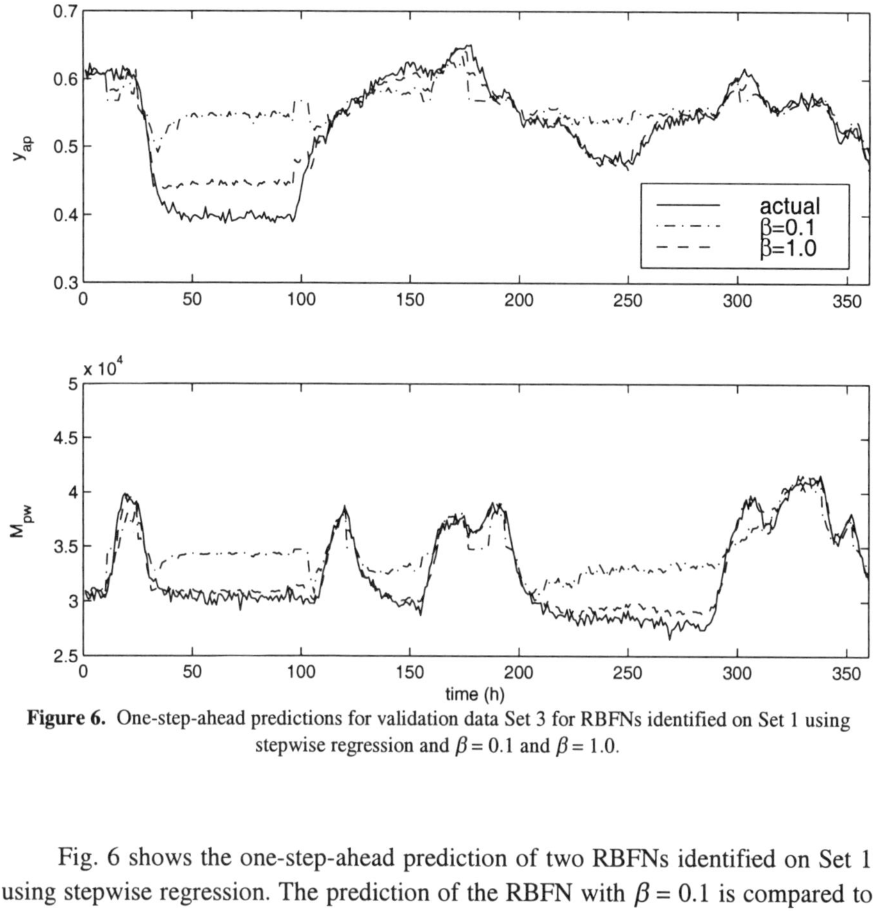

Figure 6. One-step-ahead predictions for validation data Set 3 for RBFNs identified on Set 1 using stepwise regression and $\beta = 0.1$ and $\beta = 1.0$.

Fig. 6 shows the one-step-ahead prediction of two RBFNs identified on Set 1 using stepwise regression. The prediction of the RBFN with $\beta = 0.1$ is compared to the prediction of the RBFN with $\beta = 1.0$ and the actual data for validation data Set 3. It is apparent from this Fig. 6 and the results in Table 6 that increasing β generally results in improved model accuracy on the validation data. This can be explained by the localization of the Gaussian RBFs in the network input space, and the differences between the validation data Set 3 and the identification data Set 1.

Because Gaussian RBFs decay exponentially with the squared distance from the RBF center, the RBFs are localized in the input space of the network, as shown in Eq. 3. When the RBFs have decayed to near zero values, the network may be unable to accurately predict the model output. This situation can occur when the network extrapolates from the data used for identification, as is the case here. In Fig.

7, the distribution of the data in Set 1 is shown; there is little or no data available for $y_{ap} < 0.45$ or $M_{pw} < 30,000$. In the data distribution for Set 3 shown in Fig. 8 many data points are available for $M_{pw} < 30,000$ or $y_{ap} < 0.4$. The RBFNs identified using Set 1 must therefore extrapolate to predict the outputs present in Set 3. Because RBFs with larger values of β decay to zero more slowly, they are able to extrapolate somewhat better; therefore, RBFNS with larger values of β exhibit better accuracy over the validation data in Set 3.

The effect of the data distribution is evident in other ways as well. Two RBFN models identified using stepwise regression are compared in Fig. 9. One model was identified on Set 1, and the other on Set 2. Although both models use the same β value, the model identified on Set 2 is much more accurate. The data distribution for Set 2 in Fig. 10 has many more data points for $y_{ap} < 0.45$ or $M_{pw} < 30,000$ than Set 1; this means that the RBFNs identified on Set 2 are better trained in this region than are the RBFNs identified on Set 1. In general, the RDS type of input sequence used for the identification data in Set 2 permits a more uniform distribution of data throughout the network input space than an RUS input, especially for data sets of limited length.

Model accuracy also depends on the identification method used. Given the same identification data, the two methods will not necessarily generate equally accurate models. Fig. 11 compares RBFN models on the validation data Set 3, using the most accurate models from both the stepwise regression and k-means clustering identification methods. Both models were identified on Set 2 data. While both RBFNs are fairly accurate, the RBFN identified using stepwise regression provides better predictions. This result probably occurs because stepwise regression optimizes the number and placement of RBF centers from a set of possible centers to find the best model for given model orders and width parameter, in terms of accuracy and parsimony. The k-means clustering method, on the other hand, finds a fixed number of RBF centers based on the natural clustering of data points in the input space; this ensures that RBF centers are usually located only where there is a significant amount of data. Also, in the k-means clustering algorithm, the choices for β_j are made using the p-nearest-neighbor heuristic, and are not in any way optimized to find values that might improve model accuracy for a given set of RBF centers.

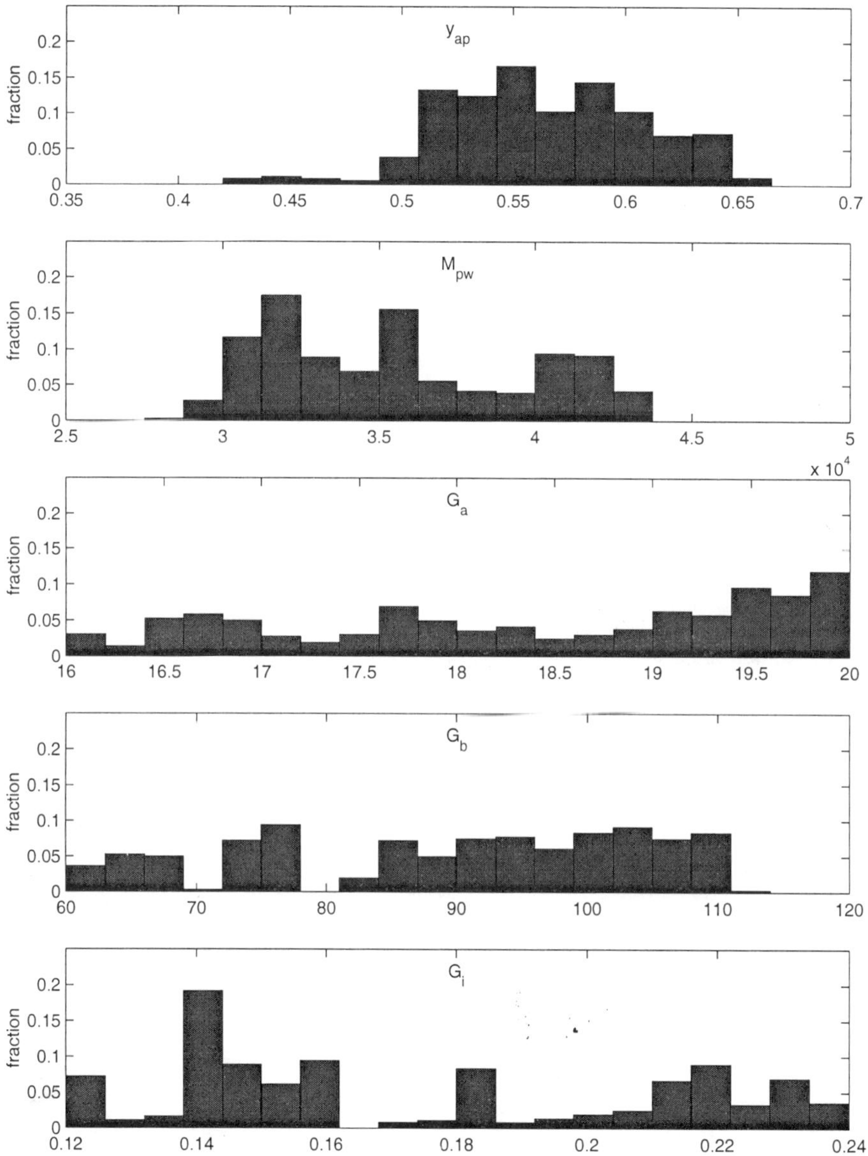

Figure 7. Normalized histogram of identification data Set 1.

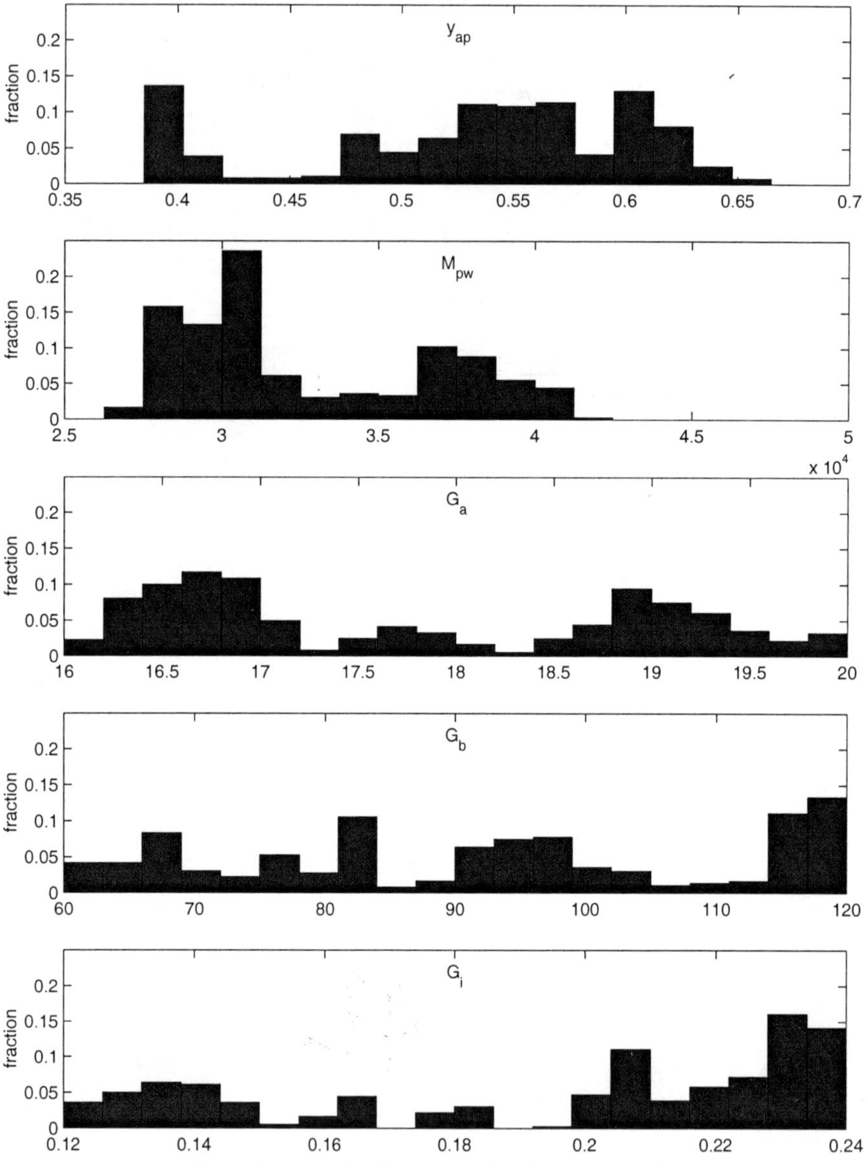

Figure 8. Normalized histogram of validation data Set 3.

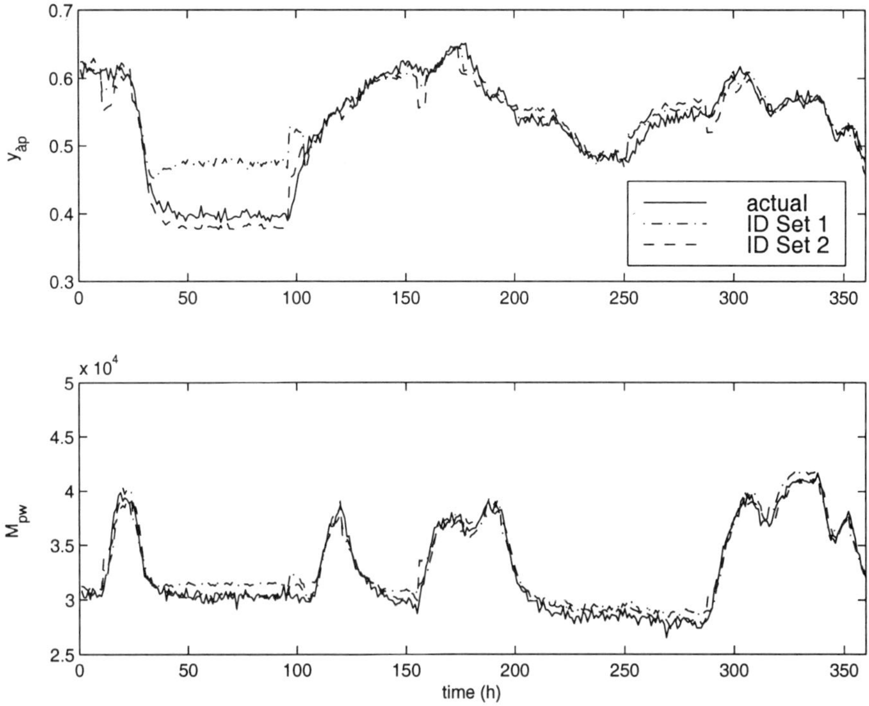

Figure 9. One-step-ahead predictions for validation data Set 3 of RBFNs identified on Set 1 and Set 3 using stepwise regression with $\beta = 0.5$.

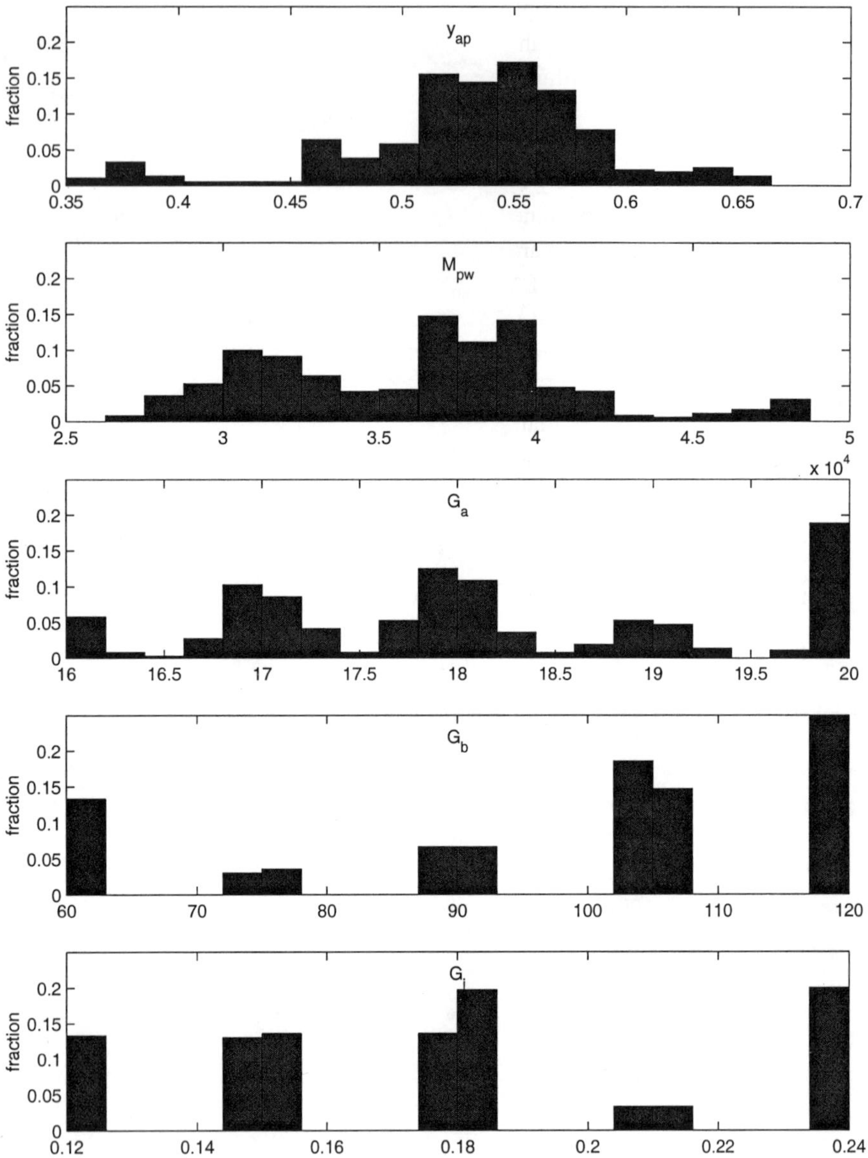

Figure 10. Normalized histogram of identification data Set 2.

As noted previously, multiple-step-ahead prediction is a more stringent test of model accuracy than one-step-ahead prediction. In Fig. 12 a comparison of one-step-ahead prediction and multiple-step-ahead prediction for the best RBFN models is made. The models were identified using stepwise regression on Set 2, with β = 0.5 for both y_{ap} and M_{pw}. While both predictions are fairly accurate, the one-step-ahead prediction is substantially better. The multiple-step-ahead prediction is worse because the networks were trained to make one-step-ahead predictions only, and because errors in the prediction are compounded in future samples.

An inadequate amount of data, or range of data, is the major source of difficulty for nonlinear model identification of multivariable processes. This especially applies to RBFN model identification, where the radial basis functions may be significantly localized in the network input space defined by the input vector $\mathbf{x}(i)$, inhibiting model extrapolation. In contrast, for linear model identification, limited data is not as important a problem. This result occurs because the superposition principle central to linear system modeling requires that the size or direction of change in a process does not alter the general process behavior. Similarly, for linear MIMO systems, changes in the process inputs act in an additive fashion and it is not necessary that all possible combinations of changes for the n_u inputs be made. Evidence is provided by the commercial success of linear multivariable control software like Aspen Technology's DMC, where linear models of large multivariable systems are identified and used for model predictive control.

It is promising, however, that the RBFN nonlinear modeling technique is able to accurately predict the nonlinear behavior of the copolymerization simulation in regions for which only limited data exists. An RBFN model may have learned little of the overall behavior of the process, but, trained only on historical data, the model is still able to predict the behavior of the process for similar operating conditions (Bomberger *et al.*, 1998). Pavilion Technologies, a commercial neural network software and engineering company, uses the ability of neural network models to easily learn nonlinear behavior by identifying steady state process models from historical operating data. These models can then be used to optimize process inputs to obtain desirable process outputs, or can be coupled with dynamic models for use in nonlinear model predictive control strategies (Keeler *et al.*, 1997).

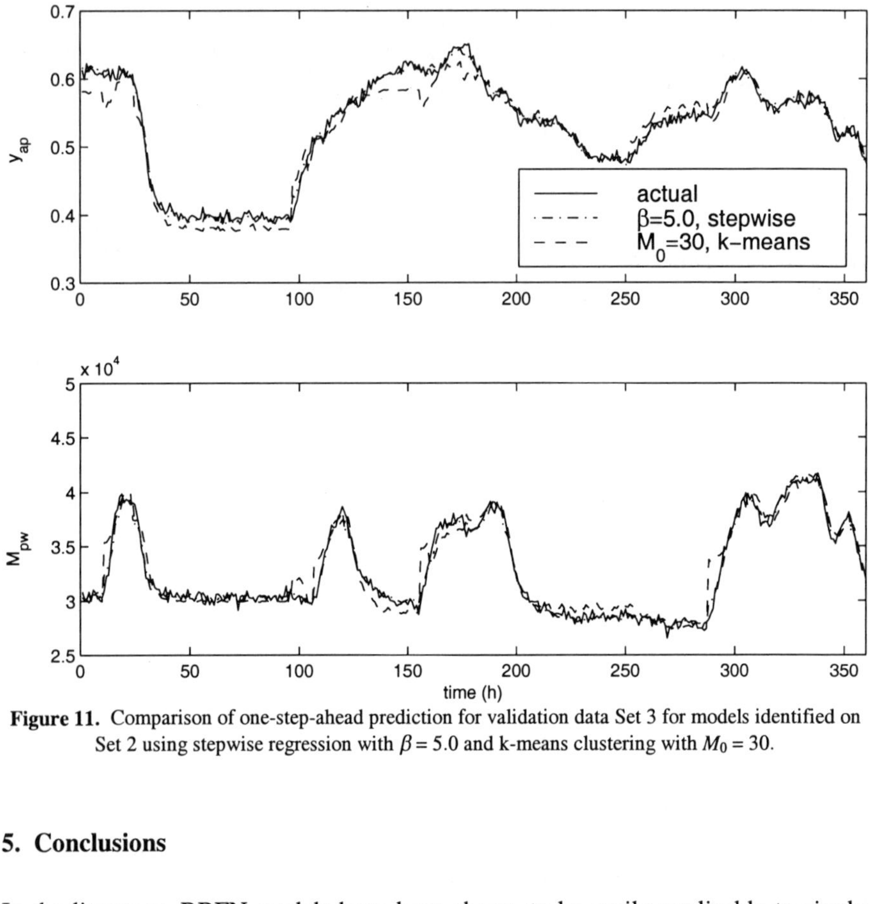

Figure 11. Comparison of one-step-ahead prediction for validation data Set 3 for models identified on Set 2 using stepwise regression with $\beta = 5.0$ and k-means clustering with $M_0 = 30$.

5. Conclusions

In the literature, RBFN models have been shown to be easily applicable to single-input, single-output and other low dimensional identification problems. This research has focused on identification of a multivariable, industrially relevant simulation of a solution copolymerization reactor. The RBFN model has three inputs and two outputs, and the dimensionality of the identification problem poses some difficulties for nonlinear empirical model identification. The primary difficulty is the amount of data required, which is an aspect of the "curse of dimensionality." The data requirement is a problem for plant testing and may cause computational difficulties for identification algorithms as well. Both the stepwise regression algorithm and the k-means clustering algorithm work robustly and quickly, even for high dimensional network input vectors. The stepwise regression algorithm exhibits

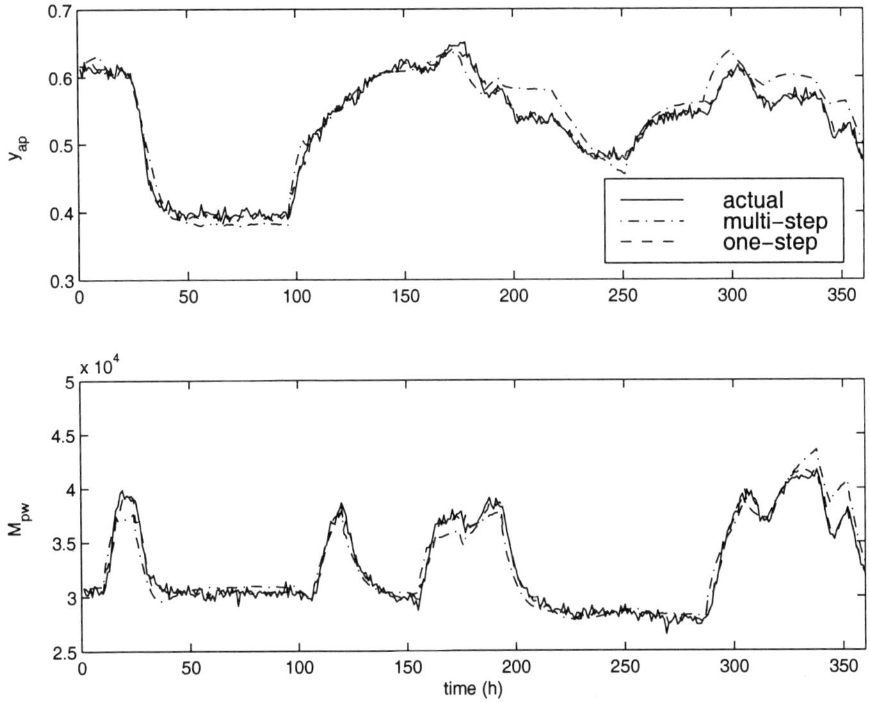

Figure 12. Comparison of one-step-ahead prediction and multiple-step-ahead prediction for validation data Set 3 for models identified using stepwise regression on Set 2 ($\beta = 5.0$ for both y_{ap} and M_{pw}).

somewhat more consistent results and has better overall accuracy at the cost of increased computational requirements which can be prohibitive for large data sets.

References

Akaike, H., In *2nd Int. Symp. on Information Theory, Tsahkador, Armenia, USSR, Sept. 2-8, 1971*, B. N. Petrov and F. Csáki, ed., Akadémiai Kiadó, Budapest, (1973), 267-281.

Banerjee, A., et al., *AIChE J.* **43** (1997), 1204-1226.

Bomberger, J. D., *Radial Basis Function Networks for Process Identification*. Ph.D. Dissertation (University of California, Santa Barbara, 1997).

Bomberger, J. D., et al., In *Chemical Process Control V*, ed. Kantor, J. C. et al., AIChE Sympos. Ser. **93**, 316, (1997), 280-283.

Bomberger, J. D. and Seborg, D. E., *J. Proc. Cont.* **8** (1998), 459-468.

Bomberger, J. D. and Seborg, D. E., Determination of the width parameter for Gaussian radial basis function networks using approximate gradient norms. Paper presented at AIChE Annual Meeting, Nov. 16-21, Los Angeles (1997).

Bomberger, J. D. et al., in *Proc. of the 1998 ACC*, June 24-26, Philadelphia, PA (1998).

Chen, S. et al., *Int. J. Control.* **55**, (1992), 1051-1070.

Chen, S. et al., *IEEE Trans. Neural Networks.* **2** (1991), 302-309.

Congalidis, J. P. et al., *AIChE J.* **35** (1989), 891-907.

Draper, N. R. and Smith, H., *Applied Regression Analysis* (3rd. Ed., John Wiley, New York 1998).

Eikens, B. and Karim, M. N., in *Preprints IFAC-ADCHEM '94*, May 25-27, Kyoto, Japan, (1994), 125-130.

Keeler, J. et al., The Process Perfector: the next step in multivariable control and optimization. Technical Report (Pavilion Technologies, Inc., Austin, TX, 1997).

Kennel, M. B. et al., *Phys. Rev. A.* **45** (1992), 3403-3411.

Keulers, M., in *Proc. of the 1993 ACC*, June 2-4, San Francisco, (1993), 2261-2265.

Leonard, J. A. and Kramer, M. A., *IEEE Control Sys. Mag.* **11**, April, (1991), 31-38.

Moody, J. and Darken, C. J., *Neural Computation.* **1** (1989), 281-294.

Pottmann, M. and Seborg, D. E., *J. Process Control.* **2** (1992), 189-203.

Rhodes, C. and Morari, M., *Computers Chem. Engng.* **21S** (1997), S1149-S1154.

Sjöberg, J. et al., *Automatica.* **31**, (1995), 1691-1724.

Zhu, Q. M. and Billings, S. A., *Int. J. Control.* **64** (1996), 871-886.

Acknowledgements

The authors gratefully acknowledge the financial support of the National Science Foundation (Grant # CTS-9424094) and DuPont. The cooperation of Dr. Yaman Arkun (Dean of the College of Engineering, KOC University, Turkey) in supplying the MATLAB® code for the copolymerization simulation is appreciated.

3. PROCESS IDENTIFICATION WITH SELF-ORGANIZING NETWORKS

B. EIKENS, M. N. KARIM, L. SIMON
Department of Chemical and Bioresource Engineering
Colorado State University
Fort Collins, Colorado 80523

This paper demonstrates the applicability of unsupervised neural networks in the form of self-organizing maps (SOM) for process visualization and modeling. The structures of SOMs and learning algorithms are summarized. Unsupervised methods are applied to identify the different physiological states which exist during a yeast fermentation. The neural network model was able to predict accurately different physiological states.

1. Introduction

Most neural networks applied in chemical engineering and in biotechnology are trained to perform a mapping $\Phi: \Re^N \rightarrow \Re^M$ by presenting input (*N*) - output (*M*) data pairs of the process. In cases, however, where output or target data is not available, the network has to extract the necessary information from the input data. Typical examples for this class of problems include clustering, dimensionality reduction and feature extraction. Neural networks designed for these problems are called self-organizing networks. They utilize *unsupervised* training algorithms, i.e., there is no "teaching signal" which indicates whether the network output is accurate. There are many types of self-organizing networks. One of the basic schemes is competitive learning as proposed by Rumelhart *et al.* (1986). Competitive learning networks are characterized by the competition process between the network nodes combined with a "winner- take- all strategy". This means that only one network output node called the best-matching unit (BMU) is allowed to fire. Only the parameters associated with this node are adjusted during 'training. A very similar network but with different emergent properties is the self-organizing map (Kohonen, 1982). Other examples of selforganizing networks are the ART networks introduced by Carpenter and Grossberg (1988) and Fukushima's cognitron (Fukushima, 1988).

The most popular unsupervised learning neural network is the self-organizing map (SOM) also known as the Kohonen's feature map (Kohonen, 1995). This network was first presented in Kohonen (1982) as a clustering and dimensionality reduction method. Kohonen linked the network architecture to the discovery of spatially ordered sensory processing areas in the brain. The SOM algorithm was presented as an example of a process which induces neighborhood relations among neurons. The result is a topology preserving mapping of the network that has the following characteristics (Murtagh and Hernandez- Paiarcs, 1995):

- Similar inputs are mapped onto identical or closely neighboring network nodes, i.e., the network nodes are ordered on themap.The mapping preserves the relative distance between the input vectors, i.e., data points which are close to each other in the input space are mapped to nearby units of the SOM. The mapping is robust against distortions due to noisy data which is an important property for real applications.
- Neighboring nodes of the self-organizing map possess similar reference vectors. This ensures the robustness of the inverse mapping. The mapping tends to reduce the dimension of the input vectors to a lower network dimension. Typically, a one- or two-dimensional network output layer is used. Although the mapping reduces the dimensionality, it usually preserves characteristic similarity relations among the input vectors.

The definition of topological neighbors modifies the "winner-take-all" strategy of classical competitive learning to a "winner-take-most" strategy for SOMs. SOMs not only modifies the parameters of the BMU, but also adjusts the vectors of topological neighbors. The area of the input space that corresponds to the BMU of a particular node is called the *Voronoi tesselation* or *region.* The topological neighborhood of the SOM is gradually decreased during training as described in the following sections.

Several modifications of the original SOM algorithm have been proposed. They include tree-structured SOMs, fuzzy-SOMs and incremental growing SOMs. SOMs have been used in many practical applications, the most common of which are pattern recognition, fault diagnosis, and robot control. The SOM partitioning may be used as a preprocessing stage. Since the Voronoi tesselation partitions the input space into disjoint sets, each of these regions may be identified by a different, local submodel. Since the SOM maps a multi-dimensional space onto a one to two dimensional surface in a nonlinear way, it is a suitable tool for visualizing and identifying the states of complex processes. In the case study presented in this paper, the task of the SOM is to identify different physiological states of a yeast fermentation. The physiological state of the fermentation depends on operating conditions. Based on these variables, which represent the inputs of the network, the SOM predicts the current mode or physiological state of the fermentation. The mode of the fermentation can be considered a "latent" variable since it is not directly measurable. Before the specific case study is presented, the structure of the SOM and its training algorithms are described in the following sections.

2. Structure of self-organizing maps

The SOM is usually represented as a two dimensional network sheet whose nodes are arranged on a grid or an array. Each node represents a vector called code-book or reference vector. The code-book vectors have the same dimension as the the input vectors $x = (x_1, x_2, \ldots, x_m)$. The input data set is assumed to consist of N vectors $\dot{x}_1, \ldots, \dot{x}_N$. All nodes of the SOM receive the same input. The basic structure of a SOM is shown in Fig. 1.

In SOMs, the network nodes are connected to adjacent nodes by a neighborhood relationship. This dictates the topology or the structure of SOMs. Two different network structures are commonly used: the rectangular and the hexagonal topology (Fig. 2). The topological relations are represented by dotted lines between the nodes. They can be defined by a distance measure, e.g., the Euclidean distance. The neighborhood set N_c, of a node $\dot{w}_c$ consists of adjacent nodes around $\dot{w}_c$ .This feature is particular to the training algorithm of SOMS.

3. Properties of self-organizing maps

SOMs have been extensively studied, in particular Kohonen's SOM algorithm. Although they are closely related to various other multivariate methods of data analysis, SOMs are difficult to analyze and their statistical properties remain unknown. So far, no quantitative analysis results have been presented for this type of network. These difficulties are caused by the heuristic choice/tuning of the neighborhood and decreasing learning rate.

Since SOMs rely on a minimal distance method, it can be argued that they are a partitioning method of the k-means type which a simultaneous ordering mechanism (Murtagh and Hernandez- Paj ares, 1995). As with the k-means clustering algorithm, convergence to the optimal solution is not guaranteed due to the heuristics of the learning algorithm although good solutions are generally approached quickly. So far, researchers have not been able to define a suitable objective function for the minimization of the clustering distance during the training process. Hence, the optimal mapping has to be determined by trial and error.

However, SOMs are able to generate interesting low dimensional representation of high dimensional input data for many applications.

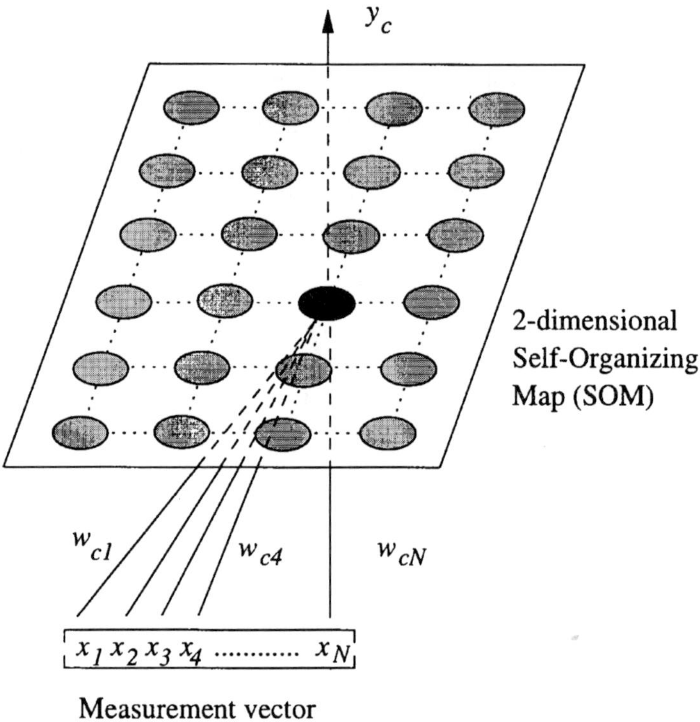

Figure 1. Mapping of the input vector through a SOM network

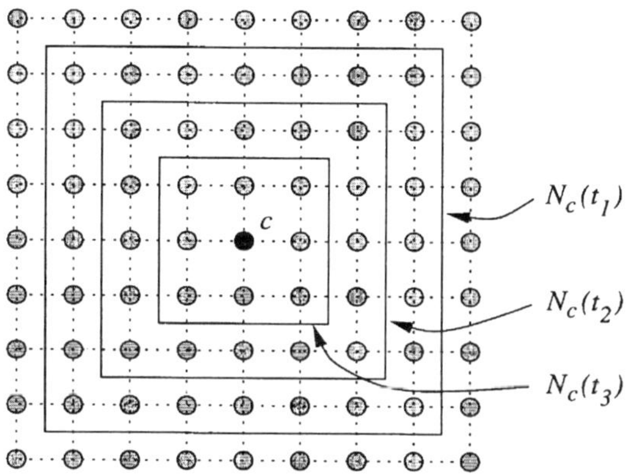

(a) Rectangular SOM topology.

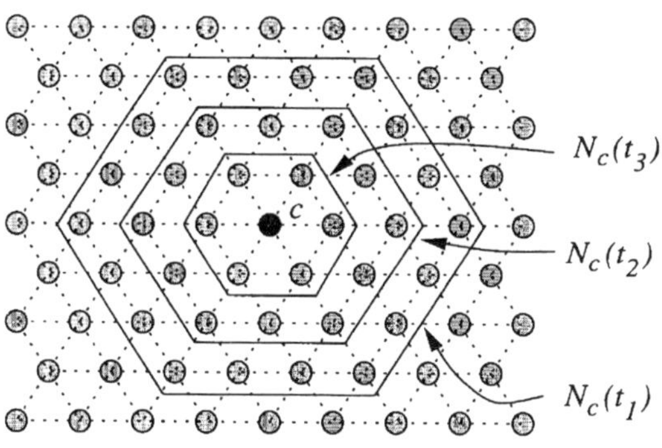

(b) Hexagonal SOM topology.

Figure 2. Typical topologies for self-organizing maps

4. Training of self-organizing maps

Three SOM initialization algorithms for determining suitable starting values for the codebook vectors w_i for i = 1,...,C are proposed by Kohonen (1995). First, the initial code-book vectors can be selected at random. Second, the code-book vectors can be selected to form the input data set. This guarantees that the reference vectors are elements of the same subset of the input domain as the data. Third, principal component analysis (PCA) can be used to determine initial code-book vectors in order to capture the linear dependency of the data. The code-book vectors are initialized to lie in one input space spanned by two eigenvectors corresponding to the largest eigenvalues. This has the effect of stretching the SOM to the same orientation as the data.

After the SOM is initialized, the training process is started. It consist of two steps that are iterated for every input vector (Kohonen, 1990).

1. **Finding the best matching units:** An input vector $\vec{x}(t)$ is selected randomly from the data set and the reference vector w_c with the greatest similarity or the "best match" is determined. The similarity is expressed in terms of a distance function. Generally, the Euclidean distance is chosen. The best matching code-book vector Z~, is calculated as

$$\|\vec{x}(t)-\vec{w}_c\| = \min\{\|\vec{x}(t)-\vec{w}_i\|\} \quad \text{for} \quad i=1,\ldots,C. \tag{1}$$

For normalized input vectors, the inner product $x^{t} W_i$ may be used.

2. **Adaptation of the weights:** After the best matching node w_c is determined, the nodes of the SOM are updated. The vector w_c, and adjacent vectors of the neighborhood set N_c, are adapted so that the similarity with the present input vector $\vec{x}(t)$ is increased. The resulting learning rule has the following form

$$\vec{w}_i(t+1) = \vec{w}_i(t) + \alpha(t)N_c(i,t)\left[\vec{x}(t)-\vec{w}_i\right] \qquad \forall i. \tag{2}$$

The learning rate is denoted by $\alpha(t)$, where $0<\alpha(t)<1$. The index c points to best matching code-book vector as determined in the previous step. $N_c(i, t)$ represents the neighborhood of node w_c at iteration t, i.e., it determines the units i that are adapted.

The neighborhood set $N_c(i, t)$ and the learning rate $\alpha(t)$ are changed dynamically during the training process. The neighborhood ($0<N_c(i, t)<1$ describes the activity of adjacent nodes. It is a decreasing function of the grid distance between unit i and unit c, such that $N_c(i, t)=1$. The grid distance between w_c, and w_i, as well as the value of $N_c(i, t)$ depend on time t. The training algorithm starts with a large range of $N_c(i, t)$ and gradually reduces the neighborhood at each iteration of the training process. Fig. 2 shows the sequences of shrinking neighborhood sets $N_c(t_1) \subseteq N_c(t_2) \subseteq N_c(t_3)$ where $t_1<t_2<t_3$ for rectangular and hexagonal topologies.

Several functions have been suggested for the description of the neighborhood. A popular choice is the Gaussian function

$$N_c(i,t) = N_o(t).\exp\left(\frac{-\|r_i - r_c\|^2}{\sigma(t)^2}\right) \tag{3}$$

where $N_o(t)$ and $\sigma(t)$ are decreasing functions of time. The neighborhood function can be defined quite freely. The overall effect of smooth mapping from the input pattern space to the output space is preserved for many functions (Kangas, 1994).

Common choices for the learning rate $\alpha(t)$ are a linearly decreasing function $\alpha(t)=\alpha(0)\ (1-t/t_{max})$ or an inverse-time function $\alpha(t) = A/(B+t)$ where t_{max} denotes the maximum number of iterations and A and B are constants. The inverse-time function is advisable for large SOMs and large t_{max} in order to allow more balanced fine-tuning. The appropriate function and its parameters have to be determined by trial and error. Some default values are given in Kohonen (1995) and Kohonen et al. (1996). The training process is typically divided in two phases. During the first phase, relatively large values for $N_o(t)$ and $\alpha(0)$ are chosen to order the code-book vectors. The reference vectors are fine-tuned during the second phase where $N_o(t)$ and $\alpha(0)$ are usually smaller and t_{max} is larger.

Once the training process is completed, the SOM can be evaluated by calculating the quantization error or the average distortion measure. The quantization error q is defined as

$$q = \frac{\|\vec{x}_i - \vec{w}_c\|}{N} \quad \text{for} \quad i = 1, \ldots, N. \tag{4}$$

The quantization error for a given input vector $\vec{x}$ is the distance between the signal and the reference vector of the BMU. Hence, the value of q may be interpreted as a measure of the quality of the prediction once the SOM is trained. For larger SOMs, the average distortion measure as described in Kohonen *et al.* (1996) should be used.

The last step of the training phase is called calibration of the SOM. A number of typical, manually analyzed data sets are projected onto the map. The corresponding BMUs are labeled in order to create a set of characteristic reference points on the map. Since the mapping is assumed to be continuous, unknown data can be interpreted by means of interpolation and extrapolation from the calibrated map (Kohonen *et al.*, 1996).

5. Case study: Visualization and identification of a fermentation

The SOM was used to identify different metabolic or physiological states which exist during a baker's yeast fermentation (Fig. 3). The measurement vector (also called feature vector) consists of measurements of the variables which have to be considered for modeling the process. The feature vector for the fermentation is made up of input variables, e.g., feed flow rate, stirrer speed, and state variables such as concentration of biomass, substrate, etc., in the reactor. This high-dimensional vector is projected on the two-dimensional array of nodes. The SOM is then used to visualize and identify the current process state. This study utilized the model of Bellgardt (1991) to simulate the fermentation process. Bellgardt's model is summarized in the next section. Other applications of SOMs in Chemical and biochemical engineering include the works of Aldrich *et. al.* (1995a, 1995b).

5.1. Model of the yeast fermentation

This section presents the dynamic process model that will be used to model the fermentation process. The model consists of a reactor model and a cell model. According to the general structure of models for biotechnological processes, the presented should describe both the reactor system, including gas and liquid phases, and the biological system, yeast. This cell model includes a kinetic model for the metabolic pathways and a dynamic model for the main regulatory systems of metabolism.

The reactor model dynamically describes concentrations in the gas and liquid phases of the reactor as governed by initial conditions, manipulating variables, and biological reactions of the yeast cells. For simplicity it is assumed that both phases are perfectly mixed. A third sub-model describes the mass exchange between gas and liquid phases.

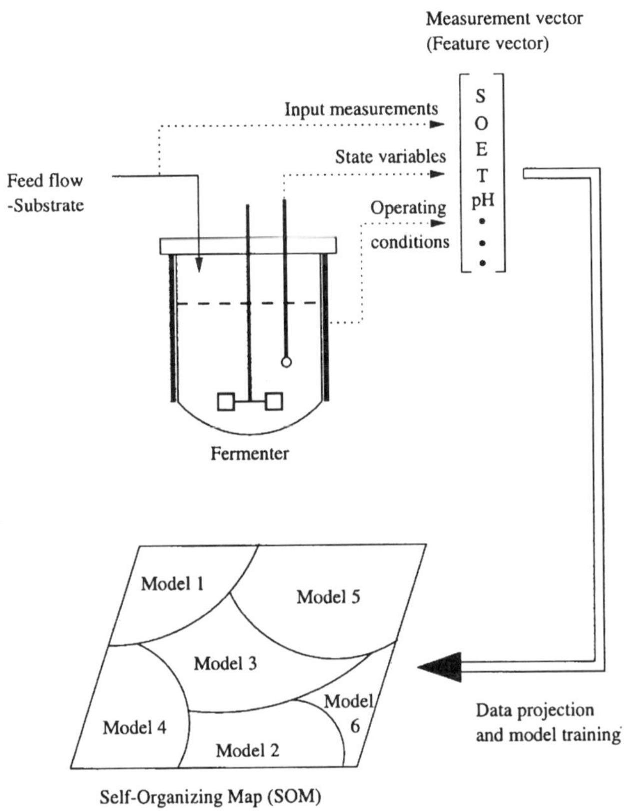

Figure 3. SOM identification of a fermentation

The liquid phase model for the main components is derived from mass balances for the fedbatch reactor. The components considered here are the cell mass *x*, substrate *s*, ethanol *e*, dissolved oxygen *o*, and dissolved carbon dioxide c. The model equations are

$$\frac{dc_x}{dt} = r_x - \frac{F}{V}.c_x \tag{5}$$

$$\frac{dc_s}{dt} = -r_s + \frac{F}{V}.\left(c_{so}^l - c_s\right) \tag{6}$$

$$\frac{dc_e}{dt} = r_e - \frac{F}{V}.c_e + ETR \tag{7}$$

$$\frac{dc_o}{dt} = -r_o + \frac{F}{V}.\left(c_{oo}^l - c_o\right) + OTR \tag{8}$$

$$\frac{dc_c}{dt} = r_c + \frac{F}{V}.\left(c_{co}^l - c_c\right) + OTR \tag{9}$$

$$\frac{dV}{dt} = F \tag{10}$$

The model includes the accumulation in the liquid phase, the biological reactions, and the mass transport which can be divided into two parts. The second term in every equation is due to the inflow of substrate, the third term in the balances for oxygen, carbon dioxide, and ethanol comes from the mass exchange with the gas phase. The substrate flow rate, *F*, is the main manipulating variable of the reactor. It determines the increase in volume of the liquid phase and the related dilution effect for the process variables. The sugar concentration in the feed c_{so}^l is an operating parameter. The reaction rates for cell growth, substrate and oxygen uptake, as well as ethanol and carbon dioxide production are calculated using the cell model. The mass transfer rates for oxygen, carbon dioxide and ethanol, are determined by the mass transfer model. The temperature is assumed to be constant.

The main components of the gas phase are oxygen, carbon dioxide, nitrogen, ethanol, and water. The model equations derived from molar balances of the gas phase components are

$$\frac{dx_o}{dt}=\frac{p_{in}T_lF_g}{pT_{in}V_g}.x_{o_{in}}-\frac{F_{go}}{V_g}.x_o-\frac{RT_lV_l}{M_opV_g}.OTR \tag{11}$$

$$\frac{dx_c}{dt}=\frac{p_{in}T_lF_g}{pT_{in}V_g}.x_{c_{in}}-\frac{F_{go}}{V_g}.x_c-\frac{RT_lV_l}{M_cpV_g}.CTR \tag{12}$$

$$\frac{dx_n}{dt}=\frac{p_{in}T_lF_g}{pT_{in}V_g}.x_{n_{in}}-\frac{F_{go}}{V_g}.x_n \tag{13}$$

$$\frac{dx_e}{dt}=-\frac{F_{go}}{V_g}.x_e-\frac{RT_lV_l}{M_epV_g}.ETR \tag{14}$$

$$\frac{dx_w}{dt}=\frac{p_{in}T_lF_g}{pT_{in}V_g}.x_{w_{in}}-\frac{F_{go}}{V_g}.x_w-\frac{RT_lV_l}{M_wpV_g}.WTR \tag{15}$$

The positive direction of the mass transfer streams, OTR, CTR, ETR, and WTR, is directed to the liquid phase. It is assumed that no nitrogen is exchanged between gas and liquid phases, and that no ethanol is present in the air flow at the inlet. The measured respiratory quotient *RQ* can be calculated under a steady state assumption. It can be used to estimate the metabolic type of growth for the yeast

- $RQ > 1$ indicates growth is fermentative
- $RQ = 1$ indicates respiratory growth on sugar substrate
- $RQ < 1$ signifies growth on ethanol

Because of these properties it is possible to use the respiratory quotient as a control variable for the substrate flow. The mass transfer rate between gas phase and liquid phase is proportional to the concentration gradient to the concentration gradient in the interfacial area and to the volumetric mass transfer coefficient.

$$OTR=\left(k_La\right)_o.\left(c^*{}_o-c_o\right) \tag{16}$$

$$CTR=\left(k_La\right)_c.\left(c^*{}_c-c_c\right) \tag{17}$$

The saturation concentrations for oxygen and carbon dioxide can be calculated according to Henry's Law:

$$c_i^* = \frac{H_i \rho_w M_i p}{M_w} . x_i \qquad \text{for} \quad i = o, \; c . \tag{18}$$

The influence of the stirrer speed on the mass transfer coefficient is modeled using the following equations. The mass transfer for a stirred bioreactor is calculated based on the following equation (Van't Ried, 1983)

$$\left(k_L a\right)_0 = 3600. \left(0.026. \left(\frac{P}{V} \right)^{0.4} . \sqrt{v_{gas}} \right) \tag{19}$$

where the linear gas velocity *vgas* is given by

$$v_{gas} = \frac{4.G_{flow}}{3600.\pi.D^2{}_r} \tag{20}$$

The gas flow rate is denoted by G_{flow}. The geometrical parameters are the diameter of the reactor D_R and the diameter of the stirrer D_s . The power consumption P for mechanical agitation is described by

$$P = P_{no}.\rho. \left(\frac{N_{stir}}{60} \right)^3 .D^5{}_s \tag{21}$$

where P_{no} denotes the power number and ρ is the density of the liquid. The following correlation between the mass transfer for oxygen $(k_L a)_o$, and the effect of temperature and biomass concentration has been suggested by Kristiansen (1994):

$$\left(k_L a\right)_0 = \left(k_L a\right)_o .(1 - 0.00176.c_x).1.022^{(T-20)} . \tag{22}$$

Based on this value, mass transfer for ethanol and carbon dioxide may be calculated according to

$$\left(k_L a\right)_e = \sqrt{\frac{1.28}{2.5} . \left(k_L a\right)_0} \tag{23}$$

and

$$\left(k_L a\right)_c = \sqrt{\frac{1.96}{2.5} \cdot \left(k_L a\right)_0} \,. \tag{24}$$

The cell model as used in this simulation consists of two parts: the metabolic model for kinetics and stoichiometry of growth and the regulation model for metabolic long-term regulation. This study uses the model of Bellgardt called metabolic regulator approach. The uptake of different substrates and the formation of primary metabolites are taken into account. The metabolic regulator is a suitable approach if the product formation depends on the growth condition in the fermenter. During the yeast fermentation, the metabolism can be directed to any mixture of fermentative growth with ethanol formation or oxidative growth with high cell yield, depending on substrate and oxygen. The stoichiometry of growth is described by the following system of equations

$$\begin{bmatrix} r_s \\ 0 \\ 0 \\ 0 \\ m_{ATP} \\ 0 \end{bmatrix} = \begin{bmatrix} -1 & 0 & 0 & 0 & K_{EG} & 0 & 1 & K_{B1} & 0 \\ -2 & 0 & -1 & 0 & -1-K_{Ad} & 1 & 2 & 0 & 0 \\ 0 & 0 & -1 & 1 & 0 & 0 & 0 & K_{B3} & 0 \\ -2 & 2 & 1 & 4 & -1-2K_{EG} & 1 & 2 & -K_{B2} & 0 \\ -3 & 2P/O & 0 & 1 & -2K_{EG} & 0 & 2 & -Y^{-1}{}_{ATP} & 0 \\ 0 & 0 & -1 & -2 & -1-K_{Ad} & 0 & 0 & 0 & 1 \end{bmatrix} \begin{bmatrix} r_g \\ r_o \\ r_{ac} \\ r_{tc} \\ r_{ep} \\ r_{ec} \\ r_s \\ r_x \\ r_c \end{bmatrix} \tag{25}$$

In *25*, K_{EG}, K_{Ad}, K_{B1}, K_{B2}, K_{B3} and are model constants (Bellgardt, *1991), P/O* denotes the effectiveness of phosphorylation, and Y_{ATP} is the yield coefficient of ATP (energy). The inherently rate-limiting steps in the model are the glucose uptake rate q_s, for which a Monod kinetics can be assumed, and the uptake steps for ethanol r_e, and oxygen r_o. The latter terms are introduced as first order kinetics. The optimal pathways for the microorganism is found by maximizing the specific growth rate

$$r_x(t) \Rightarrow \text{ maximum} \tag{26}$$

under the following constraints, which are caused by transport limitations or internal reaction mechanisms:

$$\begin{aligned}
&0 \le r_g \le r_{g\max} \\
&0 \le r_o \le \min\left(K_o.c_o, r_{o_{\max}}\right) \\
&0 \le r_{ac} \le r_{ac_{\max}} \\
&0 \le r_{tc} \le \infty \\
&0 \le r_{ep} \le \infty \\
&0 \le r_{ec} \le K_{E1}.c_e \\
&0 \le r_s \le \infty \\
&-\infty \le r_x \le \infty \\
&0 \le r_c \le \infty
\end{aligned} \tag{27}$$

The dynamic regulation model considers the lag phases of growth during phases of regulatory adaptation expressed in state space representation as

$$\frac{dy(t)}{dt} = \mathbf{F}\left(\mu\right).y(t) + \mathbf{f}(\mu).r_o(t) \tag{28}$$

The elements of the vector $y(t) = \left[r_{o\max}, E_1, E_2\right]^T$ are the maximum specific oxygen uptake rate r_{omax} and two fictitious enzymes E_1 and E_2. The matrix elements of **F** and **f** are nonlinear functions of model parameters and of the specific growth rate μ(t), and can be expressed as follows:

$$\mathbf{F}(\mu) = \begin{bmatrix} -(3.\mu + 0.63h^{-1}) & 1 & 0 \\ -\left(3.\mu^2 - 2.7h^{-1} + 7.9h^{-2}\right) & 0 & 1 \\ -\left(\mu^3 - 13.6h^{-1}.\mu^2 + 9.6h^{-2}.\mu + 1.2h^{-3}\right) & 0 & 0 \end{bmatrix} \tag{29}$$

and

$$\mathbf{f}(\mu) = 0.27.\begin{bmatrix} 1 \\ 2.\mu - 0.36h^{-1} \\ \mu^2 + 10.9h^{-2} \end{bmatrix} \tag{30}$$

This constrained optimization problem can be solved following the steps shown in Fig. 4.

The metabolic regulator approach yields set of metabolic models. Depending on the operating conditions, one of these models is utilized to describe the current growth phenomena of the yeast. The set of metabolic models consists of

- **Model 1:** oxidative growth on glucose.
- **Model 2:** aerobic fermentative growth on glucose (Crabtree effect).
- **Model 3:** anaerobic or oxygen limited growth on glucose.
- **Model 4:** oxidative growth limited by ethanol and glucose.
- **Model 5:** oxidative growth limited by ethanol and acetyl-CoA.
- **Model 6:** oxidative growth limited by glucose and oxygen.
- **Model 7:** oxidative growth limited by glucose and enzymes of gluconeogenesis.

5.2. Preparing training, testing and calibration data

During this simulation, the yeast metabolism did not choose the pathways described by model 1 and model 5. All other models were present during the simulation and had to be identified by the SOM based on the current operating condition. All simulations were carried out in MATLAB and SOM-PAK (Kohonen *et al.*, 1996).

Three network input vectors were used during this study. For the first identification study, the following input vector was used

$$\vec{x}_1 = \left(c_e, c_o, RQ, D, c^l{}_{so}, N_{stir}, G_{flow}\right) \tag{31}$$

where c_e, and c_o, denote the concentrations of ethanol and oxygen, respectively. RQ is the respiratory quotient and D represents the dilution rate. The input vector was supplemented by the substrate concentration c_s, to yield $\vec{x}_2$ for the second identification study. The third input vector $\vec{x}_3$ consisted of vector $\vec{x}_1$, the substrate concentration c_s, and the cell concentration c_x.

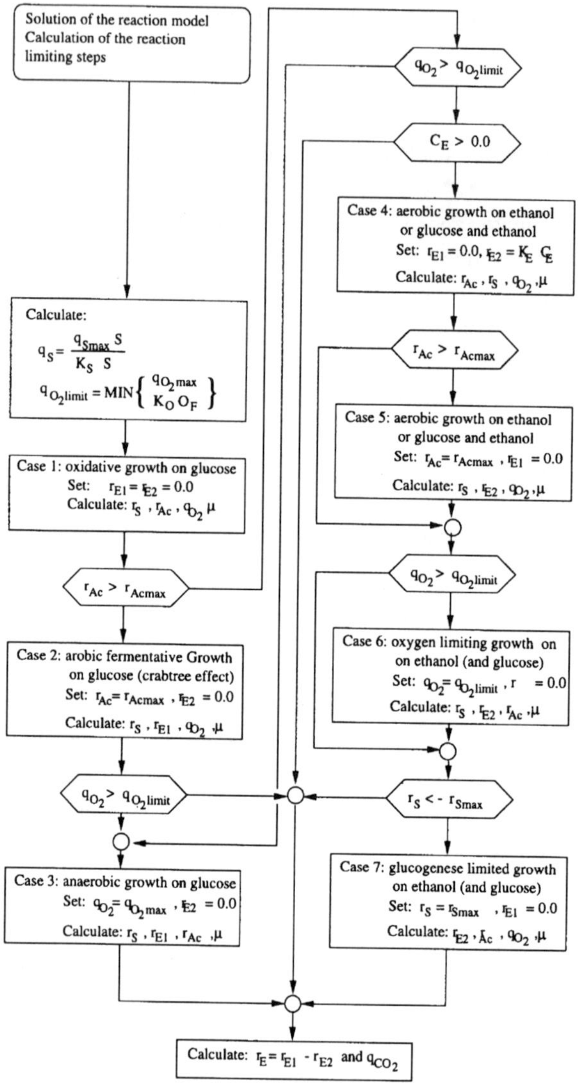

Figure 4 Diagram of the sequential steps for the solution of the reaction model

A training data set was created by simulating 70 fermentations with different operating conditions. These operating conditions were achieved by altering the manipulative variables of the fermentation. The manipulative variables were the substrate feed concentration c^{l}_{so}, the stirrer speed N_{stir}, the gas flow rate G_{flow}, and the feed flow rate F. Noise with a signal to noise ratio of 30 dB was added to the input vector. All variables were assumed to be measurable every 30 minutes. The resulting data, set of 4200 data points was split into a training set (2500 data points), a testing set (500 data points), and a calibration set (700 data points). The calibration data set consists of 140 data points for each of the metabolic models used in the simulation. This includes the data for the input vector and the class labels. These class labels indicate the membership of each data point with respect to the metabolic models. The remaining 500 data points were used to validate SOMs.

5.3. Training the SOM

The SOMs were initialized by randomly selecting data points from the training set. Initializing the model with PCA did not improve the identification result. The size node layer of the SOM was varied during this study: 8×6, 10×8, 14×10, 18×14, and 20×15 nodes were used. Since a preliminary study could not reveal a significant advantage for any particular topology, a rectangular network topology was used in this study. The learning rate decreased linearly. The Gaussian function was used to determined the neighborhood $N_c(i, t)$ at each iteration. Table 1 summarizes the quantization error for the testing set achieved during all 3 identification studies.

Table 1 Quantization error of the SOMs for the testing data set

SOM	Input vector $\vec{x}_1$	Input vector $\vec{x}_2$	Input vector $\vec{x}_3$
8x6	0.225642	0.210497	0.250972
10x8	0.187380	0.181458	0.226515
14x10	0.171176	0.156904	0.183678
18x14	0.155587	0.147505	0.160352
20x15	0.154526	0.146813	0.160125

For all input vectors, the quantization error decreases with an increasing network size. This is to be expected since more reference vectors are placed in the input domain. However, the error difference between the 18×14 SOM and the 20×15 SOM is not very significant. This suggests that a 18×14 SOM is the maximum reasonable network size and that a further increase in size will not improve the model's prediction quality.

5.4. Evaluating the SOM

The SOMs were tested for various fermentation simulations with different operating conditions. The following results correspond to the simulation study with $\vec{x}_2$ as the network input. Figs. 5-8 show the predictions for four test runs and the corresponding quantization errors. The SOM architecture with 14×10 nodes achieved the best predictions for all test runs. None of the networks tested was able to identify model 7. However, model 7 corresponds to a pathway that was rarely selected by Bellgardt's model. The 8×6 architecture did not include any reference vectors and the larger networks assigned 1-3 code-book vectors which were never activated during the testing phase.

The quantization error may be used as a credibility measure of the prediction accuracy (Kohonen, 1995). The larger the quantization error, the greater the distance between the input vector and the BMU. However, for the application presented here, this correlation is visible only for some cases. The quantization error in test run 1 is high between 12 and 18 hours. Correspondingly, all SOM networks derive less accurate predictions during this time. For most other runs, no distinct correlation can be determined.

The resulting SOMs can be visualized using the unified distance matrix representation (Umatrix) as proposed by Ultsch (1993). The calculated distance between adjacent nodes is represented by the shading of the surrounding squares. A dark color between the nodes indicates a large distance between the reference vectors in the input domain. Code-book vectors, which are close together, are separated by light coloring. The light areas can be thought of as clusters while the dark areas correspond to cluster separators. Figs. 9, 10, and 11 show the U-matrix for 3 different network sizes with $\vec{x}_2$ as the input vector.

The 8×6 SOM displays a very coarse resolution of the input domain. No reference vector corresponds to model 7 of Bellgardt's fermentation model. The resolution may be improved by increasing the network size, e.g., to a 14×10 nodes as shown in Fig. 10.

Fig. 10 reveals that some code-book vectors were not labeled while calibrating the SOM. These code-book vectors are marked as black dots. They may be assigned to a particular model by analyzing the labels of adjacent nodes and the distance to these nodes. The 14×10 map includes all 5 metabolic models present in the simulation, although model 7 is represented only by one code-book vector.

The SOM networks also offer the possibility to visualize the relative distribution of each component of the input vector. This is shown for two components of input vector $\vec{x}_2$ mapped onto the 14×10 node array. Fig. 12 and Fig. 13 show the component plane representation for oxygen concentration and RQ values, respectively. Dark colors correspond to relatively small values of the

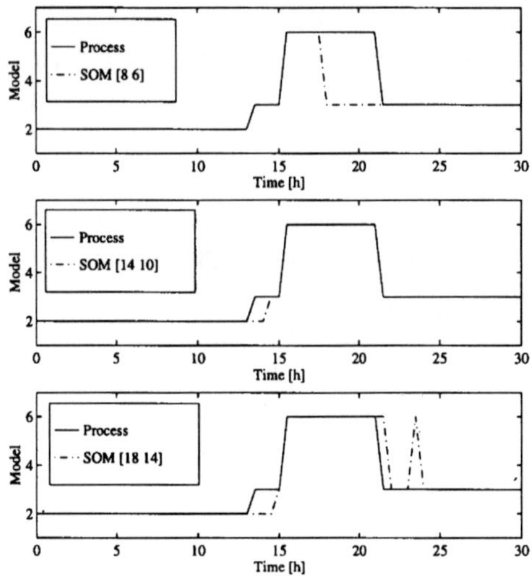

(a) Process and SOM prediction.

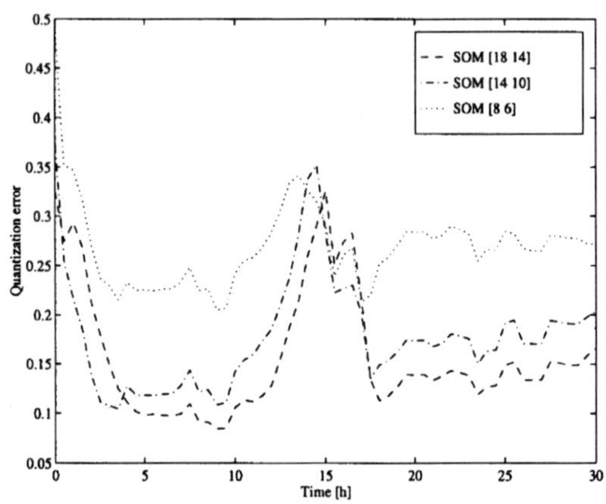

(b) Quantization error.

Figure 5. Simulation results for test run 1

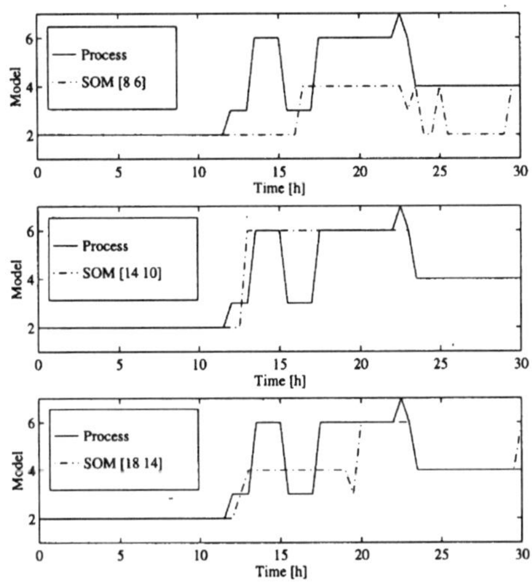

(a) Process and SOM prediction.

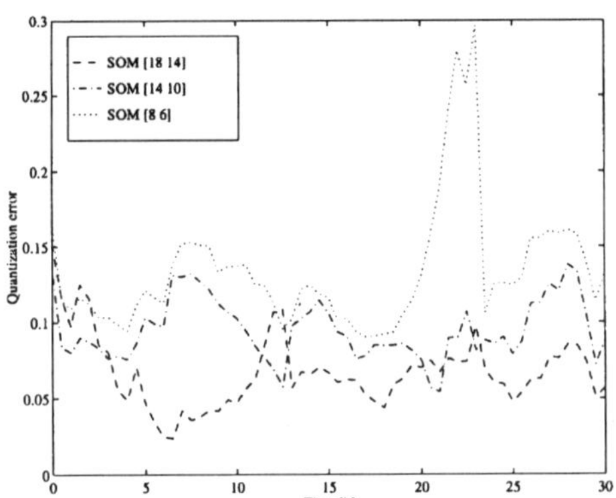

(b) Quantization error.

Figure 6. Simulation results for test run 2

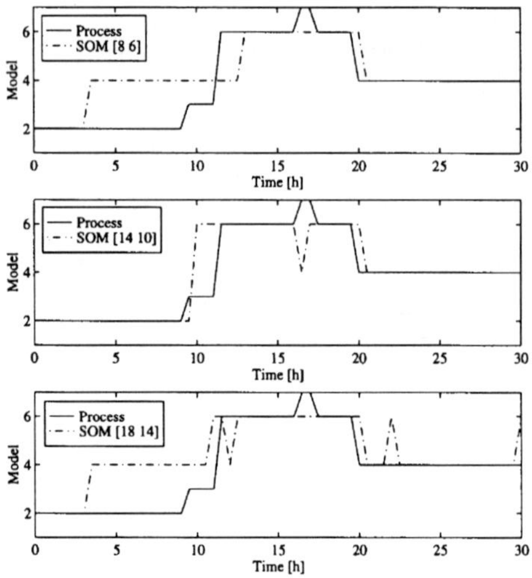

(a) Process and SOM prediction.

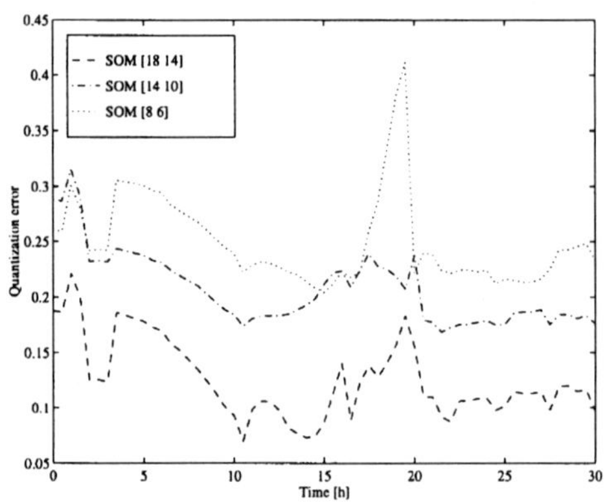

(b) Quantization error.

Figure 7. Simulation results for test run 3

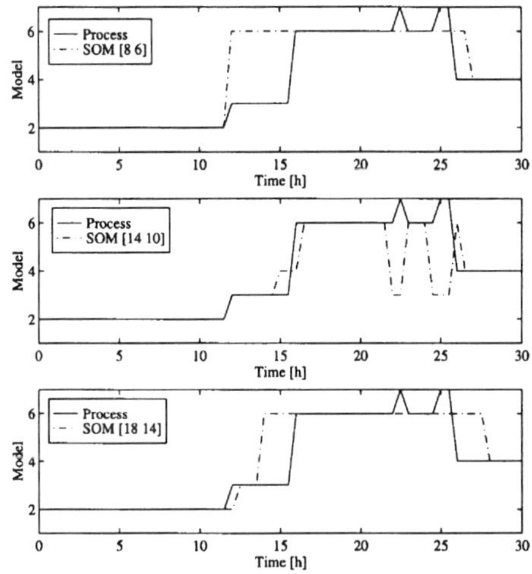

(a) Process and SOM prediction.

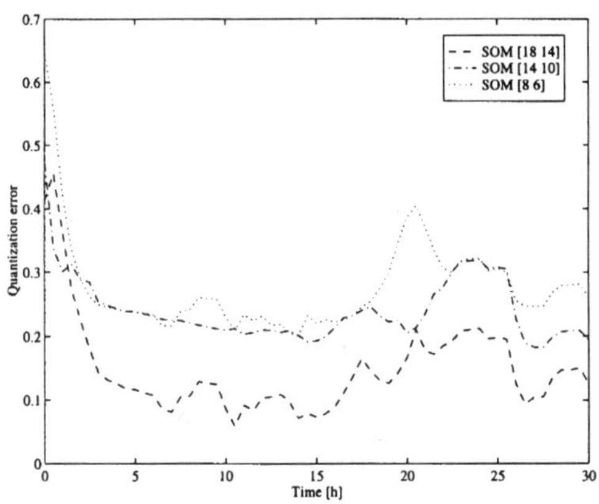

(b) Quantization error.

Figure 8. Simulation results for test run 4

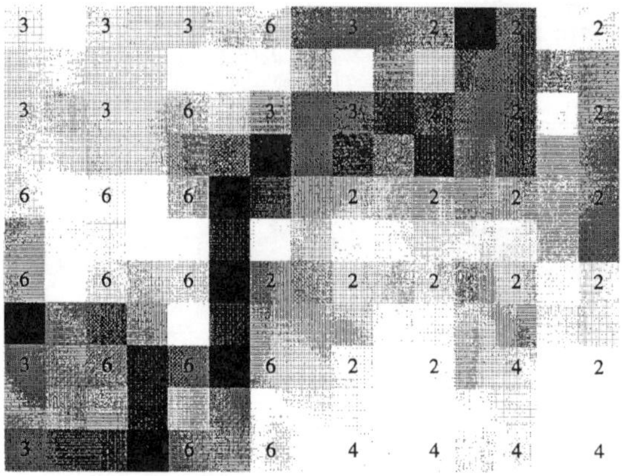

Figure 9. SOM (8×6) network for the second identification study (input vector $\vec{x}_2$)

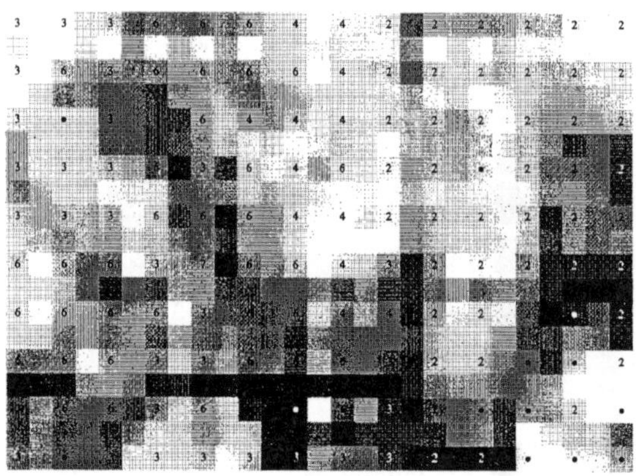

Figure 10. SOM (14×10) network for the second identification study (input vector $\dot{x}_2$)

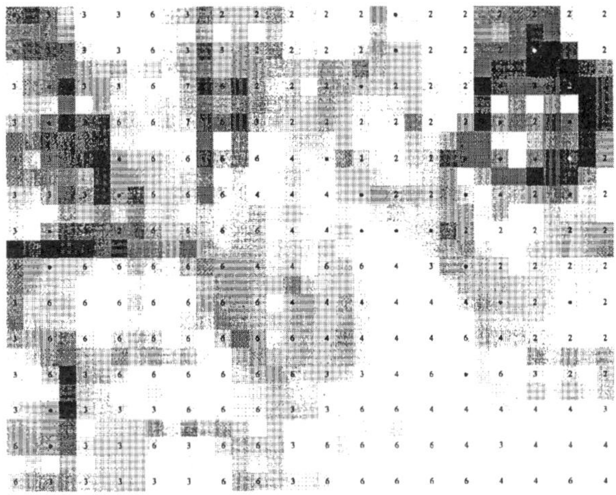

Figure 11. SOM (20×14) network for the second identification study (input vector $\dot{x}_2$)

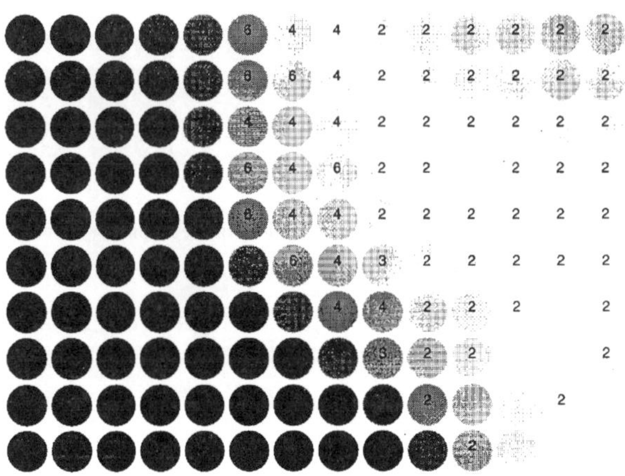

Figure 12. Component plane representation for oxygen concentration

component, while relatively large values are represented by lighter colors. Two conclusions may be derived from this representation. First, by comparing the maps for different components, it can be determined whether the components are correlated. If the sequence of dark/light is similar, then these two variables are correlated. Second, for describing the inverse model, each plane can be analyzed to determine how a desired state of a process may be reached. For instance, Fig. 12 and 13 suggest that the metabolic model 2 may be reached with high oxygen concentration and moderate RQ values.

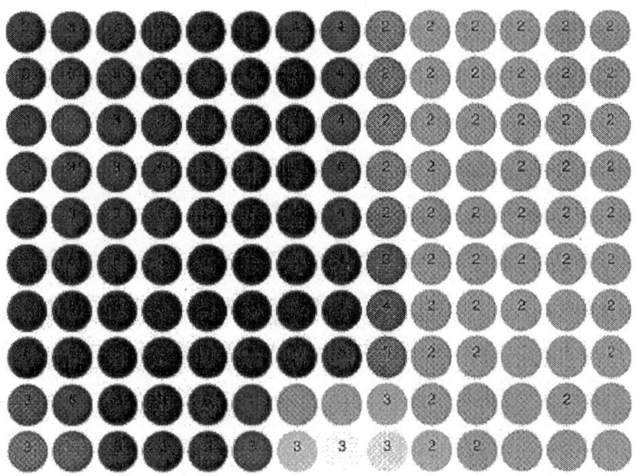

Figure 13. Component plane representation for the respiratory coefficient

6. Conclusion

In this paper, a visualization and identification method using a SOM is presented. The SOM was used to predict the metabolic pathway during a yeast fermentation. The design of appropriate SOM models can be very tedious due to the heuristics involved in the design and training of the map. During the design stage, both the network topology and the size of the map have to be selected. Generally, the appropriate network size has to be determined by trial and error. The training rate function and the neighborhood function are chosen during the training phase. No recommendations for the right choice exist; again a trial and error procedure may have to be used. As demonstrated in this case study, the quantization error achieved during training is not a suitable indicator of the quality of the map. Although the 18x14 map possessed a lower quantization error, the prediction accuracy was lower when compared with the smaller map. For this yeast fermentation simulation, the 14×10 map was found to be the best network size. The model was able to predict the different metabolic models reasonably well.

References

Aldrich C., Moolman D.W., Eksteen J.J. and VanDeventer J.S.J., *Chem. Eng. Comm.* **139** (1995a), 25-39.

Aldrich C., Moolman D.W. and VanDeventer J.S.J., *Comput. Chem. Engng.* **19** (1995b), s803-s808.

Bellgardt, K. H., in *Biotechnology* vol 4, *eds.* Rehm, H. J. and Reed, G. (VCH, Weinheim, 1991), 383-406.

Bishop, C. M., *Neural Networks for Pattern Recognition* (University Press, Oxford , 1995).

Carpenter, G. and Grossberg, S. , *Computer.* **21** (1988), 77-88.

Fukushima, K. (1988, *Neural Networks* **1** (1988), 119-130.

Kangas, J. , *On the Analysis of Pattern Sequences by Self-Organizing Maps,* Ph.D. thesis, (Helsinki University of Technology, Espoo, 1994).

Kohonen, T., *Cybernetics.* **44** (1982), 135-140.

Kohonen, T, *Proceedings of IEEE.* **78** (1990), 1464-1480.

Kohonen, T., *Self- Organizing Maps* (Springer, Heidelberg, Germany, 1995) .

Kohonen, T., Hynninen, J., Kangas, J., and Laaksonen, J., *SOM-PAK: The self-organizing map program package*, Technical report (Helsinki University of Technology, Espoo, 1996).

Kristiansen, B., *Integrated Design of a Fermentation Plant: The Production of Baker's Yeast,* (VCH ,New York, 1994).

Murtagh, F. and Hernandez-Pajares, M., *Journal of Classification.* **12** (1995), 165-189.

Rumelhart, D., Clelland, J., and the PDP Research Group , *Parallel Distributed Processing, Explorations in the Microstructure of Cognition* (MIT Press, Cambridge, MA, 1986).

Ultsch, A., in *Proceedings of the ICANN-93,* eds. Gielen, S. and Kappen, (1993), 864-867.

Van't Ried, K. '*Trends in Biotechnology.* **1**(1983), 113-119.

4. TRAINING RADIAL BASIS FUNCTION NETWORKS FOR PROCESS IDENTIFICATION WITH AN EMPHASIS ON THE BAYESIAN EVIDENCE APPROACH

L. S. KERSHENBAUM, A. R. MAGNI
Centre for Process Systems Engineering
Imperial College, London SW7 2AZ

In this paper we concern ourselves with a Bayesian approach to training radial basis function (RBF) networks, and in particular, nonlinear techniques for proper location of the basis functions in the input space. Thin plate spline networks are trained on simulated data, using both 'unsupervised' and nonlinear techniques for the basis function placements. It is observed that optimisation improves the performance of the networks by allowing the centres to spread well beyond the data space. This challenges the suitability of traditional unsupervised methods, which usually require the basis functions to be located near the input data.

1. Introduction

The radial basis function (RBF) network forms an interesting class of neural network that can be motivated from a number of seemingly disparate concepts. These include, for example, interpolation theory [10][12][13], density estimation [5][6], kernel regression, and regularisation theory [11]. From a systems identification point of view [3][4] show how the RBF network may be interpreted as a NARMAX model, and they discuss an orthogonal least squares approach for training the network. In this paper we are concerned with regression problems, in which the behaviour or output of a system is to be predicted for a given set of inputs. (This is in contrast, to say, density estimation, in which we would like to approximate an arbitrary density that has given rise to a particular distribution of observed data.) We are interested in exploring the suitability of various training schemes, particularly from a Bayesian point of view.

The paper will be divided as follows: section 2 will review the Bayesian evidence approach used for training neural network models; section 3 will then review the application of the evidence scheme specifically to radial basis function networks; section 4 will review 'unsupervised' methods for training RBFs; and in section 5 we will explore nonlinear techniques. An experimental study into the behaviour of the various training schemes is then presented in sections 6 and 7; and finally section 8 will summarise the main findings.

2. Bayesian Inference and The Evidence Scheme

In the Bayesian paradigm we concern ourselves with finding the most plausible set of parameters for a given set of observed data. If we denote the free parameters of a neural network by $\boldsymbol{w}$, the data by D, and the neural network model by H (which contains information on the type of network, and the number of basis functions, etc.), we can state our purpose thus: 'For a given neural network H what are the values of the parameters $\boldsymbol{w}$ that most plausibly explain the given data D?'. That is, we are seeking to maximise the probability $p(\boldsymbol{w}|D,H)$. This can be more conveniently expressed in the form of *Bayes rule*, which expresses the distribution $p(\boldsymbol{w}|D,H)$ in terms of other, more accessible distributions, as follows:

$$p(\boldsymbol{w} \mid \mathrm{D}, \mathrm{H}) = \frac{p(D \mid \boldsymbol{w}, H)\, p(\boldsymbol{w} \mid H)}{p(D \mid H)} \tag{1}$$

Here, $p(D|\boldsymbol{w},H)$ is known as the *likelihood function* and it expresses the probability that the data were generated for a given set of parameters $\boldsymbol{w}$. Clearly if this probability is high, the neural network solution is passing very close to the *target*, or *output*, values. The distribution $p(\boldsymbol{w}|H)$ is called a *prior* on the parameters since it summarises our prior belief/constraints about the parameters *before* the data D is seen. $p(\boldsymbol{w}|D,H)$ is therefore known as the *posterior* distribution, since it represents the belief in the parameter set *after* the arrival of information D. The final term in Eq. 1, $p(D|H)$, is called the *evidence* for the model H and it behaves as a normalisation constant that ensures that $\int p(\boldsymbol{w}|D,H)\, \mathrm{d}\boldsymbol{w} = 1$.

Equation 1 represents a particularly elegant mechanism for inference. At the outset we state our prior belief (or have an initial guess) in the final values of the parameters. As the data is presented to the network this prior belief is modified so that our *posterior* belief now reflects both the initial assumptions and the information contained in the data. Therefore, *the likelihood function is the function through which our prior knowledge or assumptions are updated by the data.*

MacKay [8] defines two levels of Bayesian inference: a) model fitting, and b) model comparison. The first level is well specified by Eq. 1 since it concerns itself with finding appropriate parameters for a given model and data set. However, it cannot address the issue of finding a 'good' model H, and if H is poor the solution will be poor irrespective of how large a posterior is generated.

The remainder of this section will be subdivided into different subsections reviewing the choice of likelihood function and prior. In the final subsection the first level of inference is explored—finding good parameters.

2.1. The Likelihood Distribution

The likelihood function $p(D|\boldsymbol{w},H)$ measures the probability that a particular neural network H with parameters $\boldsymbol{w}$ is responsible for the observed data D. The data is a fixed set of information in the form of input/output pairs $D = \{t_n, \boldsymbol{x}_n\}$, where t_n are the *target* values or *outputs* associated with the *inputs* $\boldsymbol{x}_n$. (Note that the output is assumed to be a scalar whilst the input can be a multivariate vector.)

One of the primary aims of neural computing is to find values of $\boldsymbol{w}$ such that a neural network will produce an output $y_n \equiv y(\boldsymbol{w},\boldsymbol{x}_n)$ that is similar to the target value t_n. In this respect the likelihood distribution should contain large probability in the vicinity of $\boldsymbol{w}$ when there is a good correspondence between t_n and y_n; likewise little probability around $\boldsymbol{w}$ when there is a large mismatch between t_n and y_n. It is maximised when $y_n = t_n$, $\forall\, n$. A popular choice for the likelihood is the Gaussian

$$p(D\,|\,w,H) = \frac{1}{Z_D(\beta)}\exp(-\beta E_D) \tag{2}$$

E_D is the sum-of-squares term $E_D = 0.5\,\Sigma_n\,(y_n\text{-}t_n)^2$, and $1/\beta$ is the variance between the solution and the target data. The term $Z_D(\beta)$ is a normalisation constant given by $Z_D(\beta)=(2\pi/\beta)^{N/2}$. The variable β is an unknown quantity that must be inferred from the data, and the likelihood $p(D|\boldsymbol{w},\beta,H)$ is therefore conditioned on it. Parameters like β are called *hyperparameters* because they are variables whose distribution dictates the distribution of other free parameters in the network.

The use of a Gaussian density implies that any mismatch between y_n and t_n will resemble a Gaussian noise signal, or white noise. (The likelihood is therefore sometimes referred to as a *noise model.*) If a different distribution of the errors $(y_n\text{-}t_n)$ is required then the likelihood function in Eq. 2 may be inappropriate, though for most intents and purposes it suffices well.

2.2. *Why The Likelihood Is Not Maximised*

Maximising the likelihood of a neural network can be done equivalently by minimising its negative logarithm, thus: $\max\{p(D|\mathbf{w},\beta,H)\} \equiv \min\{-\ln p(D|\mathbf{w},\beta,H)\} = \min\{0.5\ \Sigma_n (y_n - t_n)^2\}$ (where in the last equality the constants have been omitted). Maximising the likelihood is therefore equivalent to minimising a sum-of-squares objective function, which is minimised when the solution passes itself exactly through the data ($y_n = t_n$, $\forall\ n$). Since neural networks are 'semi-parametric' models 1 they will, if sufficiently complex, pass themselves very closely through the data. If the data in turn contain measurement errors, we are in danger of fitting these errors (a process known as *overfitting*), which we would otherwise rather ignore. Minimising a sum-of-squares objective function has been the mainstay of training algorithms in the traditional neural network community, and goes a long way to explaining some of the more troublesome difficulties these methods encounter.

2.3. *The Prior Distribution*

Of primary concern is the construction of neural network solutions that are representative of the data. Certainly a large portion of this responsibility lies with the likelihood distribution. However, it has been noted that use of this term alone is deficient in that it tends to capture noise from (i.e. overfit) the dataset. Overfitting is produced when excessive gradients are allowed in the solution (because noise or errors will require a solution with steeper surfaces), which are in turn encouraged by the presence of large weight magnitudes in the neural network (this is true in particular of linear weights). Therefore, the weights should be smaller rather than larger if they are to produce smooth solutions, robust to noise. A popular expression for this prior constraint is given by the exponential distribution

$$p(\mathbf{w}\,|\,\alpha, H) = \frac{1}{Z_W(\alpha)}\exp(-\alpha E_W) \qquad (3)$$

This is sometimes called a *weight-decay* prior, where E_W is sometimes known as a *weight-decay regulariser* and takes the form $E_W = 0.5\ \Sigma_i\ w_i^2$. The normalisation constant $Z_W(\alpha)$ is given by $Z_W(\alpha) = (2\pi/\alpha)^{K/2}$, where K are the number of weights in the network. The hyperparameter α is again an unknown quantity, and the prior $p(\mathbf{w}|\alpha,H)$ is therefore conditional on it. It must be stressed that not all elements of the parameter set w are included in the prior—only the weights. Other free parameters in the network, which are not weights, are excluded from the weight-decay term E_w. Practically this implies that these free parameters have a vague or uninformative prior distribution, and that they are therefore free to adopt any value in light of the data. This is discussed in greater detail in section 3, where any remaining confusion is hopefully clarified.

This form of prior has a simple and elegant interpretation. It suggests that the final weights will be normally distributed with zero mean (since we do not prefer positive or negative values) and $1/\alpha$ variance. Under the joint action of the likelihood and prior, the weights are not free to adopt arbitrary values, and provided the terms α and β are chosen appropriately, smooth network solutions can be achieved.

2.4. The First Level Of Inference: Finding Good Parameter Values

The analysis thus far has been complicated by the introduction of the unknown scale parameters α and β in the prior and likelihood distributions respectively. As such, the posterior will also be conditional on the hyperparameters, and written in the Bayes form

$$p(w \mid \alpha, \beta, D, H) = \frac{p(D \mid w, \beta, H)\, p(w \mid \alpha, H)}{p(D \mid \alpha, \beta, H)} \tag{4}$$

Inference about the parameters in this form is not very useful because of the conditioning on the unknown and variable hyperparameters. There are two approaches for dealing with this problem. The first, known as the maximum a posteriori (MAP) procedure [2][15], marginalises the hyperparameters and obtains an expression for the true posterior distribution $p(\mathbf{w}|D,H)$. That is, the true prior is determined from the integral $p(\mathbf{w}|H) = \int p(\mathbf{w}|\alpha,H)p(\alpha|H)\, d\alpha$; similarly for the likelihood. With expressions for the true prior and likelihood, inference about the parameters can be performed using the true posterior $p(\mathbf{w}|D,H) \propto p(D|\mathbf{w},H)p(\mathbf{w}|H)$. An alternative scheme, known as the evidence approach 8, assumes that the posterior hyperparameter distribution $p(\alpha,\beta|D,H)$ is sharply peaked about some most probable values α^* and β^*. In light of the data there are assumed to be a set of most probable hyperparameters, α^* and β^*, such that the true posterior can be approximated as $p(\mathbf{w}|D,H) \approx p(\mathbf{w}|\alpha^*,\beta^*,D,H)$.

It is now necessary to infer the most plausible values for the hyperparameters from the data. The posterior hyperparameter distribution is expressed in the Bayes form

$$p(\alpha,\beta \mid D,H) \propto p(D \mid \alpha,\beta,H)\, p(\alpha,\beta \mid H) \tag{5}$$

$p(\alpha,\beta|H)$ is the prior belief in the hyperparameters, which we will assume to be uninformative (no preference is specified for particular hyperparameter values *a priori*). The posterior can now be expressed as $p(\alpha,\beta|D,H) \propto p(D|\alpha,\beta,H)$. Therefore, the most plausible posterior hyperparameter values are equivalent to the hyperparameters that maximise the evidence $p(D|\alpha,\beta,H)$. Now, the evidence is the normalisation function in Eq. 4, and can be expressed as

$$p(D \mid \alpha,\beta,H) = \int_{\text{all } \boldsymbol{w}} p(D \mid \boldsymbol{w},\beta,H)\, p(\boldsymbol{w} \mid \alpha,H)\, \mathrm{d}\boldsymbol{w} \tag{6}$$

$$= \frac{1}{Z_D(\beta)\, Z_W(\alpha)} \int_{\text{all } \boldsymbol{w}} \exp(-\beta E_D - \alpha E_W)\, \mathrm{d}\boldsymbol{w} \tag{7}$$

At this point a simplifying approximation is required to solve the above integrand analytically—that the posterior parameters distribution is Gaussian distributed. This is equivalent to assuming that the function $S(\boldsymbol{w}) = \beta E_D + \alpha E_W$ is quadratic. $S(\boldsymbol{w})$ can therefore be written as a second order Taylor expansion about the most probable parameters $\boldsymbol{w}^{mp}$, thus: $S(\boldsymbol{w}) = S(\boldsymbol{w}^{mp}) + 0.5\Delta\boldsymbol{w}^{\mathrm{T}}\boldsymbol{A}\Delta\boldsymbol{w}$. $\boldsymbol{A}$ is the Hessian $\boldsymbol{A} = \nabla\nabla^{\mathrm{T}}S(\boldsymbol{w}^{mp}) = \beta\nabla\nabla^{\mathrm{T}}E_D^{\ mp} + \alpha\nabla\nabla^{\mathrm{T}}E_W^{\ mp}$; and $\Delta\boldsymbol{w} = \boldsymbol{w} — \boldsymbol{w}^{mp}$. Eq. 7 can therefore be expressed in the analytical form

$$p(D \mid \alpha,\beta,H) = \frac{\exp(-S(\boldsymbol{w}^{mp}))}{Z_D(\beta)\, Z_W(\alpha)} \int_{\text{all } \boldsymbol{w}} \exp(-0.5\Delta\boldsymbol{w}^{\mathrm{T}}\boldsymbol{A}\Delta\boldsymbol{w})\, \mathrm{d}\boldsymbol{w} \tag{8}$$

$$= \frac{\exp(-S(\boldsymbol{w}^{mp}))}{Z_D(\beta)\, Z_W(\alpha)} (2\pi)^{K+M/2} \mid \boldsymbol{A} \mid^{-1/2} \tag{9}$$

The integrand in Eq. 8 is a multivariate Gaussian distribution with covariance matrix $\boldsymbol{A}^{-1}$, whose solution is $(2\pi)^{K+M/2}|\boldsymbol{A}|^{-1/2}$ (multivariate version of the normalisation constant $(2\pi\sigma^2)^{1/2}$ in one dimension). It is raised to a factor of $K+M$ because the Hessian $\boldsymbol{A}$ contains second-order details of the K weights and M possible free parameters that are not weights. The vagueness of this last statement will be clarified in section 3. For now the general framework is presented, and when the specifics of radial basis function networks are considered the interpretation of M will be clearer.

We are seeking to locate the maximum of the evidence, which can be solved equivalently by minimising its negative logarithm, $-\ln p(D|\alpha,\beta,H)$. By setting the derivatives of this negative logarithm with respect to the hyperparameters to zero, one finds the following result [1,9,14]

$$\alpha = \frac{\gamma}{2E_W^{mp}} \tag{10}$$

$$\beta = \frac{N - \gamma - M}{2E_D^{mp}} \tag{11}$$

where

$$\gamma = \sum_{j=1}^{K+M} \frac{\eta_j - \alpha}{\eta_j} [\boldsymbol{U}^{\mathrm{T}} \boldsymbol{I}_1 \boldsymbol{U}]_{jj} \tag{12}$$

η_j are the eigenvalues of the Hessian $\boldsymbol{A}$ and $\boldsymbol{U}$ is a matrix containing the eigenvectors of $\boldsymbol{A}$; $\boldsymbol{I}_1$ is a zero matrix except for possible diagonal entries of 1 if the diagonal corresponds to a weights that is included in the prior of Eq. 3. Each diagonal element in $\boldsymbol{I}_1$ refers to a particular parameter in $\boldsymbol{w}$, and takes a value of 1 if the parameter is a weight that is included in the prior. For example, imagine that $\boldsymbol{w}$ is a vector of $K+M$ entries where the first K elements are the network weights (and which are included in the prior). The matrix $\boldsymbol{I}_1$ will have its first K diagonal entries equal to 1, and the remainder being 0. N are the number of samples in the training data, and E_D^{mp} and E_W^{mp} are the sum-of-squares term and weight-decay regulariser evaluated at the most probable weights $\boldsymbol{w}^{mp}$ respectively. The term γ has an intuitive interpretation as the number of well determined parameters in the network. It represents the number of weights that can be determined from the information in the data. If there is copious data that is uncorrupted by noise then we would expect γ to approach K. If, on the other hand, there is significant measurement error or relatively small amounts of data, we would not expect there to be sufficient information to determine all of the weights fully, and $\gamma < K$ [1][8].

Eqs. (10)—(12) represent part of an iterative process between hyperparameter updates and weight updates, for which the following algorithm can be implemented:

1. Initialise the hyperparameter α to some small value (e.g. 1×10^{-2}), and β to some moderate value (say 10 or 100). The motivation for these settings is that the

parameters are initially unconstrained, and some initial solution is found close to the maximum likelihood solution.

2. Hold the hyperparameters fixed, and update the posterior parameters. Since in RBF networks the weights are linear, they can be optimised in a single step using standard pseudo-inverse methods. Other free parameters are not necessarily linearly related to the output and may require nonlinear methods to update. In the case of nonlinear parameters, several updates (of the order of 10—100) are usually performed before proceeding to the next step.
3. Using Eqs. (10)—(12), update the hyperparameters using the parameters calculated in step 2.
4. Repeat steps 2 and 3 until there are no further changes in the parameters and hyperparameters.

3. The Evidence Framework As Applied To Radial Basis Function Networks

This section deals primarily with two issues: 1) which parameters of a radial basis function network are included in the Bayesian inferencing scheme, and 2) a more precise definition of the prior presented in section 2. In particular, the meaning of 'parameters that are not weights', and the mechanism of 'including' and 'excluding' parameters from the prior.

3.1. Which Parameters Require A Posterior Description?

Put simply, any parameter that is being inferred from observation (data) should be described by a posterior belief. The general form of a radial basis function network can be written

$$y(x) = \sum_{k=1}^{K} w_k \phi(\| \boldsymbol{x} - \mu_k \|) + w_0 \tag{13}$$

w_k are the final layer weights and w_0 is the final layer bias. The basis functions $\phi(\cdot)$ depend on the Euclidean distance between the input $\boldsymbol{x}$ and some *location* in the input space μ_k. Now, in the RBF literature there are a plethora of techniques for finding appropriate locations μ_k, and not all of them can be regarded as inference from observation. For example, finding basis function locations using random subsets of data, the orthogonal least squares algorithm [3][4], or random locations in the input space cannot be considered inference. They do not consider the plausibility or utility of small changes in the locations. The basis functions are fixed at certain points, considered no further, and the weights w_k are then optimised. In these types of training algorithms the only free parameters about which inference is being made are the weights and the final layer bias. In this case the number of parameters that are not weights is $M = 1$ (the bias term). The bias ensures that the mean neural network solution is the same as the mean target data, and we do not believe that subjecting it to an informative prior is useful. The prior prefers smaller rather than larger values, and would compromise the bias' ability to compensate for the difference between the mean network solution and target data. Hence the dimension of the network Hessian is $K+1$.

Alternative approaches to basis function locations permit free movement in the input space, and find their 'optimal' locations by continual assessment of the utility of small changes in the locations. These include clustering algorithms, Expectation Maximisation (EM) algorithms, and nonlinear approaches. In these cases inference is clearly being made about the suitability of different basis function locations, and as such μ_k should form part of the posterior parameters w (in which case $M = K \times D_{in}+1$ where D_{in} is the dimension of the input space, and the dimension of the Hessian A is $K(D_{in}+1)+1$). However, μ_k do not have an obvious consequence to overfitting, and as such are not subjected to an informative prior of the form in Eq. 3.

3.2. A Closer Look At The Prior

It has been argued that different parameters are included in the posterior inference depending on the training strategy employed. Irrespective of this, however, the prior of Eq. 3 is informative only for the weights w_i, and uninformative for all other parameters. By 'uninformative' it is meant that no prior bias is applied to these parameters, and that they can plausibly adopt any value in light of the data. For present purposes an uninformative prior is flat or constant. Eq. 3 is then strictly written

$$p(w \mid \alpha, H) \propto \frac{1}{Z_W(\alpha)} \exp(-\alpha E_W) \tag{14}$$

The $\propto$ arises from the uninformative (constant) prior associated with parameters that are not weights (the bias term, and possibly the basis function locations).

4. Unsupervised Methods For Training RBF Networks

Most of the popular training methods for RBF networks rely on the centres being located near the data points in the input space. There are a number of techniques that aim to distribute the centres according to the input distribution, so regions of greater data density are more likely to have basis function centres located within them. Since these methods locate basis functions according to the input distribution, and without direct reference to the target values, they are usually called *unsupervised* methods. For example, in density estimation, RBF centres are commonly located using either the *means clustering* or *expectation-maximisation* (EM) algorithms [5]. Means clustering looks for convenient clusters of data and places an RBF centre in the middle of each cluster. The EM algorithm determines the parameters of the network by maximising the likelihood that the observed data were generated from the network. It is an iterative procedure that is a popular alternative to nonlinear methods. Unfortunately, it is restricted to RBFs that are proper densities only (such as the Gaussian). Neither method makes any reference to the target values. Fig. 1 shows typical solutions of networks trained using the k-*means* clustering technique. In all, 10 Gaussian and 10 thin plate spline networks were selected from larger sets of 100 networks, and the best and worst cases (in a sum-of-squares sense) are plotted. Similar results were achieved for the EM algorithm. Clearly these unsupervised methods are not well suited to regression problems because of their disregard for the output data.

An unsupervised version of the evidence approach can be performed by training a set of randomly selected networks and choosing those with the largest evidences. This often leads to an improvement over purely unsupervised methods (such as clustering methods and the EM algorithm) because the evidence takes account of the model fit (i.e. the target values) via the likelihood function. So although the evidence does not use the target values to decide the basis function locations directly, it does use them in ascertaining the goodness-of-fit of the model, and we can subsequently expect more reliable modelling than purely unsupervised methods would allow. However, the method is still *ad hoc* and scales poorly with dimension, because the probability of finding good centre placements randomly reduces exponentially with the dimension of the input space.

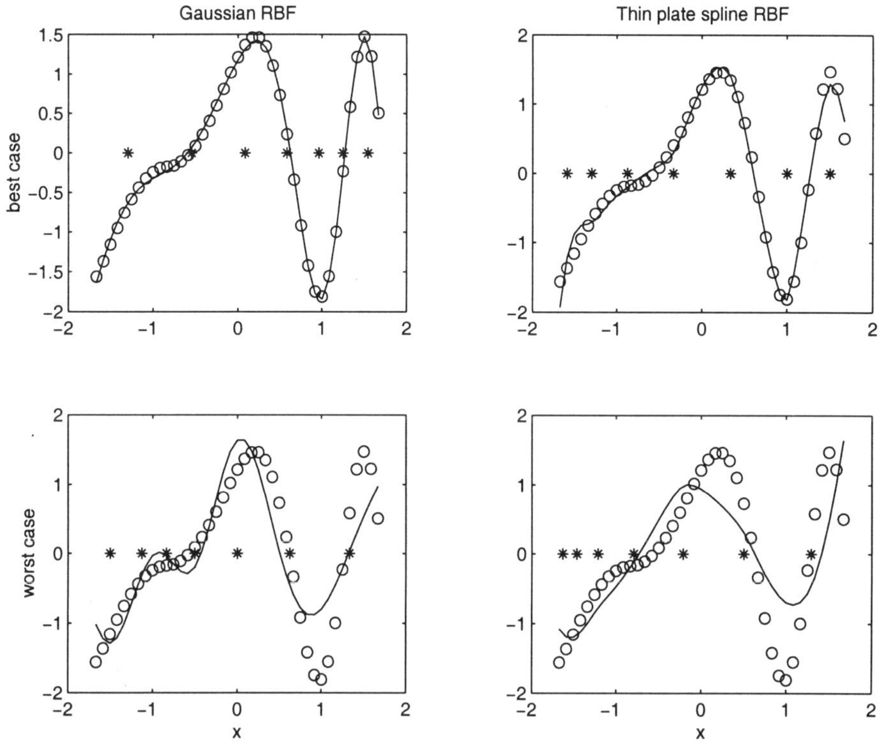

Figure 1. Examples of different neural networks trained using an unsupervsied clustering algorithm. The disregard for the target values makes these schemes unreliable. The target data are plotted as °, the location of the network centres are depicted as *, and the network solution is plotted as the solid line.

In the regression community, researchers have been quick to observe that RBFs trained via these unsupervised approaches do not perform as well as the popular multi-layer perceptron (MLP) network. This suggests that RBFs ought to be trained using gradient-based methods.

5. Supervised Methods For Training RBF Networks

The conditional evidence $p(D|\alpha,\beta,H)$ for a model H is given by Eq. 9. Expressing this as its negative logarithm (so it may be minimised like a regular error function)

$$-\ln p(D\,|\,\alpha,\beta,H) = \beta E_D^{mp} + \alpha E_W^{mp} + \frac{1}{2}\ln|A| - \frac{N}{2}\ln\beta - \frac{K}{2}\ln\alpha + \frac{N-M}{2}\ln 2\pi \quad (15)$$

The gradient of Eq. 15 with respect to a single location parameter in a single dimension, μ_{jk}, (denoted for simplicity by ∇_μ) can be approximated as

$$-\nabla_\mu \ln p(D\,|\,\alpha,\beta,H) \approx \beta\nabla_\mu E_D^{mp} \quad (16)$$

where it has been assumed that the gradients ∇_μ ln $|A|$ are negligible. It can be argued that when the centre locations are ill matched to the data the sum-of-squares gradient $\beta\nabla_\mu E_D^{mp}$ is expected to strongly dominate the location updates. As the solution approaches its optimum the volume measure ln $|A|$ is not expected to change much[1]. The gradient of a sum-of-squares objective function with respect to a location parameter in a single dimension, μ_{jk}, is given as

$$\frac{\partial E_D}{\partial \mu_{jk}} = \sum_{n=1}^{N} w_j(y_n - t_n)(\mu_{jk} - x_{nk})\frac{1}{r_{jn}}\frac{\partial\phi(r_{jn})}{\partial r_{jn}} \quad (17)$$

The identity $r_{jn} = \|\mu_j - x_n\|$ has been used where x_{nk} is the kth dimension of the nth input. The term $\frac{\partial\phi(r_{jn})}{\partial r_{jn}}$ depends on the basis function being used, which, for a thin plate spline function ($\phi(r) = r^2\ln r$), is given by

$$\frac{\partial\varphi(r_{jn})}{\partial r_{jn}} = r_{jn}(2\ln r_{jn} + 1) \quad (18)$$

6. Application To Simulated Datasets

We are now interested in applying the results of sections 2 and 3 to some typical identification problems. Performance on two datasets will be investigated:

A simulated continuous stirred-tank reactor (cstr) system. A first order, liquid phase, irreversible, exothermic reaction occurs in a constant volume stirred-tank

---

[1] The 1-dimensional version of ln $|A|$ is -ln σ, which is a measure of the total volume encompassed by the (unnormalised) Gaussian distribution.

reactor. The objective of this identification is to predict the one-step-ahead temperature profile, given measurements of the temperature, concentration of reactants, and temperature of the coolant in the surrounding cooling jacket.

The simulated forward kinematics of an 8-link all-revolute robot arm. The task associated with these datasets consists of predicting the distance of the end-effector from a target, given the angular positions of the joints. The dataset comes from the DELVE archive, maintained at the University of Toronto and freely available at http://www.cs.utoronto.ca/~delve/.

The remainder of this section will be divided into 2 subsections, where the cstr and robot arm (or *kin*, for *kinematic*) datasets will be discussed in greater detail. Section 7 will then discuss the results obtained for the subsequent optimisation of the radial basis function networks.

6.1. The Cstr System

The cstr system is a model of a simple first-order reaction inside a continuous-stirred-tank reactor system. The reaction is in the liquid-phase, exothermic and irreversible; the volume of liquid inside the reactor is assumed constant and perfectly mixed at all times; and the temperature inside the reactor is moderated by a cooling jacket surrounding the tank through which water of a certain flowrate passes.

The reaction kinetics for the first-order reaction $A \xrightarrow{k0} B$ (where A is the reactant and B is the product) can be expressed as,

$$\dot{C}_A = -k(T)C_A, \quad k(T) = k0\exp\left(-\frac{E}{RT}\right) \tag{19}$$

C_A is the concentration of reactant A (mol.m^{-3}); $k(T)$ is the rate of reaction where $k0$ is the Arrhenius pre-exponential factor (min^{-1}); E is the activation energy (J.mol^{-1}); R the universal gas constant (J.mol^{-1}.K^{-1}); T is the temperature (K). It follows that the overall mole balance for A in the reactor is given by

$$V\dot{C}_A = -Vk(T)C_A + Q(C_{A,f} - C_A) \tag{20}$$

V is the volume (m^3), Q the volumetric flowrate into and out of the reactor ($m^3.min^{-1}$), and C_{Af} is the concentration of A in the feed stream. A similar heat balance yields

$$\dot{T} = -\frac{\Delta H}{\rho C_p} k(T) C_A + \frac{Q}{V}(T_f - T) + \frac{U_r A_r}{\rho C_p V}(T_c - T) \tag{21}$$

ΔH is the heat of reaction ($J.mol^{-1}$), ρ the reactant density ($kg.m^{-3}$), C_p the specific heat capacity ($J.kg^{-1}.K^{-1}$), T_f the temperature of the feed stream, T_c the temperature of the coolant stream in the jacket, U_r the heat transfer coefficient ($J.min^{-1}.m^{-2}.K^{-1}$), and A_r the heat transfer area (m^2).

Table 1. Relationships used to reduce the cstr model into dimensionless form.

Dimensionless Quantity	Relationship	Nominal Value
Activation Energy	$\boldsymbol{e} = E/RT_{f0}$	20
Damkohler Number	$\boldsymbol{D} = Vk_0\exp(-\boldsymbol{e})/Q_o$	0.11
Heat of Reaction	$\boldsymbol{h} = -\Delta HC_f\boldsymbol{e}/(\rho C_p T_{fo})$	7
Heat Transfer Coef.	$\boldsymbol{c} = U_r A_r/(\rho C_p Q_o)$	0.5
Volumetric flowrate	$\boldsymbol{q} = Q/Q_o$	1

Hussain [7] reduces the above model to dimensionless form via the relationships in Table 1. The quantities T_{fo} and Q_o are the nominal values of the feed temperature and volumetric flowrate respectively. The operating points for the dimensionless quantities were chosen in [7] to represent a reasonably nonlinear operating region for the system. Under these transformations, Eqs. (20) and (21) can now be written in dimensionless form

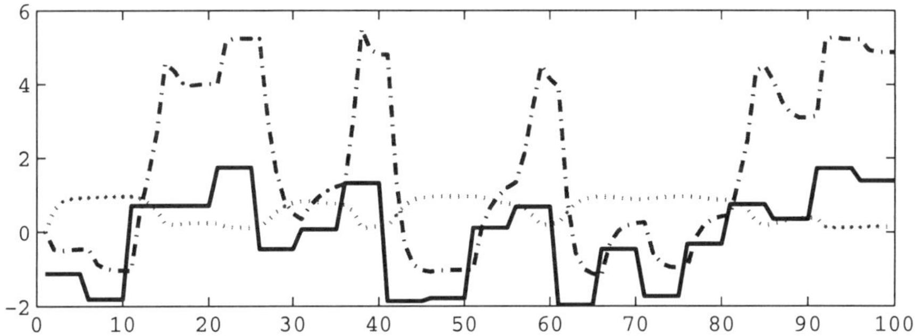

Figure 2. Training data for the cstr system. The dimensionless coolant temperature u is plotted as the solid line, the dimensionless reactor temperature x_2 as (– ·), and the dimensionless reactor concentration x_1 as (…).

$$\dot{x}_1 = -\boldsymbol{D}x_1K(x_2) + \boldsymbol{q}(1 - x_1) \tag{23}$$

$$\dot{x}_2 = \boldsymbol{hD}x_1K(x_2) - (\boldsymbol{q} + \boldsymbol{c})x_2 + u + v \tag{24}$$

$K(x_2) = \exp(\mathbf{e}x_2/(\mathbf{e}+x_2))$. Here $x_2 = \mathbf{e}(T-T_{fo})/T_{fo}$ is the dimensionless reactor temperature, $x_1 = C/C_o$ the dimensionless concentration, $u = \mathbf{ec}(T_c-T_{fo})$ is the dimensionless temperature of the cooling medium and $v = \mathbf{eq}(T_f-T_{fo})/T_{fo}$ is the dimensionless feed temperature. Finally, the derivatives of Eqs. (22) and (23) are with respect to the dimensionless time $\tau = Q_0t/V$.

The objective of this study is to model the temperature profile of the reactor. For simplicity it is assumed that the feed temperature to the plant is always the nominal temperature T_{fo}, and therefore $v = 0$.

6.1.1. The Training Data

The training data was generated by subjecting the dimensionless temperature of the cooling medium, u, to random step changes every 5 dimensionless time units (dtu), and within the range $u \in [-2,2]$. The system was sampled every dtu for a total of 300 samples, where the concentration x_1, and the temperatures x_2 and u were measured. A portion of these results is plotted in Fig. 2.

6.1.2. The Test Data

In order to test the models 6 test data sets were generated as follows:

1. Random step changes in u every 5 dtu and within the range $u \in [-2,2]$. This is to test whether the information in the training set is representative of the system as a whole when u changes as a random step in the range [-2,2] every 5 dtu.
2. Random step changes in u every 3 dtu and within the range $u \in [-2,2]$. This is to test whether the model has captured higher frequency behaviour, which is essentially completely dynamic with no settling or steady states.
3. Random step changes in u every 8 dtu and within the range $u \in [-2,2]$. Conversely, this tests the steady-state information gathered by the networks from the training data.

The remaining three test sets were generated precisely as above, except that the step changes are replaced by random *ramp* changes. Each test set contains 500 examples, samples of which are depicted in Fig. 3.

6.1.3. The Addition Of Noise

In itself the identification of the cstr system described above is reasonably straightforward. Though this form of the system will be used as an initial assessment tool, for a more challenging and realistic problem noise is added to the datasets. In each case uniformly distributed noise in the range [-1,1] has been added to the training and test measurements. (The development of the Bayesian evidence scheme assumed a Gaussian noise model (likelihood distribution), and is in contradiction with the uniform noise added to this dataset. Whilst a uniform noise model would be more consistent with this system, it would negate the analytical tractability of the evidence approach, and an alternate scheme would be necessary to fulfill the Bayesian inference. It is therefore of interest to observe the evidence scheme in precisely such scenarios where the true noise model is not Gaussian. Many real-world applications have noise structures that are zero-mean but not normally distributed, and we are interested in the validity and performance of our analytic solution in these systems.)

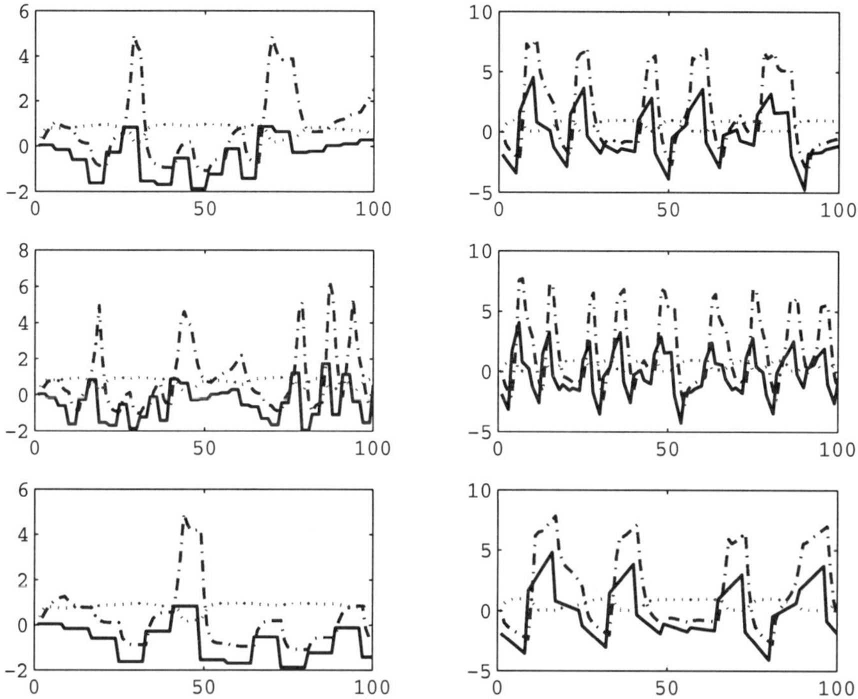

Figure 3. Samples of test data for the cstr system. Plots [1], [3] and [5] are data generated by random step changes in u, with frequencies of 5 dtu, 3 dtu and 8 dtu respectively. Plots [2], [4], and [6] are data generated by random ramp changes in u, with frequencies of 5 dtu, 3 dtu and 8 dtu respectively. The coolant temperature u is plotted as the solid line, the reactor temperature x_2 as (–.), and the reactor concentration x_1 as (...).

6.1.4. The Input And Output Spaces

An important issue in constructing a neural network representation of this system is in the choice of inputs for the model. In particular how many past samples of each of x_1, x_2 and u should be used in order to predict the next value of reactor temperature. In keeping with the model of [7], we arbitrarily choose to use the current and one past sample of each of x_1, x_2 and u. The input data to the neural network will subsequently look like $\boldsymbol{x}_n = [x_2(n)\ x_2(n-1)\ x_1(n)\ x_1(n-1)\ u(n)\ u(n-1)]$, where the required output is $y_n = x_2(n+1)$. Using the current and one past sample of each variable permit for a limited degree of gradient information. I.e. we are given an idea of the recent rates of changes of each variable, which is useful in light of dynamic systems, though it is by no means clear that this particular choice is 'optimal'.

6.2. *The Kin Robot Arm*

The kin datasets are a family of synthetically generated datasets from a realistic simulation of the forward kinematics of an 8-link all-revolute robot arm. The task associated with these data sets consists of predicting the distance of the end-effector from a target (the Cartesian coordinate [*0.1 0.1 0.1*]), given the angular positions of the joints. For this 8-dimensional input family there are 4 possible systems to consider: high nonlinearity vs. fairly linear, and medium vs. high noise. In this paper we have used the high nonlinearity, medium noise set (the kin-8nm set in the DELVE archive).

The inputs are sampled from a uniform distribution in the range [$-0.5\pi, 0.5\pi$], and uniform noise in the range [*-0.2,0.2*] is then added; the output is perfectly measured.

6.2.1. The Training And Test Data

In this paper we use training sets of 4 different sizes: 64, 128, 256 and 512 samples respectively. For each size there are 8 different training and test sets, where the inputs are the 8 angular positions of the robot arm and the output is the distance of the end-effector from a target.

7. Results and Discussions

To demonstrate the effect of the alternative training strategies, different neural networks were trained on the cstr and kin data sets using thin plate spline basis functions. For the cstr system, each neural network for the different training sets has 20 basis functions, and for the kin system each network has 26 basis functions. In the unsupervised training, each network was selected by comparing the evidences of 10 randomly generated networks. The locations of the basis functions for the candidate models were drawn from a multivariate Gaussian distribution (this is justified assuming the input space has been properly normalised). The optimised networks were trained using the unsupervised networks as starting points for the optimisation. For each of these schemes the weights were trained using both a traditional sum-of-squares objective function, and the Bayesian evidence scheme. The results are summarised in Tables 2 and 3.

In both Tables 2 and 3, $E_{\mathrm{nms}}^{\mathrm{test}}$ is the normalised mean square error defined on the test data: $\frac{\|\mathbf{y}-\mathbf{t}\|^2}{\|\mathbf{t}-\langle\mathbf{t}\rangle\|^2}$ (where y are the network predictions and t are the target values of the system).

Table 2. Results for thin plate spline networks trained on the cstr dataset, where the optimised networks are compared with those trained using unsupervised techniques.

	Unsupervised (SSE)	Unsupervised (evidence)	Optimised (SSE)	Optimised (evidence)
	Noiseless data			
Evidence	-	404.2	-	934.3
E_D^{train}	1.752	1.763	0.0318	0.0324
$E_{\mathrm{nms}}^{\mathrm{test}}$	0.0155	0.0157	0.0109	0.0109
$\langle r_j^0 \rangle$	2.298	2.298	2.749	2.749
Var[r_j^0]	0.6320	0.6320	1.179	1.179
	Noisy data			
Evidence	-	-278.0	-	-253.8
E_D^{train}	251.2	284.9	264.8	280.6
$E_{\mathrm{nms}}^{\mathrm{test}}$	0.1985	0.2340	0.2094	0.2230
$\langle r_j^0 \rangle$	2.561	2.561	129.6	129.6
Var[r_j^0]	0.6829	0.6829	3.13×10^4	3.13×10^4

Table 3. Results for thin plate spline networks trained on the kin dataset, where the optimised networks are compared with those trained using unsupervised techniques.

	Unsupervised (SSE)	Unsupervised (evidence)	Optimised (SSE)	Optimised (evidence)
	Training set with 64 samples			
Evidence	-	-81.55	-	-64.57
E_D^{train}	13.36	22.94	18.31	24.82
$E_{\text{nms}}^{\text{test}}$	0.8857	0.7266	0.7372	0.6459
$\langle r_j^0 \rangle$	2.800	2.800	214.8	214.8
$\text{Var}[r_j^0]$	0.4335	0.4335	5.51×10^4	5.51×10^4
	Training set with 128 samples			
Evidence	-	-161.0	-	-146.0
E_D^{train}	39.22	49.70	40.77	54.73
$E_{\text{nms}}^{\text{test}}$	0.5933	0.6086	0.6186	0.6038
$\langle r_j^0 \rangle$	2.738	2.738	117.3	117.3
$\text{Var}[r_j^0]$	0.5204	0.5204	2.67×10^4	2.67×10^4
	Training set with 256 samples			
Evidence	-	-316.6	-	-272.6
E_D^{train}	87.02	101.5	66.32	63.37
$E_{\text{nms}}^{\text{test}}$	0.5349	0.5541	0.5012	0.5175
$\langle r_j^0 \rangle$	2.745	2.745	41.51	41.51
$\text{Var}[r_j^0]$	0.4946	0.4946	5.65×10^3	5.65×10^3
	Training set with 512 samples			
Evidence	-	-602.1	-	-334.0
E_D^{train}	179.6	184.6	45.76	45.82
$E_{\text{nms}}^{\text{test}}$	0.4911	0.4950	0.3072	0.3023
$\langle r_j^0 \rangle$	2.802	2.802	5.474	5.474
$\text{Var}[r_j^0]$	0.5196	0.5196	6.458	6.458

7.1. Comparing Bayesian Regularisation With The Sum-Of-Squares Solution

In data sets where the data/weights ratio is small, there is a greater potential to overfit. The sum-of-squares solution invariably produces smaller training errors E_D^{train} and greater testing errors $E_{\mathrm{nms}}^{\mathrm{test}}$ than the Bayesian scheme; which is indicative of overfitting. This is particularly well illustrated for the kin datasets with 64 samples (Table 2). For the remainder of the experiments performed in this paper the data/weights ratios are reasonably large, there is more information from which the weights may be inferred, and the Bayesian and sum-of-squares solutions become more similar[2]. We observe throughout Tables 2 and 3 that the Bayesian scheme and sum-of-squares solutions produce similar training and testing errors.

7.2. Performance Of Optimised Basis Functions Relative To Unsupervised Training

Tables 2 and 3 show that the optimisation of thin plate spline centre locations has yielded better neural network models (i.e. supervised sum-of-squares solutions perform better than unsupervised ones, and similarly for the Bayesian trained networks). The mechanism for this improvement is identified as the distance between the basis function centres and the data space. This is measured crudely by the term $\left\langle r_j^0 \right\rangle$, which is the mean Euclidean distance between the location and the data space origin. Var[r_j^0] is the variance of this mean. Nonlinear optimisation has, in all cases, increased the mean distance (and variance) of the centres from the data space origin. In this sense, *common unsupervised methods of training are fundamentally restricted in that they do not search a large enough region in the data space.*

8. Summary

A Bayesian paradigm for the process of inferring the weights and locations of the basis functions in an RBF network has been presented. By expressing prior belief in the final state of the weights, the Bayesian scheme allows a natural and plausible

---

2 In the presence of large datasets the prior becomes more vague and the weights are less constrained, and therefore approach the sum-of-squares solution, whose weights are completely unconstrained.

mechanism for regularising the network weights, and thus avoids overfitting. In most experiments in this paper the potential of this scheme is not fully explored because the data/weights ratios are not small enough to warrant much regularisation (they are usually >10). In fact, it is only in the case of the kin datasets of 64 samples that the distinction between the Bayesian and sum-of-squares solution is apparent. It is nonetheless encouraging to observe that the Bayesian scheme produces at least similar results to the sum-of-squares solution when the data/weights ratio is large.

The main idea of this paper is a humble one: that optimisation of the basis function locations yields notably better models than those trained using conventional unsupervised methods. In a sense the choice of training scheme for the weights is irrelevant. The Bayesian paradigm offers a natural means for avoiding overfitting, but if the method of choice is minimisation of a sum-of-squares objective then the idea that the basis functions should be located by nonlinear means is still preserved. This calls into question the suitability of unsupervised approaches to regression modelling. Optimisation invariably causes the basis functions to be located further from the data space origin than the unsupervised method would usually allow. We conclude that unsupervised training methods for regression modelling are fundamentally restricted in that they do not explore a sufficiently large space for the basis function locations.

References

1. Bishop, C. M., *Neural Networks for Pattern Recognition* (Oxford Press, 1995).
2. Buntine, W. and Weigend, A., *Complex Systems.* **5** (1991), 603-643.
3. Chen, S. et al., *International Journal of Control.* **50** (1989),1873-1896.
4. Chen, S. et al., *IEEE Transactions on Neural Networks.* **2** (1991), 302-309.
5. Dempster, P. et al., *Journal of the Royal Statistical Society.* **39** (1977), 1-38.
6. Hand, D. J. and Batchelor, B. G., *Information Sciences.* **14** (1978), 171-180.
7. Hussain, M. A., *Inverse-Model Control Strategies Using Neural Networks: Analysis, Simulation and On-Line Implementation.* Ph.D. thesis (Imperial College of Science, Technology and Medicine, 1996).
8. MacKay, D. J. C., *Neural Computation.* **4** (1992), 415-447.
9. Magni, R., Bayesian Methods for Neural Networks in Process Identification. Ph.D. thesis (Imperial College of Science, Technology and Medicine, 2000).
10. Michelli, C., *Constructive Approximation.* **2** (1986), 11-22.
11. Poggio, T. and Girosi, F., *Proceedings of the IEEE.* **78** (1990), 1481-1497.
12. Schoenberg, J., *Annals of Mathematics,* **39** (1938), 811-841.

5. PROCESS IDENTIFICATION OF A FED-BATCH PENICILLIN PRODUCTION PROCESS - TRAINING WITH THE EXTENDED KALMAN FILTER

R. SCHEFFER, R. MACIEL FILHO

LOPCA/DPQ, Faculty of Chemical Engineering, State University of Campinas (UNICAMP), Cidade Universitária Zeferino Vaz, CP 6066, , Campinas – SP, Brazil, CEP 13081-970,

In this work a recurrent neural network is applied as a non-linear identification tool of a fed-batch penicillin process. The recurrent neural network is trained by a multiple-stream extended Kalman filter, which allows the process to be identified in real-time. It is shown that the fed batch process can be estimated very accurately by a recurrent network and that the extended Kalman filter is a very efficient and rapid training algorithm for non-linear process identification. The recurrent neural network could be trained to give an one-step ahead prediction for a sample-time of six minutes. This ensures that the neural network can be used as an identification model in a model predictive control loop for calculation of the optimal feeding strategy in real-time.

1. Introduction

Control of non-linear chemical processes relies on a good dynamical model, which usually is linear and has to be continuously updated to follow the process behaviour. In the past decade, this has led to an increased interest for non-linear process control, which can be divided into optimisation and transformation methods. Only recently, artificial neural networks (ANN) have attracted a great deal of attention, providing a simple way to describe non-linear processes at low computational cost. If trained by appropriate input/output data, an ANN is able to approximate any function mapping of the process input/output behaviour inside its training area. So, an ANN can be viewed as a black-box model of the process.

The classical approach to identify a non-linear dynamical process by an ANN, is to augment the number of inputs of the ANN by past values of the input data. For simple systems, which can be described by a one-dimensional state space system, this will result in a small and workable ANN. But this approach can result in too large ANNs suitable for training more complicated systems, such as fixed-bed reactors, fluidised beds and bubble columns, and for systems with slow dynamics.

Biochemical processes are normally more complex than chemical processes, making modelling their behaviour and dynamics a very hard task. This is true, since often the reaction kinetics are complicated and small variations in temperature and pH deteriorate the biochemical catalysts, such as its enzymes. Thus biochemical processes are ideal candidates to be identified by recurrent neural networks where the recurrent neural network is trained on-line to be able to learn the process changes.

In this work the production of penicillin was chosen fir a case-study. The feeding strategy of the penicillin process is one of the key factors in obtaining a high final penicillin concentration. The recurrent neural network will be used in a model predictive control algorithm, which determines the best feeding strategy from the current operating point to maximise the final penicillin concentration.

The first step in creating such a control algorithm is a good process identification and therefore the main objective of this work is the development of a dynamical ANN trained by a recursive algorithm such as the Kalman filter, which can be used to identify the penicillin process on-line and afterwards be used in an advanced control scheme.

2. Penicillin Production process

The production of antibiotics for clinical applications is an important industrial process and about 60-65% is produced by means of fermentation. The penicillin is produced by bacteria or fungi and in this work the process studied is the production of Penicillin G, which is done by the fungi *Penicillium chrysogenum*.

The process is initiated by growing an inoculate which is injected in the pre-fermentation batch reactor, where a high concentration of the carbon source, such as lactose or glucose, is maintained. When enough fungi is produced the production phase is started in the fermentation tank. In this phase the feeding strategy is vital in obtaining a high final penicillin concentration and the carbon source concentration has to be kept low to promote the penicillin production but not too low to starve the fungi. After the production phase the penicillin has to be purified. The flow diagram of the penicillin production process is shown in Fig. 1.

The emphasis is put on the production phase to determine an optimal on-line feeding strategy and in this work more specifically on the on-line identification of the process, which is done by a recurrent neural network trained by Kalman filter.

One input, the substrate feed flow, and four state variables, the bio-mass concentration, the substrate concentration, the penicillin concentration and the dissolved oxygen concentration, were taken to identify the process.

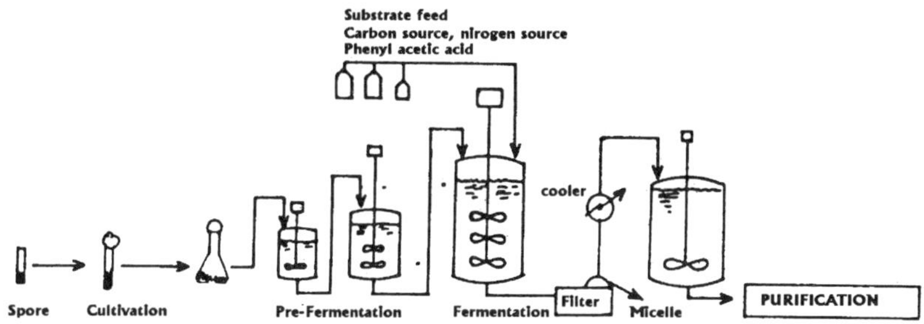

Figure 1. Flow diagram of the fed-batch production process (Crueger and Crueger, 1984)

The substrate feed influences directly the cellular concentration, the substrate concentration and the penicillin concentration. The dissolved oxygen concentration is also influenced, because of the cellular growth which requires oxygen for respiration.

The dissolved oxygen concentration is a vital variable for a good process operation. Rodrigues (1999) mentioned that a dissolved oxygen concentration lower than 30% causes deterioration of the fungi. The dissolved oxygen concentration can be controlled by aeration and stirring, but the latter will destroy the fungi at high rotation speeds. This shows the need for an optimisation algorithm which takes into account such constraints.

In the identification of the process the stirring and aeration were not accounted for as they were kept constant, but can be done so if necessary.

Rodrigues (1999) determined an optimal feeding strategy for this studied penicillin process off-line, but will be optimal only if the exact specifications can be followed. If the system is subjected to changes this strategy will become sub-optimal and there will be another strategy which will be optimal.

To determine this new optimal feeding strategy, the process model has to be updated to incorporate the changes which were not accounted for in the model. If the model is updated, it can be used in an optimisation criterion to calculate the new best control actions to be taken. A schematic of this type of the proposed control system is shown in Fig. 2.

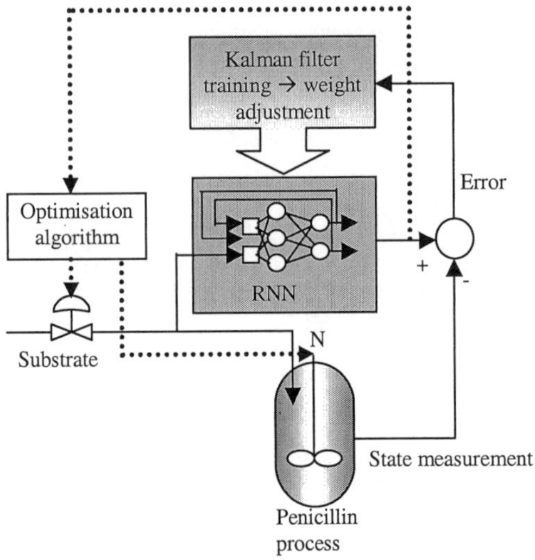

Figure 2. Schematic of the proposed control scheme

3. Training Neural Networks with the Extended Kalman Filter

The training of a neural network with the Kalman filter is done in conjunction with the back-propagation algorithm. The back-propagation pass is used to calculate the derivatives and the errors of every neuron in the network. The weight adjustment done with the Extended Kalman Filter will give a much faster convergence than with the back-propagation method as it is a second order method based on the least squares principal. But the storage and computational requirements may become too high as the network size is increased. In the next subsection a short review is given for the standard calculation of the common neural network and the way the recurrent neural network is implemented. Afterwards, the differences are shown when applying the Kalman filter theory to the training of neural networks. A derivation of the Kalman filter can be found in for example, Goodwin and Sin (1984).

3.1. Principle Calculations Of A Neural Network

The most popular ANN is the multiple layer perceptron trained by the bac k-propagation method of Rumelhart et al. (1986), which consists of an input layer, a number of hidden layers and an output layer. The output of a neuron j, y_j in the hidden or output layer k is calculated as a function of the outputs of the former layer as follows:

$$y_{k,j} = f\left(v_{k,j}\right) = f\left(\sum_{i=0}^{N_{k-1}} w_{k,ji} y_{k-1,i}\right) \tag{1}$$

where $w_{k,ji}$ is the weight for the connection of input y_i with neuron j in layer k and N_{k-1} is the number of inputs or the number of neurons in the former layer as the network is fully connected. The bias or threshold is defined by putting input $y_{k-1,0}$ equal to 1. k ranges from 0 to the number of chosen layers, where layer 0 is the input layer. The function $f(\cdot)$ is typically a sigmoidal or tangent hyperbolic function for a neuron in the hidden layer and linear for the output layer in case of function approximation. If the data are appropriately scaled, the latter can be a non-linear function also. A signal flow presentation of a neuron is shown in Fig. 3.

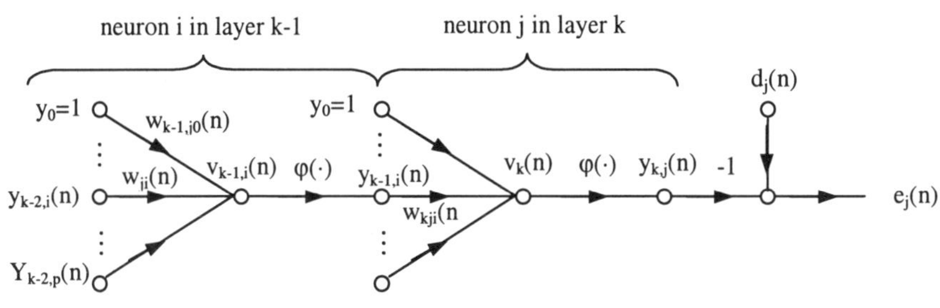

Figure 3. Signal flow diagram of neurons in subsequent layers in a neural network

The ANN's output and the desired output define an error, $e = d_j - y_j$, which can be propagated back through the system. The error for intermediate or hidden neurons is calculated by:

$$e_{k,j} = \sum_{j=1}^{p} w_{k+1,ji} \delta_{k+1,j} \tag{2}$$

$\delta_{k,j}$ is the local error gradient for neuron j in layer k and is calculated by:

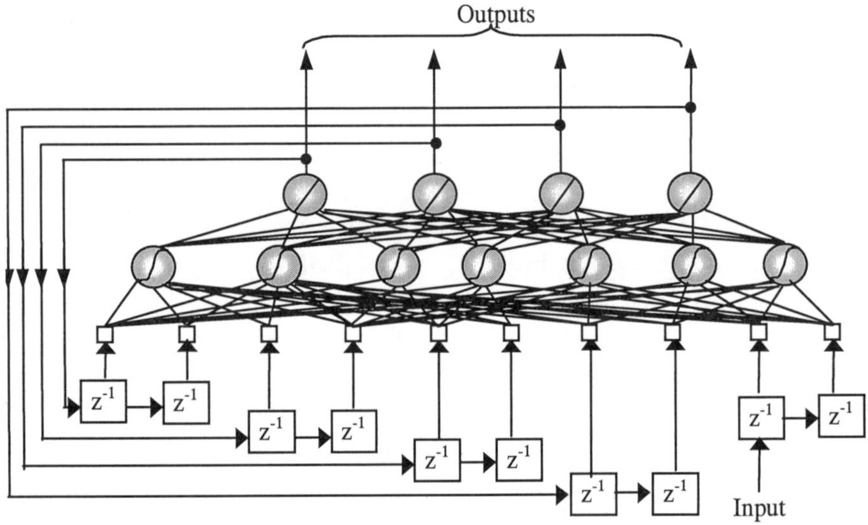

Figure 4. A schematic model of a NARX recurrent neural network of order 1

$$\delta_{k,j} = \frac{df(v_{k,j})}{dv_{k,j}} e(k,j) \tag{3}$$

In the case of the back-propagation algorithm, this leads to a parameter adjustment based on the steepest descent by:

$$w_{k,ji}(n+1) = w_{k,ji}(n) + \eta\partial_{k,j} y(k-1,i) \tag{4}$$

where η is the learning parameter. A faster convergence can be obtained by adding a momentum term (Rumelhart et al., 1986). Optimisation methods, such as the conjugate gradients (Fletcher et al., 1964) and the method of Levenberg-Marquardt can be used to obtain a much faster convergence using second-order gradient information. A line search is conducted in the calculated direction by a quadratic approximation. These methods need a good estimate of the gradient, and therefore can only be used in a batch training mode, where an average gradient to the weights is calculated over the whole training set.

A dynamical ANN, also known as a recurrent neural network (RNN) is obtained when some of the neurons in the layer k have feedback connections with the neurons in layer l, where $l < k$. In this work only external feedback connections were chosen, which lead the outputs from the output layer back to the input layer. The advantage of this type of RNN is that during the training phase the target values instead of the RNN's outputs can be fed to the input layer, so-called *teacher forcing*, which leads to a faster convergence. When the error of the output is small enough, the network outputs are fed to the input layer.

If re-feeding of the outputs is not sufficient, then more memory has to be built into the ANN. This can be done by applying a tap-delay filter of order q to the inputs and the re-fed outputs. Figure 4 shows a RNN with one input and four outputs with a tap-delay filter of order 1 for both input and re-fed output. The present and past values of the input represent exogenous inputs, while the delayed values of the output form the regressive inputs of the recurrent neural network. This non-linear auto regression model with exogenous inputs (NARX model) is fed to a multi-layer perceptron (MLP) which calculates the new output of the RNN. If this results in a large NARX model order, then the network might become too large and a slow down of training occurs. In this case it might be necessary to make the network fully recurrent, which is more powerful in acquiring the system dynamics.

For on-line applications, the optimisation methods usually do not have a recursive calculation scheme and cannot be used, but the back-propagation algorithm is typically slow and forgets the past data. A system identification method such as the Kalman filter could be used to update the network parameters, which has the advantage over the back-propagation that it takes into account the past data when it calculates a new optimal estimate with the newly arrived data.

3.2. Neural Networks and the Kalman Filter

One of the first attempts of training neural networks with the Kalman filter was conducted by Singhal and Wu (1989). They used a Global Extended Kalman Filter (GEKF) to train feed-forward neural networks which had an excellent performance on training the network weights but at the cost of a large increase in storage and computational requirements. Shah et al. (1992) proposed a Multiple Extended Kalman Filter Algorithm (MEKA) to train feed-forward neural networks on a classification problem. With this algorithm a local Kalman filter is designed for every neuron present in the network. They compared their algorithm with the global extended Kalman filter algorithm and concluded that the MEKA algorithm has similar convergence properties but is computationally less expensive. Though the last algorithm is adopted in this work, both will be shown to give a more complete overview of the training of neural networks with the Kalman filter.

The Kalman filter identifies a linear stochastic dynamical system. To be able to estimate parameters with the Kalman filter, the weights of the network have to be written as a dynamical system. The weights for neuron j in layer k can be written as the following dynamical system:

$$w_{k,ji}(n+1) = w_{k,ji}(n) + q_{k,ji}(n) \qquad i = 0 \ldots N_{k-1}$$
$$y_{k,j}(n) = f\left(\sum_{i=0}^{N_{k-1}} w_{k,ji}(n)\, y_{k-1,i}(n)\right) + r_{k,j}(n) \tag{5}$$

where $q_{k,ji}$ and $r_{k,j}$ are stochastic variables with a Normal random Gaussian distribution, N(0, Q) and N(0,R), respectively.

The stochastic process noise, q, would be in fact zero as a parameter has in its definition no stochastic noise compound. But in training neural networks it was pointed out by Puskorius and Feldkamp (1994) that adding process noise stabilises the Kalman filter and also prevents the algorithm getting stuck in poor local minima. A larger process noise also speeds up the training process.

The Kalman filter provides an elegant and simple solution to the problem of estimating the states of a linear stochastic dynamical system. For non-linear problems, such as in Eq. 5, the Kalman filter is not strictly applicable as the linearity plays an important role in its derivation. The extended Kalman filter tries to overcome this problem by linearising the stochastic dynamical system about its current state estimation, which first seems to be suggested by Kopp and Orford (1963) and Cox (1964). It should be noted that the extended Kalman filter will not be optimal in general. Moreover, due to the linear approximation, it is quite possible that the filter may diverge and therefore care has to be taken in applying this method (Goodwin and Sin, 1984).

Expansion of the non-linear dynamical system of the neural network weight parameters around the estimate of the weight state vector w(n-1) at time t leads to:

$$w_{k,ji}(n+1) = w_{k,ji}(n) + q_{k,ji}(n) \qquad i = 0 \ldots N_{k-1}$$
$$y_{k,j}(n) = f\left(\sum_{i=0}^{N_{k-1}} w_{k,ji}(n)\, y_{k-1,i}(n)\right)\Bigg|_{w=\hat{w},q=0} + C(n)^T\left[\underline{w}(n) - \underline{\hat{w}}(n)\right] + r_{k,j}(n) \tag{6}$$

where C(n) is the Jacobian matrix resulting from the Taylor expansion about the state at time n and is recalculated at every sampling instance. w is the real parameter, while $\hat{w}$ is the estimated weight based on the information at time n-1, and C(n) is given by:

$$C(n) = \left.\frac{\partial f\left(w(n), q(n)\right)}{\partial w(n)}\right|_{w=\hat{w}(n), q=0} \tag{7}$$

As for the linear stochastic system shown in the appendix, a Kalman filter can be set up to estimate the non-linear system and is (Goodwin and Sin, 1984):

$$\begin{aligned} w_{k,ji}(n+1) &= w_{k,ji}(n) + K_{k,ji}(n) e_{k,j}(n) \qquad i = 0 \dots N_{k-1} \\ K_{k,ji}(n) &= P_{k,j}(n)\, C(n)^T \Big/ \left(C(n)\, P_{k,j}(n)\, C(n)^T + R(n)\right) \\ P_{k,j}(n+1) &= \left(I - K(n)\, C(n)\right) P_{k,j}(n) \left(I - K(n)\, C(n)\right)^T + K(n)\, R(n)\, K(n)^T \end{aligned} \tag{8}$$

where K is the Kalman gain vector, R is measurement noise covariance matrix, I is the identity matrix and P is the error covariance matrix. The update equation for the covariance matrix P as written in most textbooks is:

$$P_{k,j}(n+1) = \left(I - K(n)\, C(n)\right) P_{k,j}(n) \tag{9}$$

which is a simplification of the update equation for matrix P in Eq. 8 after substituting the formula of the Kalman gain. But care has to be taken with this substitution, because the covariance matrix of Eq. 9 is not positive definitive any more by definition and can lead to numerical instability. This was one of the main reasons for the non-popularity of the filter some decades ago.

The Kalman filter of Eq. 8 is used together with the forward pass of the back-propagation when there is no process noise present. The forward pass calculates the new estimates of the observations y, after which the innovation (= output – desired output) or error is propagated back to the network calculating the local errors and the weight derivatives of the observation matrix C. Then the Kalman filter is used to adjust the weights.

If there is process noise present ($Q \neq 0$) then the error covariance matrix has to be updated during the forward pass by:

$$P_{k,j}(n+1) = P_{k,j}(n) + Q(n) \tag{10}$$

The Kalman filter has to be initiated with initial values for the states and the covariance matrixes, while the matrixes Q and R are tuning parameters and can be chosen to obtain a certain convergence behaviour (see paragraph 3.2). The initial value for the states or ANN weight parameters are chosen at random from a normal or uniform distribution resulting in weights ranging from –0.2 to 0.2. The error covariance matrix is set initially to a matrix with large values, like 100, on its diagonal.

The Jacobian matrix C(n) is calculated conveniently by the back-propagation method, but differs for the two different Kalman training algorithms.

The MEKA algorithm uses a Kalman filter for every neuron, having its own Kalman filter gain vector, error covariance matrix, process noise variance matrix, Q, and measurement noise covariance matrix, R. The advantage of using a Kalman filter per neuron is that the denumerator of the Kalman filter equation becomes a scalar and no matrix inversion is needed any more.

The GEKF algorithm adjusts all the weights of the neural network by *one* extended Kalman filter. To do so, the weights have to be positioned in W x p matrix, where W is the total number of weights present in the network and p is the number of outputs of the network. The derivatives in matrix C have to be calculated for a weight with respect to the networks output and not to its neurons output. This can be done by propagating back the output instead of the error and calculating the derivative with respect to the back-propagated output.

3.3. The Tuning Parameters of the Kalman Filter

The process noise covariance matrix Q and the measurement noise covariance matrix R are normally regarded as the tuning parameters of the Kalman filter. It was already mentioned that the process noise for a parameter would be zero, but accelerates the learning process and helps to avoid local minima. From Eq. 10 it can be seen that the error of the state estimation will start to grow after the sampling point until the new measurement arrives. The measurement noise determines how much the new measurement can be trusted and according to this a correction is made by the Kalman filter gain. So a larger value for R will result in a smaller adjustment for the weights. The matrix R cannot have any zeros on its diagonal, otherwise this may lead to a division by zero.

It can also be shown that if the matrixes Q and R are constant, then the Kalman gain will converge to a constant value. This is exactly what is not wanted for an on-line process identification tool. Therefore a way has to be found to keep the filter excited to new data.

Shah et al. (1992) used a similar formulation as is used for the method of recursive minimum least squares, where a forgetting factor is introduced in the minimisation criterion and thus in the Kalman filter equations.

A more elegant way is to make the process noise covariance matrix and/or the measurement noise covariance matrix a function of time. Rivals and Personnaz (1998) made the measurement noise covariance matrix a function of the number of epochs, while maintaining the process covariance matrix 0. The function used for R is an exponential function and is:

$$r(n) = (r_0 - r_f)exp\{-\alpha i\} + r_f \tag{11}$$

where r_0 was chosen in the order of one (about the order of the initial mean squared error averaged over the number of data), r_f a small value like 10^{-10}, α 0.5 to 1 and i is the number of epochs.

Though their function is a function of the number of epochs, it is not in a form suitable for on-line training. Therefore it is proposed to make the matrices a function of the error, in such way the training of a neural network can be controlled nicely and a certain convergence behaviour can be obtained depending on the characteristics needed for the problem. For example, the larger the elements of Q, the more adjustment is made by the filter, which can result in fitting noise as well.

Therefore both the process noise covariance matrix and the measurement noise covariance matrix can be made a function of the error. The functions are linear or exponential. The process noise covariance matrix will be larger in the beginning and go to zero when the error decreases. The measurement noise covariance matrix can be made larger in the beginning also to make the weight changes less severe, which is normally desired in the beginning of neural network training as all weights are non-optimal.

It was chosen to make the process noise covariance matrix, Q, a function of error only, as both have similar effects. The measurement noise covariance matrix, R, was kept constant at a value of 50. Q was made a function of the total Summed Square Error (SSE) calculated over all outputs and over the whole training set.

$$Q(n) = Q_0\left[\exp\left(1.0 * 10^{-2} * SSE\right) - 1.0\right] \tag{12}$$

Observe that Q will become zero when the SSE reaches zero. Q_0 was set to 0.01, and shows how strong the weight adjustment responds to an error. If Q_0 is set higher the weight change will be larger. However, if the measurements contain much noise and the Q_0 is set high, then the noise will be fitted also.

For on-line training it would be better to make the error a function of the absolute local error, in such a way the weight change will be higher for those neurons which have a high local error.

It was already shown by Scheffer and Maciel Filho (2000) that the Kalman filter was a potential candidate for the training of recurrent neural networks, but in the way the algorithm was implemented, the memory requirements became too large. To diminish the memory requirements, the weight matrix of the network was re-structured into a vector by subsequent numbering of the weights. The Kalman filter matrices were defined from this vector, which reduced the required memory and made the algorithm have a calculation time comparable to the conjugate gradient method.

4. Results

The data of the penicillin fed-batch process were obtained from the optimal batch run, which was determined by Rodrigues (1999). The batch run was obtained with a sampling interval of 6 minutes, which is quite large and should be enough when the neural network is used in an optimisation scheme. Only a training set was made, as the main objective is to create an on-line identification tool.

Various recurrent neural networks were trained with variable sizes and different architectures. It was noted that some of the states have a linear behaviour, while others exhibit a highly non-linear response. To account for both linear and non-linear behaviour a specific network architecture was created, which is as follows:

- The activation function of the output layer is a linear function, while the activation functions of the hidden layer are non-linear and were chosen as the tangent hyperbolic function.
- All the inputs of the recurrent neural network; the re-fed outputs, the input and the time-delayed inputs and delayed outputs, are directly connected to the output layer.

The direct connection from the input layer to the output layer models the linear behaviour of the inputs and states, because of the linear activation function in the output layer. This kind of network structure will be called "RRNlin" from now on. A representation of a RNNlin is shown in Fig. 5.

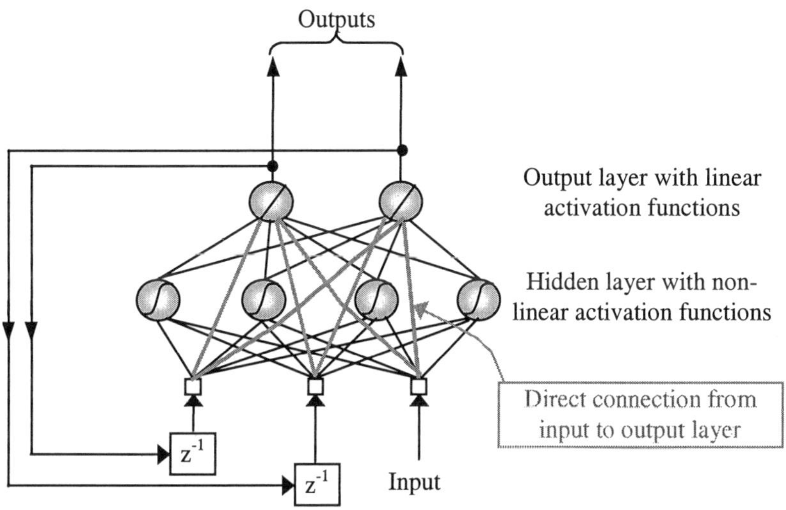

Figure 5. An example of a RNNlin with an order of the NARX model of 0

Only the results are shown from the training of a RNNlin consisting of a hidden layer with 15 neurons with a tangent hyperbolic activation function and a output layer with 4 neurons with a linear activation function. The order of the NARX model was chosen to be 1, so the present values of input and outputs and 1 past value of the input and outputs are taken into consideration. The conventional recurrent neural networks were only able to describe the penicillin process with teacher forcing, when the outputs were re-fed it resulted in high values for the error. It should be mentioned that the error of the Kalman filter was small during the training due to the filtering done during the sequential mode. When the final weights were used for network simulation, it resulted in high errors.

The RNNlin networks proposed here were much better in modelling the penicillin process, because of their specific architecture. The conjugate gradients method used in training the network, only converges to a small error, when the network is pre-trained to a very small error in the teacher forcing mode. Otherwise the conjugate gradient method gets stuck in a local minimum with a much higher error.

In Figs. 6 and 7 the training of this RNNlin is shown for three different implemented training algorithms. One epoch is one presentation of every training sample of the whole training set. Actually, the method of Levenberg-Marquardt was implemented also, but it showed itself inferior to the method of conjugate gradients. This is probably due to the approximation of the Hessian used or because it is more dependent on the initial shoot. Therefore only the conjugate gradients method is shown here.

The error for the sequential mode is much lower, because for every presented sample an adjustment is made for every weight while in the batch training mode an average adjustment is made over the whole training set.

It can be seen that in the teacher forcing mode the training algorithms behave in the same way (Fig. 6), giving a rapid adjustment in the beginning and afterwards a slower fine-tuning of the weights. The back-propagation algorithm with momentum is slower as the learning parameter is varied with a fixed step-size.

The Kalman filter converges in the same way as the conjugate gradient algorithm, which shows that it is a second order method.

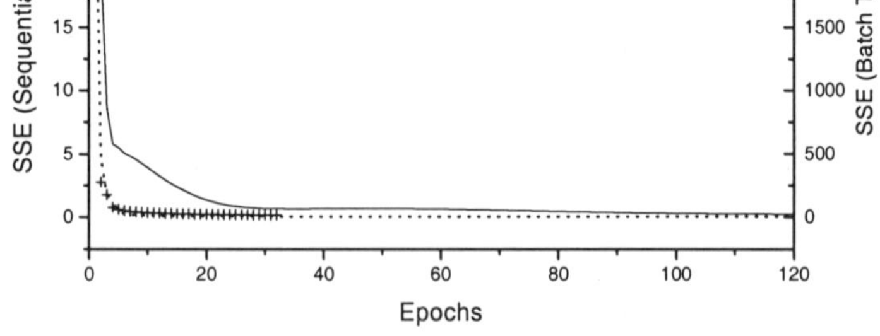

Figure 6. Comparison of the training of a MA1lin-15 network in teacher forcing mode

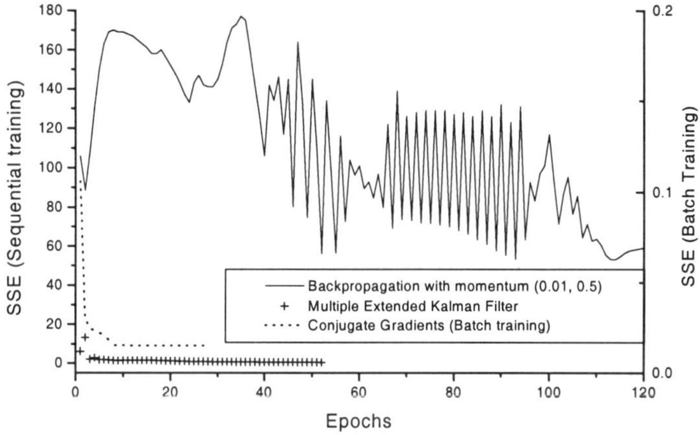

Figure 7. Comparison of the training of a MA1lin-15 network with the network outputs re-fed to the input

Figure 7 shows that when the outputs are re-fed, the behaviour of the sequential algorithms is totally different. The neural network has become dynamical which makes the system much more complex for its weight change. A weight change will affect both the neural network´s input as the network´s output. The oscillation for the back-propagation algorithm with momentum is probably due to the momentum term, but it can also be that the learning parameter was too big. Though the MEKA algorithm takes more calculation time, it converges very rapidly, making it very suitable for on-line training of recurrent neural networks as a system identification tool.

In Figs. 8-11 the trained RRNlin networks are shown after several presentations of the training set. The outputs of the network were re-fed.

Very good approximation of the penicillin process is obtained with the RRNlin network, which was trained with the Kalman filter. The experimental data were modelled perfectly and shows the potential of the multiple Kalman filter training algorithm. It should be mentioned that the Linear RNN, which is a network with *no* hidden layer, the same order of the NARX model and linear activation functions in the output neurons, also describes part of the outputs quite reasonably. The biomass concentration and the penicillin concentration are described more accurately by the linear RNN, because they exhibit a more linear behaviour.

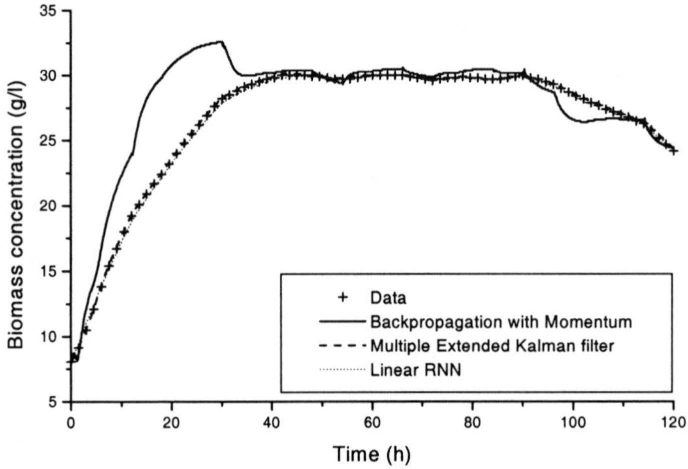

Figure 8. Prediction of the biomass concentration A RNNlin with 15 hidden tanh neurons and 4 output linear neurons trained the different training algorithms with re-feeding of the outputs after several presentations

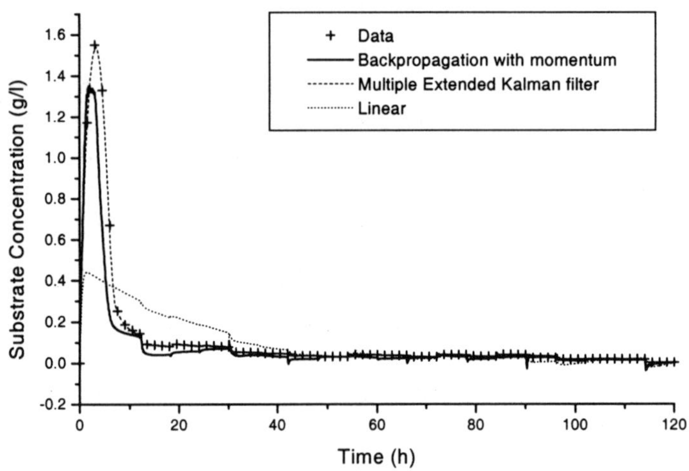

Figure 9. Prediction of the substrate concentration A RNNlin with 15 hidden tanh neurons and 4 output linear neurons trained the different training algorithms with re-feeding of the outputs after several presentations

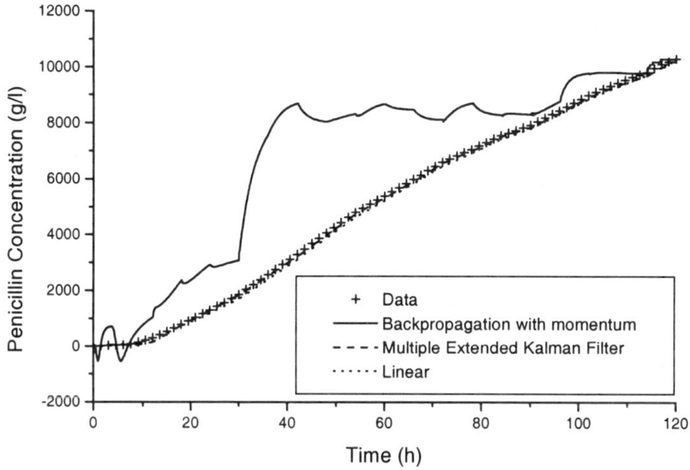

Figure 10. Prediction of the penicillin concentration - A RNNlin with 15 hidden tanh neurons and 4 output linear neurons trained the different training algorithms with re-feeding of the outputs after several presentations

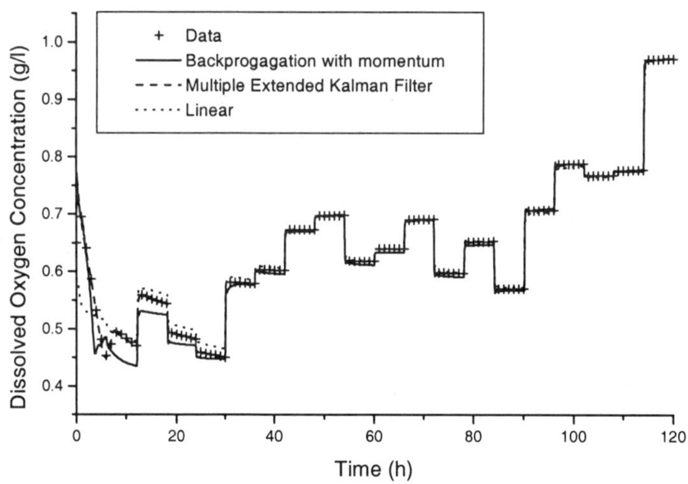

Figure 11. Prediction of the dissolved oxygen concentration - A RNNlin with 15 hidden tanh neurons and 4 output linear neurons trained the different training algorithms with re-feeding of the outputs after several presentations

However the RNNlin trained with the back-propagation with momentum algorithm did not well describe the penicillin process. The RNNlin could be trained with the back-propagation algorithm to a lower error.

Finally an on-line training test was conducted with the RRNlin and is shown in Figs. 12 and 13 Both sequential training algorithms were used to learn the dynamics of the fed-batch penicillin process. The RNNlin was not pre-trained at all, but directly subjected to training the process on-line. Only the prediction of the biomass concentration and the dissolved oxygen concentration are shown. The other two state variables give a similar response.

For training with the Kalman filter on-line, it can be seen that the RNNlin weights are rapidly adjusted by the filter and a reasonable estimation of the penicillin fed-batch process is obtained taken into account that the RNNlin was never trained before. The peaks could be caused by the steps in the dissolved oxygen concentration, which excites the filter to give a rapid adjustment when identifying the dissolved oxygen concentration. Probably the peaks can be prevented by lowering the parameter Q_0, giving a less severe change to the weights, but this will also result in slower training and a worse process estimation in the initial phase of the process.

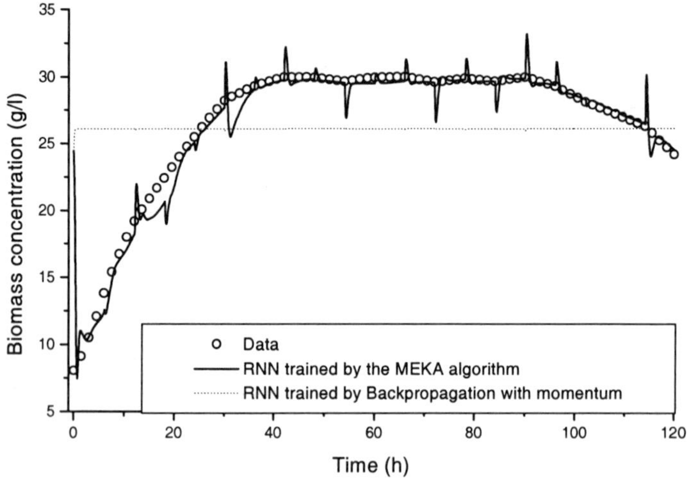

Figure 12. Prediction of the biomass concentration - On-line training of a batch run of the RRNlin with the sequential training algorithms

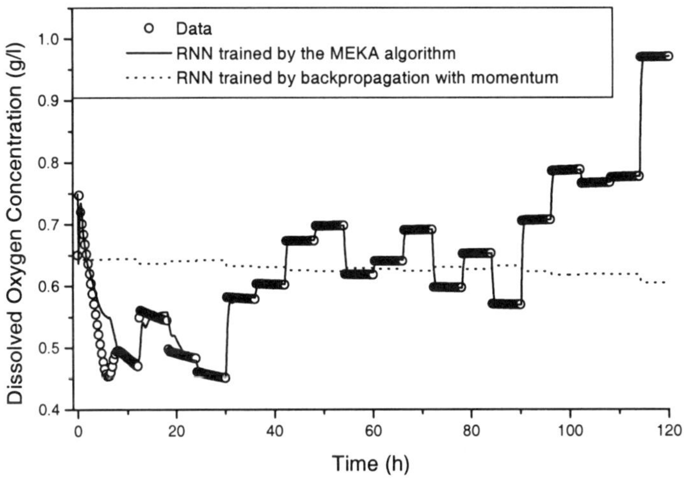

Figure 13. Prediction of the dissolved oxygen concentration - On-line training of the RRNlin with the sequential training algorithms

It shows that when the Kalman filter is used as a training algorithm no off-line training is necessary if small deviations are allowed for the process to operate appropriately. But in biochemical processes it is vital to have little or no deviation, as a small variation affects the organism and the activity of enzymes, making it necessary to pre-train the neural network.

The back-propagation algorithm cannot cope with on-line training of recurrent neural networks, as the dynamics of the process are not followed and the prediction is not good. If the back-propagation is used to train a neural network on-line, then there has to be implemented two networks, one used to predict the process, while the other is trained off-line to learn the changes. Switching of the networks will maintain a good process estimation.

5. Conclusions

It was shown in this study that the Multiple Extended Kalman filter is a very powerful training algorithm, especially for the training of recurrent neural networks. Good process descriptions were obtained with a recurrent neural network which has direct connections from the inputs to the outputs. The extended Kalman filter can be used in on-line training schemes, giving reasonable process estimations throughout the process, even when the network was never trained before. The ability of predicting the process behaviour over a sample time of six minutes demonstrates its possible use in a model predictive control algorithm.

References

Cox, H., *IEEE Trans. Autom. Control.* **AC-9** (1964), 5-12.

Crueger, W. and A. Crueger, *Biotechnology: a Textbook of Industrial Microbiology*, (Suderland, Sinawer Associates, Inc., 1984)

Goodwin, G.C. and K.S. Sin, *Adaptive Filtering Prediction and Control* (Prentice-Hall, Inc., Englewood Cliffs, New Jersey, 1984), 284.

Kopp, R.E. and R.J Orford, *AIAA J.* **1**(10) (1963), 2300-2306.

Puskorius, G.V. and L.A. Feldkamp, *IEEE Transactions on Neural Networks*, **5**(2) (1994), 279-297.

Rivals, I. and L. Personnaz, *Neurocomputing*, **20**(1-3) (1998), 279-294.

Rodrigues, J.A.D. and R. Maciel Filho, *Chem. Eng. Sci.* **54**(1999), 2745-2751.

Scheffer, R. and R. Maciel Filho, in *Escape-10 symposium proceedings* (Elsevier, Amsterdam, The Netherlands, 2000) 223-228.

Shah, S., F. Palmieri and M. Datum, *Neural Networks*, **5** (1992), 779-787.

Singhal, S. and L. Wu, *Proceedings of IEEE International Conference on Acoustics, Speech and Signal Processing* (IEEE Press, 1989) 1187-1190. Glasgow, Scotland.

Acknowledgements

The authors would like to thank CAPES for the financial support in the form of a scholarship.

PART II HYBRID SCHEMES

6. COMBINING NEURAL NETWORKS AND FIRST PRINCIPLE MODELS FOR BIOPROCESS MODELING

B. EIKENS, M. N. KARIM, L. SIMON
Department of Chemical and Bioresource Engineering
Colorado State University, Fort Collins, Colorado 80523

This paper analyzes the combination of prior knowledge in the form of first principle models (parametric models) and neural networks. These models are called hybrid models. Neural networks and hybrid models were used to identify a fed-batch fermentation. Different neural networks were integrated into the hybrid model structure. The performance of these hybrid models is compared with "traditional" neural networks.

1. Introduction

A parametric model consists of equations obtained through theoretical analysis and experimental testing. The parameters of the resulting parametric model can be associated with specific physical characteristics of the system. Most processes encountered in chemical engineering can be described by a parametric model in the form of a first principle model (FPM). The FPM may be based on mass, momentum, and energy balances as well as empirical correlations. FPMs used for system identification and modeling have to be simple enough for real-time evaluation. Hence, only major characteristics and trends of the process are described by the FPM and not all potential variables may be included in the input vector of the model. Additionally, FPMs do not incorporate unmeasured disturbances which are present in many real systems. However, the parametric modeling approach guarantees plausible predictions since it is based on fundamental principles that have to be fulfilled at all times. Empirical models on the other hand, are computationally efficient, data-driven models. They can represent a non-linear process accurately in the domain reflected by the data even if unmeasured disturbances are present. Shortcomings of this approach surface when the model has to extrapolate. This is true especially for models with localized receptive fields, e.g., radial basis function networks.

Several researchers have suggested to synthesize *hybrid models* which overcome the drawbacks of each approach while combining the advantages. The expression "hybrid model" is used in this context for a model which consists of both an empirical and a parametric submodel. These two submodels can be arranged in series or in parallel (Fig. 1). In the serial approach, the neural network estimates unmeasured process parameters such that the first principle constraints are satisfied. The FPM then specifies process variable interactions based on physical considerations.

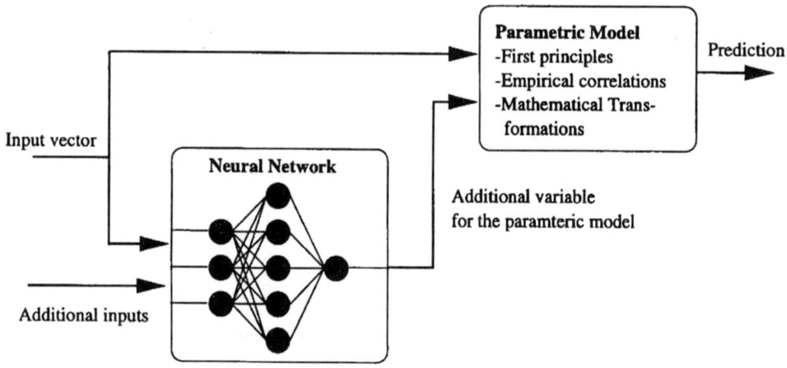

Figure1. Serial combination of a parametric model and a neural network

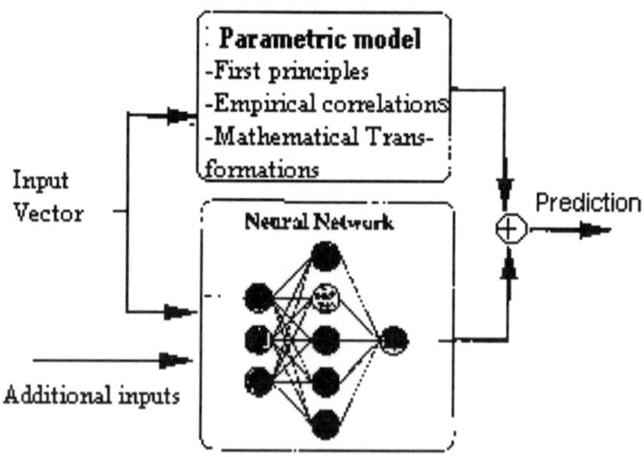

Figure 2. Parallel combination of a parametric model and a neural network

A serial hybrid model was implemented by Psichogios and Ungar (1992) to identify a biochemical reactor. The neural network in the form of a multilayer perceptron approximates the unknown kinetic (cell growth rate), which is used as an input parameter of the FPM. The FPM predicts the concentration of substrate and biomass based on the growth rate and the remaining input variables. Compared to standard neural network models, the prediction of the serial hybrid model was found to be more accurate. Further advantages reported are: 1.) better generalization (extrapolation and interpolation) and 2.) fewer training data requirements.

The parallel hybrid model was presented by Su *et al.* (1992) and Thompson and Kramer (1994). In the parallel approach, the hybrid model prediction is an additive combination of the output of the parametric model and the output of the neural network. The neural network compensates for the residuals between the process and the FPM caused by inherent process complexity or unmeasured disturbances.

A parallel hybrid model was applied to a polymerization process by Su *et al.* (1992). The model consists of a multilayer perceptron and a simplified first principle model that describes the polymerization mechanism. The same model structure was also used to model an activated sludge wastewater treatment process (Zhao and McAvoy, 1996). In both cases, the parallel hybrid model structure resulted in improved prediction accuracy.

Kramer *et* al. (1992) employed the parallel hybrid model to predict the behavior of a vinyl acetate polymerization reactor. A radial basis function network was trained to predict the residual between the first principle model and the process. In a second study, Thompson and Kramer (1994) extended the model structure by using a parametric output model in series to the parallel hybrid model. The task of the output model is to guarantee that the predictions are consistent with the physical process. The FPM serves as default estimate of the process if training data are missing. The resulting model was applied to predict the cell biomass and secondary metabolite in a fed-batch penicillin fermentation. Prediction accuracy was found to improve for the hybrid model.

The accuracy of the hybrid modeling approach depends on the quality of the prior knowledge. This is especially true for the serial structure since it is based on the assumption that all essential process behaviors are present in the FPM. The performance of serial and parallel hybrid models was compared in a recent study by Tsen *et al.* (1996) for emulsion polymerization of vinyl acetate in a batch reactor. That comparison, however, could not establish the superiority of either approach.

This study tests the parallel hybrid model with different neural network types. Both local and global neural networks are used in connection with first principle models. While local neural networks ensure that the hybrid model extrapolates according to the first principle model, global networks influence the output of the hybrid model throughout the entire regime. In addition to these "static" networks, a continuously adapted local network was implemented. The weight parameters of this network are continuously adjusted according to the process characteristics. The extrapolation quality of the hybrid model with an on-line adapted neural network depends on both submodels, the first principle model and the neural network.

The hybrid models were evaluated for the identification of a fermentation process. A yeast fed-batch fermentation which was simulated with Bellgardt's model (Bellgardt, 1991) was modeled with different parallel hybrid models. The model proposed by Fukuda et al. (1978) was used as the first principle model in the hybrid framework. Opposite to previous studies with hybrid models, the process and the first principle model are based on modeling approaches derived independent of each other.

2. Neural Network Models

The following paragraph will briefly review the different neural networks incorporated in the parallel hybrid modeling approach. For a more detailed description, please refer to Bishop (1995) and Ripley (1996). Three neural networks were employed in this study: a multilayer perceptron (MLP), a radial basis function network (RBFN), and ail adaptive radial basis function network (ARBFN). The MLP and RBFN were trained using off-line training algorithm, hence the resulting networks are "static" since they have fixed parameters. In the hybrid framework, the neural networks were trained to predict the residuals between the FPM and the process. The input vector of the neural network consists of delayed residuals and process inputs. Inputs not included in the FPM may also be added to improve prediction accuracy.

2.1. Multilayer Perceptron

The multilayer perceptron belongs to class of global mapping neural networks. Their basis functions influence a large part of the input space. Typical basis functions are the nonlinear sigmoid function and the hyperbolic tangent function which was chosen in this case study. The parameters of the basis function were adjusted using the Levenberg-Marquardt algorithm which is a well known nonlinear least-squares optimization procedure. It is a second order Gauss-Newton type method. As shown in Hagan and Menhaj (1994), the Levenberg-Marquardt algorithm is a very efficient and robust learning algorithm for neural networks with up to a few hundred weights. The application of this method to larger networks, however, is restricted by computational requirements. The update of the weights Δw is calculated according to

$$\Delta w = \left[J^T(w) J(w) + \lambda I \right]^{-1} J^T(w) e(w) \tag{1}$$

where I denotes the identity matrix, λ is a parameter and $J(w)$ is the Jacobian matrix with respect to the weights

$$J(w) = \begin{bmatrix} \frac{\partial e_1(w)}{\partial w_1} & \frac{\partial e_1(w)}{\partial w_2} & \cdots & \frac{\partial e_1(w)}{\partial w_n} \\ \frac{\partial e_2(w)}{\partial w_1} & \frac{\partial e_2(w)}{\partial w_2} & \cdots & \frac{\partial e_2(w)}{\partial w_n} \\ \vdots & \vdots & \ddots & \vdots \\ \frac{\partial e_N(w)}{\partial w_1} & \frac{\partial e_N(w)}{\partial w_2} & \cdots & \frac{\partial e_N(w)}{\partial w_n} \end{bmatrix} \tag{2}$$

The learning algorithm can be summarized as follows:

1. Initialize the weights *w(0)* at random. Set the parameter λ to an initial value (e.g. $\lambda = 0.01$).
2. Present the training data pairs and calculate the error *E* between the network output *Y* and the target values *T*. Compute the performance index.
3. Calculate the Jacobian matrix *J(w)* and the update of the weights Δw according to equation 1.
4. Calculate the performance index for the new weights $w+\Delta w$. If the index is lower than that for the previous weights *w*, then reduce the parameter λ to $\lambda_{new} = \lambda_{old}/10$. If the performance index is greater, than apply $\lambda_{new} = 10\ \lambda_{old}$ .Go back to step 1.
5. Reiterate until a stopping criterion (e.g., SSE is smaller than a preset value) is satisfied.

2.2. *Radial Basis Function Network*

The RBFN was trained in two steps. First, the centers of the Gaussian basis functions were learned using the adaptive k-means clustering algorithm. The adaptive k-means clustering algorithm developed by Chinrungrueng (1993) determines an optimal clustering solution with an efficient adaptive learning rate. The structure of the algorithm is as follows:

1. Initialize the K centers by randomly selecting K vectors $\vec{c}_1,...,\vec{c}_k$ from the input domain consisting of N vectors $\vec{x}_1,...\vec{x}_N$
2. Determine the membership function $M_i\left(\vec{x}_j\right)$, $1 \le i$, $k \le K$ and $1 \le j \le N$ according to:

$$M_k\left(\vec{x}_j\right)=\begin{cases}1 & \text{if} \quad v_k(t)\left(\left\|\vec{x}_j-\vec{c}_k\right\|^2\right) \le v_i(t)\left(\left\|\vec{x}_j-\vec{c}_i\right\|^2\right) \quad \text{for all} \quad i \ne k \\ 0 & \text{otherwise}\end{cases}$$

(3)

where $v_i(t)$ denotes the variance of center *i* and *t*, the iteration.

3. Update the variance $v_k(t)$ using the equation:

$$v_k\left(k+1\right)=\alpha v_k\left(t\right)+\left(1-\alpha\right).\left[M_k\left(\vec{x}_j(t)\right)\left\|\vec{x}_j\left(t\right)-\vec{c}_k\left(t\right)\right\|^2\right] \tag{4}$$

where α is a constant.

4. Calculate the adaptive learning rate η based on

$$\eta=1-\frac{H\left(\overline{v}_1,\overline{v}_2,\ldots,\overline{v}_K\right)}{\ln K} \tag{5}$$

where

$$H(\overline{v}_1,\overline{v}_2,\ldots,\overline{v}_K)=\sum_{i=1}^{K}-\overline{v}_i \ln \overline{v}_i \tag{6}$$

with

$$\bar{v}_i = \frac{v_i}{\sum_{l=1}^{K} v_l}$$

The learning rate η depends only on the values of the variations vi and is limited to a range between 0 and 1. It is close to 1 when the current partition is far from the optimal solution, and close to 0 when it is close to a final optimal partition with all equal within-region variances.

5. Calculate the new center $\vec{c}_k(t+1)$ according to

$$\vec{c}_k(t+1) = \vec{c}_k(t) + M_k(\vec{x}_j(t))\left[\eta(\vec{x}_j(t) - \vec{c}_k(t))\right] \tag{7}$$

6. Go to step 2 until the clustering process has converged, i.e., until there is no change in within-region variances.

2.3. Adaptive Radial Basis Function Network

The training algorithms presented in the previous sections are based on batch-mode learning, i.e. the network parameters are calculated off-line based on the information contained in the training data. Adaptive RBFN learning, however continuously adjust their topology to the complexity of the process dynamics. Several methods for training a RBFN on-line can be found in the literature. Chen *et al.* (1992) present a recursive hybrid algorithm that allows on-line adaptation of the network. However, the fact that the number of centers is fixed at the start of the computation restricts the ability of these on-line methods to process time-varying systems. A Resource Allocating Network (RAN) that dynamically increases the number of activation functions is presented by Platt (1991) and Lowe and McLachlan (1995). An extended version of the original algorithm for NARX (Nonlinear AutoRegressive with eXogenous inputs) is implemented by Sargantanis (1996). Luo *et al.* (1996) present an algorithm for on-line adaptation of NARX and RBFN models called GFEX (Givens rotation with Forward selection and exponential windowing). The GFEX algorithm provides a possible solution to the on-line structure modification and parameter updating of a RBFN. In the present study, we follow the method presented by Luo *et al.* (1996).

The implementation of an adaptive RBFN consists of three parts: generation of candidate RBF centers, recursive orthogonal transformations, and on-line structure detection. The centers used at each time point are selected from all candidate centers. In the GFEX algorithm, the first data point is defined to be the first candidate center. With each new data point collected, the distances between the new data point and the existing candidate centers, d_i, $i = 1, 2, \ldots$, are computed. If the minimum d_i is larger than d_c a new candidate center located at this point is created, where d_c is a tolerance limit. The candidate centers are related to a set of weight factors, w_i. The weights of the newly created candidate center and all previously selected centers are set to unity for the current time point. The weight factors for candidate centers that are not selected for use at the present time point are multiplied by $\sqrt{\lambda}$, where λ is the forgetting factor in the exponential windowing algorithm. If the weight factors of a candidate center is less than the tolerance, it is eliminated since it has not been used for a period defined by the asymptotic memory length.

While many candidate regressors may emerge in the initial stage of selection, most of them are insignificant or linearly dependent. The linearly independent regressors may be decomposed and the significant regressors determined by a preset tolerance limit for the residual error. The number of selected regressors will typically be less than the number of candidate variables. Therefore, the selection is continued for m_s steps until the normalized residual error reaches the preset tolerance level. Generally, the selected regressors differ at each computational interval and the number of selected regressors m_s is time-varying. The contribution of each candidate regressor can then be computed (Luo et al., 1994). Since the number of candidate centers is variable, the dimension of the augmented matrix needs to be adjusted on-line. All elements in the new column of the augmented matrix are initialized to very small values at time t. The data associated with these new variables are added successively and then computed.

The initialization of the- adaptive RBFN is limited to the calculation of' initial centers based on the adaptive k-means clustering. The neural networks implemented in the hybrid modeling architecture are denoted as MLP-HM, RBFN-HM and ARBFN-HM.

3. Case Study: Modeling A Fed-Batch Fermentation

The hybrid modeling approach was applied to a fed-batch fermentation of Saccharomyccs cerevisiae. Several FPMs of the fermentation process have been presented (Fukuda et al., 1978; Barford, 1990; Coppella and Dhurjati, 1990; Bellgardt, 1991; Dantigny et al., 1992; Kristiansen, 1994). This study uses a detailed first principle model (Bellgardt, 1991) for the simulation of the process. Bellgardt's model consists of a reactor model and a kinetic regulator model. It simulates the process with ten ordinary differential equations. Compared to Bellgardt's model, the structured model developed by Fukuda et al. (1978) is a simpler description of the process. His approach was implemented in the hybrid modeling framework as the FPM.

3.1. Simulating The Yeast Fermentation

The fermentation reactor is described by Bellgardt's model (Bellgardt, 1984; Bellgardt, 1991). This model consists of a reactor model describing both gas and liquid phases and a cell model describing the kinetics. The reactor model describes the concentration of each ingredient in the gas and liquid phase. The following model equations are used for the cell mass c_x, substrate molasses c_s, ethanol c_e, dissolved oxygen c_o, dissolved carbon dioxide c_c, and the liquid volume V:

$$\frac{dc_x}{dt} = r_x - \frac{F}{V}.c_x \tag{8}$$

$$\frac{dc_s}{dt} = -r_s + \frac{F}{V}.\left(c_{so}^l - c_s\right) \tag{9}$$

$$\frac{dc_e}{dt} = r_e - \frac{F}{V}.c_e + ETR \tag{10}$$

$$\frac{dc_o}{dt} = -r_o + \frac{F}{V}.\left(c_{oo}^l - c_o\right) + OTR \tag{11}$$

$$\frac{dc_c}{dt} = r_c + \frac{F}{V}.\left(c_{co}^l - c_c\right) + CTR \tag{12}$$

$$\frac{dV}{dt} = F \tag{13}$$

ETR, CTR, and *OTR* denote the mass transfer between gas and liquid phases. The molasses flow rate, F, is the main manipulating variable of the reactor. It determines the increase in volume of the liquid phase and the related dilution effect for the process variables. The sugar concentration in the feed c_{so}^l is an operating parameter. The reactions rates for cell growth, substrate and oxygen uptake, as well as ethanol and carbon dioxide production are calculated using the cell model. The mass transfer rates for oxygen, carbon dioxide and ethanol are determined by the mass transfer model. The temperature is assumed to be constant. The main components of the gas phase are oxygen, carbon dioxide, nitrogen, ethanol, and water. The model equations derived from molar balances of the gas phase components are

$$\frac{dx_o}{dt} = \frac{p_{in}T_lF_g}{pT_{in}V_g}.x_{o_{in}} - \frac{F_{go}}{V_g}.x_o - \frac{RT_lV_l}{M_opV_g}.OTR \tag{14}$$

$$\frac{dx_c}{dt} = \frac{p_{in}T_lF_g}{pT_{in}V_g}.x_{c_{in}} - \frac{F_{go}}{V_g}.x_c - \frac{RT_lV_l}{M_cpV_g}.CTR \tag{15}$$

$$\frac{dx_n}{dt} = \frac{p_{in}T_lF_g}{pT_{in}V_g}.x_{n_{in}} - \frac{F_{go}}{V_g}.x_n \tag{16}$$

$$\frac{dx_e}{dt} = -\frac{F_{go}}{V_g}.x_e - \frac{RT_lV_l}{M_epV_g}.ETR \tag{17}$$

$$\frac{dx_w}{dt} = \frac{p_{in}T_lF_g}{pT_{in}V_g}.x_{w_{in}} - \frac{F_{go}}{V_g}.x_w - \frac{RT_lV_l}{M_wpV_g}.WTR \tag{18}$$

The positive direction of the mass transfer streams, OTR, CTR, ETR, and WTR, is directed to the liquid phase. It is assumed that no nitrogen is exchanged between gas and liquid phases, and that no ethanol is present in the air flow at the inlet. The mass transfer rate between gas phase and liquid phase is proportional to the concentration gradient in the interfacial area and to the volumetric mass transfer coefficient.

$$OTR = (k_La)_o.(c^*{}_o - c_o) \tag{19}$$

$$CTR = (k_La)_c.(c^*{}_c - c_c) \tag{20}$$

The saturation concentrations for oxygen and carbon dioxide can be calculated according to

$$c_i^* = \frac{H_i \rho_w M_i p}{M_w} . x_i \qquad \text{for} \quad i = o, \; c . \tag{21}$$

The influence of the stirrer speed on the mass transfer coefficient is modelled using the following set of equations. The mass transfer for a stirred bioreactor is calculated as (Van't Ried, *1983)*

$$(k_L a)_0 = 3600. \left(0.026. \left(\frac{P}{V} \right)^{0.4} . \sqrt{v_{gas}} \right) \tag{22}$$

where the linear gas velocity v_{gas} is given by

$$v_{gas} = \frac{4.G_{flow}}{3600\pi.D^2{}_r} \tag{23}$$

The gas flow rate is denoted by G_{flow}. The geometrical parameters are the diameter of the reactor D_R and the diameter of the stirrer D_s . The power consumption P for mechanical agitation is described by

$$P = P_{no}.\rho. \left(\frac{N_{stir}}{60} \right)^3 .D_s^5 \tag{24}$$

where P_{no} denotes the power number and ρ is the density of the liquid. The following correlation between the mass transfer for oxygen $(k_L a)_o$ and the effect of temperature and biomass concentration was suggested by Kristiansen:

$$(k_L a)_0 = (k_L a)_o .(1 - 0.00176.c_x).1.022^{(T-20)} . \tag{25}$$

Based on this value, mass transfer for ethanol and carbon dioxide may be calculated according to

$$\left(k_L a\right)_e = \sqrt{\frac{1.28}{2.5}\cdot\left(k_L a\right)_0} \tag{26}$$

and

$$\left(k_L a\right)_c = \sqrt{\frac{1.96}{2.5}\cdot\left(k_L a\right)_0} \tag{27}$$

The cell model as used in this simulation consists of two parts: the metabolic model for kinetics and stoichiometry of growth and, the regulation model for metabolic long-term regulation. This study uses the model of Bellgardt called metabolic regulator approach. The uptake of different substrates and the formation, of primary metabolites is taken into account. The metabolic regulator is a suitable approach if the product formation depends on the growth condition in the fermenter. During the yeast fermentation, the metabolism can be directed to any mixture of fermentative growth with ethanol formation or oxidative growth with high cell yield, depending on substrate and oxygen. The stoichiometry of growth is described by Eq. 28.

$$\begin{bmatrix} r_s \\ 0 \\ 0 \\ 0 \\ m_{ATP} \\ 0 \end{bmatrix} = \begin{bmatrix} -1 & 0 & 0 & 0 & K_{EG} & 0 & 1 & K_{B1} & 0 \\ -2 & 0 & -1 & 0 & -1-K_{Ad} & 1 & 2 & 0 & 0 \\ 0 & 0 & -1 & 1 & 0 & 0 & 0 & K_{B3} & 0 \\ -2 & 2 & 1 & 4 & -1-2K_{EG} & 1 & 2 & -K_{B2} & 0 \\ -3 & 2P/O & 0 & 1 & -2K_{EG} & 0 & 2 & -Y^{-1}{}_{ATP} & 0 \\ 0 & 0 & -1 & -2 & -1-K_{Ad} & 0 & 0 & 0 & 1 \end{bmatrix} \begin{bmatrix} r_g \\ r_o \\ r_{ac} \\ r_{tc} \\ r_{ep} \\ r_{ec} \\ r_s \\ r_x \\ r_c \end{bmatrix} \tag{28}$$

The inherently rate-limiting steps in the model are the glucose uptake rate qs, for which a Monod kinetic can be assumed, and the uptake steps for ethanol r_e, and oxygen r_o. The latter terms are introduced as first, order kinetics. The optimal pathways for the microorganism is found by maximizing the specific growth rate (Eq. 29) under the constraints, which are caused by transport limitations or internal reaction mechanisms (Eq. 30).

$$r_x(t) \Rightarrow \quad \text{maximum} \tag{29}$$

$$\begin{aligned}
&0 \le r_g \le r_{g\max} \\
&0 \le r_o \le \min\left(K_o.c_o, r_{o_{\max}}\right) \\
&0 \le r_{ac} \le r_{ac_{\max}} \\
&0 \le r_{tc} \le \infty \\
&0 \le r_{ep} \le \infty \\
&0 \le r_{ec} \le K_{E1}.c_e \\
&0 \le r_s \le \infty \\
&-\infty \le r_x \le \infty \\
&0 \le r_c \le \infty
\end{aligned} \tag{30}$$

The metabolic regulator approach yields set of metabolic models. Depending on the operating conditions, one of these models is utilized to describe the current growth phenomena of the yeast. The set of metabolic models consists of

- Model 1: oxidative growth on glucose.
- Model 2: aerobic fermentative growth on glucose (Crabtree effect).
- Model 3: anaerobic or oxygen limited growth on glucose.
- Model 4: oxidative growth limited by ethanol and glucose.
- Model 5: oxidative growth limited by ethanol and acetyl-CoA.
- Model 6: oxidative growth limited by glucose and oxygen.
- Model 7: oxidative growth limited by glucose and enzymes of gluconeogenesis.

3.2. FPM for Hybrid Modeling

A simple mathematical description of the fed-batch culture in a well mixed bioreactor was presented by Fukuda et al. (1978). The growth of baker's yeast is assumed to be limited by the inhibitory substances produced by microorganisms. The relationships describing baker's yeast cultivation are expressed as:

$$\frac{d(Vc_x)}{dt} = \mu.Vc_x - a_1 Y_{x/e} \nu_e.Vc_x \tag{31}$$

$$\frac{d(Vc_s)}{dt} = Fc^l{}_{so} - \mu.\frac{Vc_x}{Y} - m.Vc_x - \frac{a_2 \pi_e}{Y_{e/s}}.Vc_x \tag{32}$$

$$\frac{d(Vc_e)}{dt} = a_2 \pi_e.Vc_x - a_1 \nu_e.Vc_x \tag{33}$$

$$\frac{d(Vc_p)}{dt} = k_1.Vc_x - k_2 \mu.Vc_x \tag{34}$$

$$\frac{d(V)}{dt} = F \tag{35}$$

where

$$\pi_e = 0.155 - 0.123.\log c_s \tag{36}$$

$$\nu_e = 0.138 - 0.062.c_p - \frac{0.0028}{(S - 0.28)} \tag{37}$$

$$\mu = \mu_{\max} \cdot \frac{c_s}{(k_s - c_s)(1 - c^2{}_p)}. \tag{38}$$

The parameter π_e is the specific ethanol production rate and ν_e is the specific growth rate for the assimilation of ethanol. Whenever π_e and ν_e are negative values, they should be taken as zero. If c_e is zero, then ν_e is zero as well. In the above equation, c_x, c_s, and c_p represent the concentrations of biomass, substrate, inhibiting substance, and ethanol respectively. F represents the flow rate, c^i_{so} is the concentration of feed (molasses) and V is the working volume of the fermenter.

Constant parameters are taken as $k_1 = 0.0023$, $k_2 = 0.0070$, $k_s = 0.025$, $m = 0.03$, $Y = 0.5$, $\mu_{max} = 0.42$, $Y_{x/e} = 0.48$, and $Y_{e/s} = 0.51$ in the numerical calculation. The following relationships are required for constants a_i:

$$\begin{aligned} a_1 = 0, \quad a_2 = 1 \quad \text{for} \quad S > 0.28 \\ a_1 = 1, \quad a_2 = 0 \quad \text{for} \quad S \leq 0.28 \end{aligned} \tag{39}$$

The initial concentration of the inhibitor is assumed to be zero, $c_p(0) = 0$.

3.3. Creating Training And Testing Data

During the training phase, the fermentation process described by Bellgardt's model was simulated for 25 hours. A sampling interval of 30 minutes was assumed. The training and testing data were created by pseudo-randomly varying the input variables in the following ranges:

- Feed flow rate F: 13 l/h $\leq F \leq$ 27 1/h,
- Substrate feed concentration c^l_{so} : 230 kg/m^3 $\leq c^l_{so} \leq$ 420 kg/m^3,
- Gas flow rate G_{flow}: 18 m^3/h $\leq G_{flow} \leq$ 40 m^3/h,
- Stirrer speed N_{stir}: 18 rpm $\leq N_{stir} \leq$ 22 rpm.

The following variables were assumed to be measurable during the fermentation: the concentration of biomass c_x, the concentration of substrate c_s, the concentration of ethanol c_e, the concentration of oxygen c_o, the concentration of carbon dioxide c_c, the respiratory quotient RQ, and the dilution rate D. Additional noise with a signal to noise ratio of 30 dB was superimposed on these variables. The data of 32 fermentation simulations were collected for the training process. This data set was split in a training set containing data from 26 batches and a testing set consisting from 6 fermentation simulations. The input variables were kept within their ranges when the data were created. In addition to these two data sets, 5 batches were created to test the interpolation and extrapolation capabilities of the models. For the extrapolation tests, the input variables were allowed to vary beyond the limits given above.

3.4. Identification And Modeling

The different types of neural networks and hybrid models were used to predict future values of the biomass concentration c_x. The input vector to the FPM consisted of the input variables F and c^l_{so} and the state variables ($c_x(t)$, $c_s(t)$, $c_e(t)$, $c_p(t)$). The neural networks used in the hybrid model served two purposes:

1. The neural networks were trained to capture the influence of the input variables not included in Fukuda's model. Theses variables are the gas flow rate G_{flow} and the stirrer speed N_{stir}.
2. The neural network compensated for the difference between the biomass prediction of the First Principle Model, c_x^{FPM} and the process concentration c_x.

The input vector of the neural networks was chosen as

$$\vec{x}(t) = \left[x_x(t), c_s(t), c_e(t), c_o(t), RQ(t), D(t), F(t), c^l{}_{so}(t), G_{flow}(t), N_{stir}(t) \right]^T$$

where $x_x = c_x$ for the neural network models, and $x_x = c_x - c_x^{FPM}$ for the hybrid models. In addition to the one-step-ahead prediction, the models were evaluated for long-term predictions as k-step-ahead predictor, i.e., $c_x(t+k) = \Phi(\S(\int))$ where k = 1,...,6. For long range predictions, the neural network was linked to itself. This approach was proposed by SaintDonat et al. (1991) and proved to reflect the process dynamics more accurately than a single network mapping. The weight parameters were kept constant at each iteration while the missing variables of the input vector, i.e., $y(t+1)$ and $y(t+2)$ were substituted by their predicted values $\hat{y}(t+1)$ and $\hat{y}(t+2)$. This "chaining" method was implemented for the neural network models and the hybrid models.

After the neural network and the hybrid model were trained on the data from 26 batches as described above, the interpolation and extrapolation capabilities of the models were tested using the data of 3 fermentation simulations. The input variables were altered as follows:

- Fermentation simulation 1: c^l_{so} = 100 kg/m^3, N_{stir} = 10 rpm, F = 25 1/h, and G_{flow} = 10 m^3 /h.
- Fermentation simulation 2: c^l_{so} = 100 kg/m^3, N_{stir} = 35 rpm, F = 25 1/h, and G_{flow} = 60 m^3 /h.
- Fermentation simulation 3: c^l_{so} = 500 kg/m^3, N_{stir} = 35 rpm, F = 30 1/h, and G_{flow} = 60 m^3 /h.

• Fermentation simulation 4: c^l_{so} = 280 kg/m^3, N_{stir} = 30 rpm, F = 18 1/h, and G_{flow} = 25 m^3 /h.
• Fermentation simulation 5: c^l_{so} = 330 kg/m^3, N_{stir} = 25 rpm, F = 22 1/h, and G_{flow} = 37 m^3 /h.

The extrapolation qualities of the models were evaluated based on the first three runs which include high and low values, for the gas flow rate and the substrate feed concentration. The data for batch four and five are contained in the range of the training data and evaluate how good the models interpolate between the "learned" data points. In addition to the changes in the input variables, these five fermentations were simulated over a period of 36 hours, while the batches included in the training set lasted 25 hours. Hence, the last 11 simulation hours require the model to extrapolate.

3.4.1. Multilayer Perceptron Models

In this subsection, the results for MLPs with and without the support of the FPM are compared. The optimal number of hidden nodes was found to be 12 for the MLP model and 10 for the MLP-HM. The hyperbolic tangent activation function was chosen for both networks.

The modeling results for the MLP model and the MLP-HM are summarised in Table 1. The prediction accuracies are shown for the k-step ahead prediction (k = l,...,6). The data in these tables are based on the root- mean- squared (RMS) error between the prediction and the process values.

These results are illustrated for the 1-step ahead predictor for the fermentation runs 1, 3, and 4 in Fig. 3, Fig. 4, and Fig. 5. For fermentation run 1 (Fig. 3), Fukuda's model is very accurate during the first 15 hours of the fermentation, but it can not describe the process during the second half of the fermentation. The MLP model describes the process dynamics fairly well for the first 20 hours. A constant bias is present for the last 16 hours of the process simulation. The MLP-HM approach improved the prediction accuracy. The output hybrid model seems to follow Fukuda's model for the first phase of the fermentation. During the second phase, the MLP-HM compensates for the error between Fukuda's model and the process model. This transition takes place after 20 hours and is characterized by a temporary model inaccuracy that lasts for approximately four hours. The results for the k-step ahead prediction arc also included in Table 1. The accuracy of the pure MLP model decreases significantly for 3 and 6-stepahead predictions.

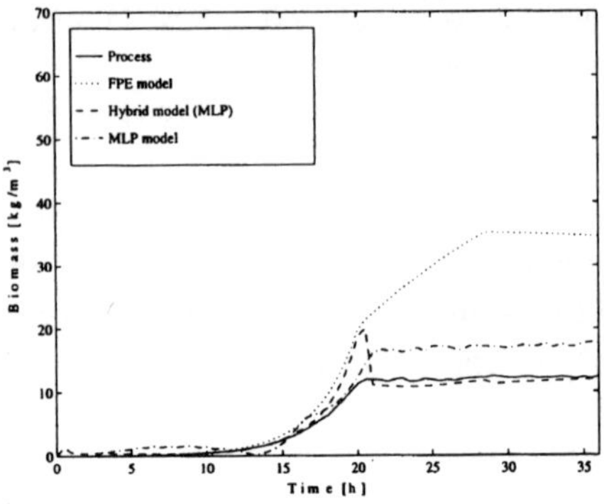

Figure 3. Comparison of the MLP based modeling approaches for run 1

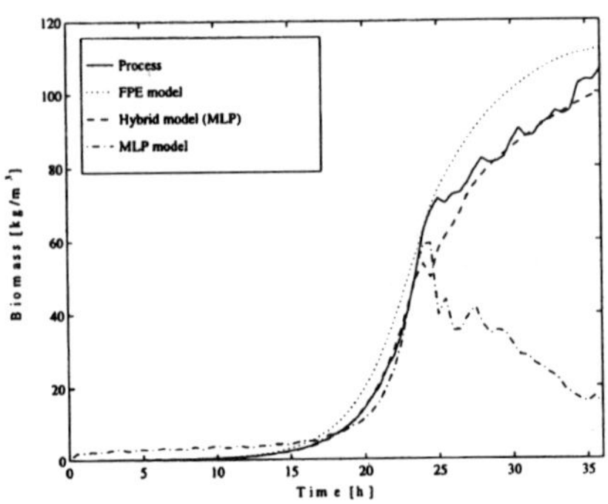

Figure 4. Comparison of the MLP based modeling approaches for run 3

The fermentation run 3 shown in Fig. 4 shows a completely different behavior than the previous batch. The biomass concentration is significantly higher due to existing operating conditions. In this case, the FPM closely describes the process dynamics for the complete simulation time. However, the MLP-HM achieves improved prediction accuracy, especially for the last 12 simulation hours. During this period, the "pure" MLP model shows a very poor performance. The MLP model is not able to extrapolate and predict the data for this run accurately. This also holds true for multistep ahead predictions. On the other hand, the MLP-HM produces reasonable results even for long term predictions. Its performance is very consistent for 1-through 6-step ahead predictions, with the RMS error increasing only slightly.

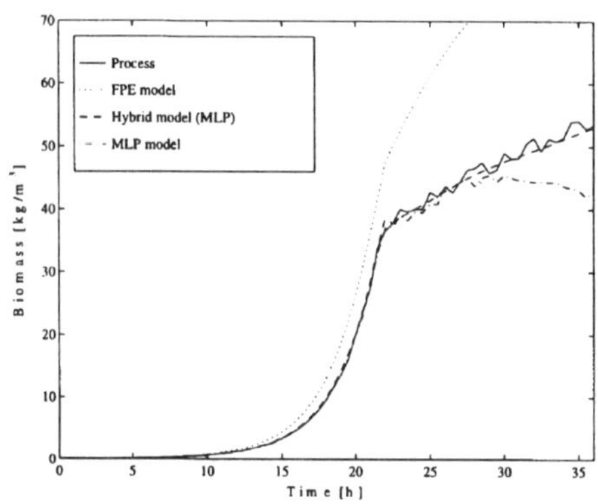

Figure 5. Comparison of the MLP based modeling approaches for run 4

The interpolation capabilities of the models are reflected in the figure for simulation 4 (Fig. 5). In this case, Fukuda's model overestimates the biomass concentration during the final stages of the fermentation. The MLP-HM, however, leads to accurate predictions of the cell mass. The hybrid model as a 1-step ahead predictor achieves a very close approximation of the real process data with a RMS error of less than 0.75. The model accuracy deteriorates for long term predictions. The MLP model interpolates very well as a one-step ahead predictor during the first 28 hours of the simulation. However, it is not able to extrapolate during the last 8 hours and the predicted cell mass concentration is much lower than the values given by the process model. Employing the MLP for long term prediction leads to higher modeling errors.

Table 1. Modeling results for MLP based models (RMS error).

	k-step-ahead prediction (MLP)					
Run	1	2	3	4	5	6
1	3.3867	6.5599	8.7144	9.3571	9.5316	9.7071
2	3.0347	3.7779	5.0240	6.0931	6.2191	7.1087
3	34-5559	21.4110	20.1394	17.3198	21.0982	18.8572
4	3.3978	3.8728	4.1940	4.6442	5.1247	5.6018
5	1.6980	1.6925	2.0776	2.9865	3.8468	4.8285
	k-step-ahead prediction (MLP-HM)					
1	1.6301	1.2350	2.5901	2.5538	3.8666	4.6246
2	1.4248	1.8388	2.0080	2.0658	2.1362	2.2505
3	3.6674	3.5911	3.5738	3.7268	3.9419	4.1866

3.4.2. Radial basis function models

A RBFN model with Gaussian basis function was designed based on the training data set. The optimal number of hidden nodes for this application was determined to be 18. The center vectors corresponding to these hidden nodes were calculated using the adaptive k-means algorithm.

The RBFN model and the hybrid RBFN model were tested on the same 5 fermentation simulations. The results in form of the RMS error are summarized in Table 2. Three runs are shown in Figs. 6, 7, and 8 which demonstrate the model characteristics.

Fermentation run 4 is represented accurately by the RBFN model and the hybrid RBFN model. This is especially true for the one-step ahead prediction. The simulation results are shown in the Fig. 8. The long term behavior during these fermentations is more closely described by the hybrid RBFN model.

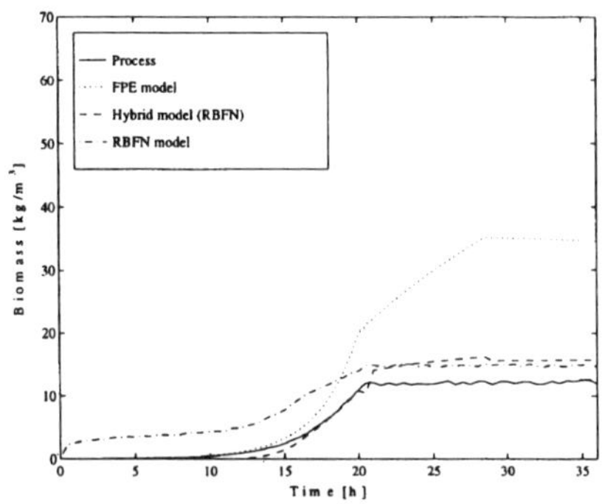

Figure 6. Comparison of the RBFN based modeling approaches for run 1

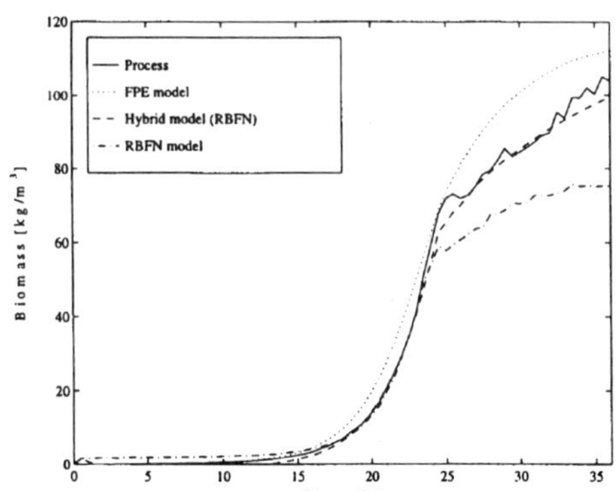

Figure 7. Comparison of the RBFN based modeling approaches for run 3

As shown in Fig. 6, fermentation run 1 is not predicted accurately by the RBFN model. The model predicts higher biomass values for the entire simulation.

The hybrid model produces proper prediction for the first 20 hours of the simulation. During the remaining 16 hours, the offset of the hybrid RBFN model is similar to the error of the "pure" RBFN model. As shown in Table 2, none of the RBFN models can be employed as a multistep predictor for this fermentation run. All models overestimate the cell mass concentration.

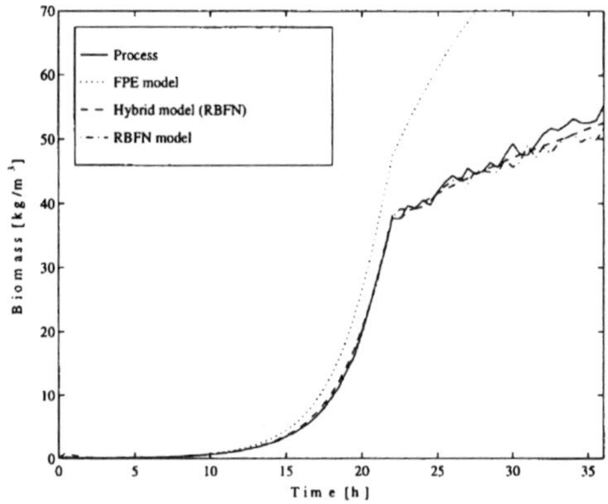

Figure 8. Comparison of the RBFN based modeling approaches for run 4

The hybrid RBFN modeling approach shows very good extrapolation results for the higher biomass values of the third simulation (Fig. 7). The modeling error of the hybrid model is very small throughout the entire simulation. The long term behavior of the process is also approximated properly as indicated by the RMS error. The RBFN model, on the other hand, shows very good results for the first 25 hours, but falls to predict the remainder of the run accurately.

3.4.3 Adaptive radial basis function models

The learning phase of the adaptive radial basis function network (ARBFN) is reduced to determine an optimal set of initial center vectors and to find a pool of suitable candidate centers which might have to be added during the first stage of a fermentation. In this case study, the initial centers were found by using the adaptive k-means clustering algorithm. During the learning phase, 8 centers were selected. The pool of suitable candidate vectors was also determined through the adaptive clustering procedure. In this case, only the data vectors collected during the first 5 hours of the fermentation were clustered since they represent potential center vectors which might have to be added during the initial stage of the fermentation.

Table 2. Modeling results for the RBFN models (RMS error)

	k-step-ahead prediction (RBFN)					
Run	1	2	3	4	5	6
1	3.5693	5.0832	5.7093	5.9115	5.9050	5.7910
2	4.7523	7.0113	8.0533	8.4777	8.5279	8.6975
3	12.1297	17.8413	21.8321	24.8809	27.3911	29-5565
4	1.4384	2.1789	2.9226	3.7137	4.7171	5.6747
5	0.9765	1.3181	1.7783	2.7721	3.8693	5.1422
	k-step-ahead prediction (RBFN-HM)					
1	2.2895	4.9469	6.4685	7.3481	7.8550	8.1440
2	1.4226	2.0179	2.4729	2.7889	3.0419	3.1965
3	1.5942	2.2245	3.0599	3.6714	4.0793	4.3714
4	1.1113	1.0855	1.9428	2.6457	3.1614	3.5708
5	0.8699	1.1585	1.8699	2.4082	2.8145	3.0833

The modeling results for the ARBFN model and the hybrid ARBFN model are listed in Table 3. The results show that the prediction errors increase only marginally for long-term forecasts. This is the case for all fermentation simulations.

It can also be concluded that the combination of a first principle model and the ARBFN does not yield a significant increase in model accuracy. This is valid for all test runs shown in Figs. 9—11.

Table 3. Modeling results for the ARBFN model (RMS error).

	k-step-ahead prediction					
Run	1	2	3	4	5	6
1	0.4577	0.4599	0.5076	0.5662	0.5676	0.6726
2	0.5623	0.5581	0.6296	0.5588	0.5723	0.7379
3	1.3661	1.3383	1.3925	1.3925	1.3949	1.6555
4	0.6687	0.6665	0.6811	0.6920	0.6870	0.6805
5	0.8012	0.8487	0.8994	0.8761	0.8801	1.0523
	k-step-ahead prediction					
Run	1	2	3	4	5	6
1	0.4556	0.4449	0.4452	0.4427	0.4754	0.4755
2	0.5623	0.5576	0.5539	0.5589	0.5712	0.5723
3	1.3657	1.3181	1.2924	1.3924	1.3924	1.3152
4	0.6484	0.6464	0.6604	0.6717	0.6767	0.6803
5	0.7905	0.8385	0.8691	0.8751	0.8599	0.8544

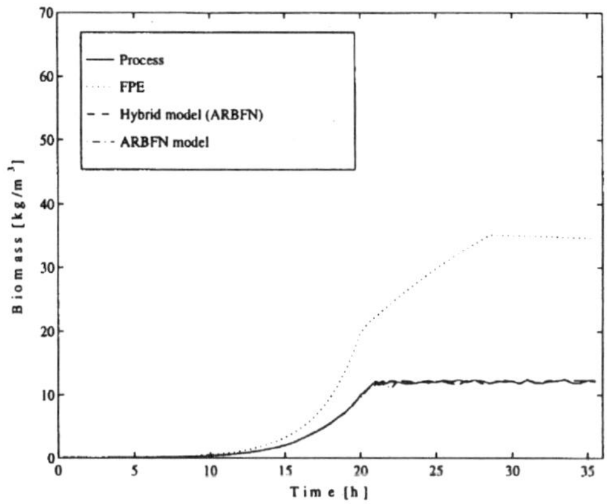

Figure 9. Comparison of the ARBFN based modeling approaches for run 1

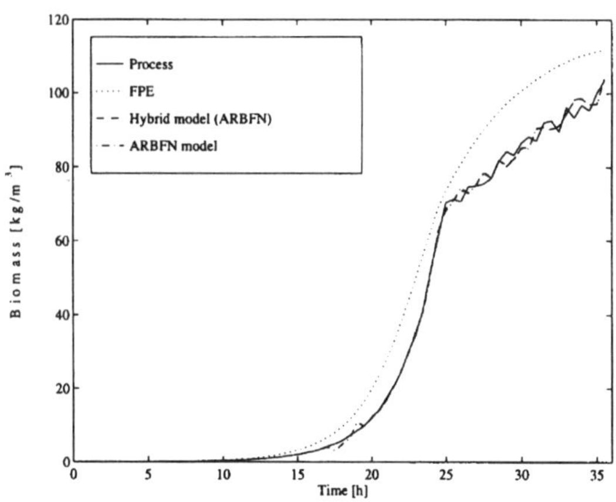

Figure 10. Comparison of the ARBFN based modeling approaches for run 3

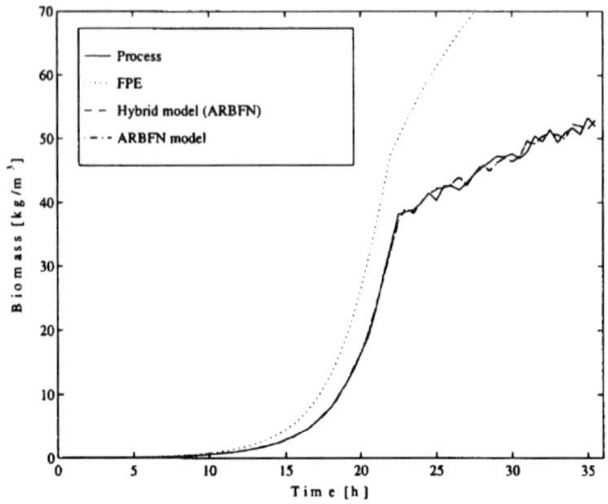

Figure 11. Comparison of the ARBFN based modeling approaches for run 4

4. Conclusion

This paper demonstrates how neural networks and FPMs can be combined to form hybrid models. In particular, the parallel combination of FPMs and neural networks was analyzed. Three different neural network types were applied to the hybrid modeling approach: a RBFN, a MLP, and an adaptive RBFN. Global and local neural networks should affect the extrapolation capabilities of the hybrid model differently. The MLP is generally expected to extrapolate better than the R.BFN. However, for the application presented here, the extrapolations of the RBFN are superior to the predictions of the MLP. The hybrid modeling approach resulted in improved prediction results compared to both the RBFN model and the MLP model. An adaptive radial basis function was also tested in the hybrid modeling framework. Here, no significant improvement in model accuracy was detectable. However, one might argue that, the hybrid approach improves the stability of all on-line adaptive network since the network is trained only on the error between the FPM and the process. The quality of the FPM is critical for the success of hybrid modeling approaches. This is true in particular for the serial combination of FPM and neural network. The case study presented here showed that neural networks are able to compensate for a mismatch between the FPM and the process.

References

Barford, J. P., *Biotechnol. Bioeng.* **35** (1990), 907-920.
Bellgardt, K., *Modellbildung des Wachstums von Saccharomyces Cerevisae in Rührkesselreaktoren*, Ph.D. thesis, (Universitä Hannover, Germany, 1984).
Bellgardt, K. H., in *Biotechnology* vol 4, *eds.* Rehm, H. J. and Reed, G. (VCH, Weinheim, 1991), 383-406.
Bishop, C. M., *Neural Networks for Pattern Recognition* (University Press, Oxford , 1995).
Chen, S., Billings, S., and Grant, P., *International Journal of Control.* **55**(1992), 1051-1070.
Chinrungrueng, C., *Evaluation of Heterogeneous Architectures for Artificial Neural Networks,* PhD thesis, (University of California, Berkeley, 1993).
Coppella, S. J. and Dhurjati, P., *Biotechnol. Bioeng.* **35**(1990), 356-374.
Dantigny, P., Ziouras, K., and Howell, J. A., in *Modeling and Control of Biotechnical Processes,* eds. Karim M.N and Stephanopoulos G. (Pergamon Press, Oxford, 1992), 223-226.
Fukuda, H., Shiotani, T., Okada, W., and Morikawa, H., *Journal of Fermentation Technology.* **56**(1978), 361-368.
Hagan, and Menhaj, M. B., *IEEE Transaction on Neural Networks.* **5**(1994), *989-993.*
Kramer, M. A., Thompson, M. L., and Bhagat, P. M., *Proc. of the American Control Conference* (1992), 475-479.
Kristiansen, B., *Integrated Design of a Fermentation Plant: The Production of Baker's Yeast,* (VCH ,New York, 1994).
Lowe, D. and McLachlan, A., in *Fourth IEE International Conference on Artificial Neural Networks* (1995).
Luo, W., Billings, S. A., and Tsang, K. M, *Technical Report 503*, (University of Sheffield, UK, 1994).
Luo, W., Karim, M. N., Morris, A. J., and Martin, E. B, in *ESCAPE-6,* (Rhodes, Greece, 1996).
Platt, J., *Neural Computation.* **3**(1991), 213-225.
Psichogios, D. C. and Ungar, L. H. (1992), *AIChE Journal.* **38**(1992), 1499-1511.
Ripley, B. D., *Pattern Recognition and Neural Networks,* (Cambridge University Press, Cambridge, 1996).
Saint-Donat, J., Bhat, N., and McAvoy, T. J., *International Journal of Control.* **54**(1991), 1453-1468.
Sargantanis, I, *Model based control with variable structure: Application to DO control for β-Lactamase production,* Ph.D. thesis, (Colorado State University, Colorado, 1996).

Su, H. T., Bhat, N., Minderman, P.A., and McAvoy, T. J., in *IFAC Symp. on Dynamics and Control of Chemical Reactors, Distillation Columns, and Batch Processes (DYCORD)*, (1992).
Thompson, M. L. and Kramer, M. A. , *AIChE Journal,* **40**(1994), 1328-1340.
Tsen, A. Y., Jang, S. S., Wong, D. S. H., and Joseph, B, *AIChE Journal.* **42**(1996), 455-465.
Van't Ried, K. , *Trends in Biotechnology.* **1**(1983), 113-119.
Zhao, H. and McAvoy, T. J., in *Proceedings of the 13th Triennial World Congress, IFAC* , 1996, 455-459.

7. NEURAL NETWORKS IN A HYBRID SCHEME FOR OPTIMISATION OF DYNAMIC PROCESSES: APPLICATION TO BATCH DISTILLATION

M. A. GREAVES, I. M. MUJTABA
Computational Process Engineering Group, Department of Chemical Engineering, University of Bradford, West Yorkshire BD7 1DP, UK.

M. A. HUSSAIN
Department of Chemical Engineering, University of Malaya Kuala Lumpur 59100, Malaysia.

It is well understood that the optimal control policies can be significantly different with and without due consideration to the *plant-model mismatches*. In our previous work, the detailed dynamic model was assumed to be the exact representation of the plant while the difference in predictions of the plant behaviour using a simple model and the detailed model was assumed to be the dynamic *plant-model mismatches*. Theses dynamic mismatches were modelled using neural network techniques and were added to a simple model to produce a hybrid model. Previously, we developed a general optimisation framework based on the hybrid model for dynamic plants.
In this work, a hybrid model for an actual pilot plant batch distillation column is developed. However, taking advantage of some of the inherent properties of batch distillation process a simpler version (new algorithm) of the general optimisation framework is developed to find optimal reflux ratio policies which minimises the batch time for a given *separation task*. Finally, discrete reflux ratio used in most pilot plant batch distillation columns, including those used in industrial R&D Departments, does not allow a direct implementation of the optimum reflux ratio (treated as a continuous variable) obtained using a model based technique. Here a relationship between the continuous and the discrete reflux ratio is developed. This allows easy communication between the model and the plant and comparison on a common basis.

1. Introduction

Continuous processes operating at steady state become dynamic because of external disturbances or during start-up operation (Barolo et al., 1994; Henry et al., 1997). Batch processes, on the other hand, are inherently dynamic and remain dynamic until the end of their operation. Optimal operation (also known as optimal control) of such processes has been the subject of many researchers in the past (Cuthrell and Biegler, 1989; Farhat et al., 1990; Logsdon et al., 1990; Vassiliadis et al., 1994;

Luss,1994; Sorensen and Skogestad, 1996; Mujtaba and Macchietto, 1996, 1998). In most cases, models of these processes (as described by DAEs) are considered to be the exact representative of the system. However, accurate modelling of such processes is often very difficult due to complex non-linearity of the thermophysical properties in addition to basic mass and energy balances. For example, modelling of vapour-liquid equilibrium calculations is often difficult for many non-ideal and azeotropic mixtures. Availability of faster computers and sophisticated numerical methods although allow development of complex models, these models are not completely free from *plant-model mismatches*. Therefore, optimal control policy of dynamic processes can be significantly different with and without due consideration to the *plant-model mismatches*. The nature of the mismatches of a dynamic system is also dynamic. The magnitude of the error in predicting the dynamic behaviour of the actual process using a model depends on the extent of the *plant-model mismatches*. While modelling of steady state mismatches can be simple, the modelling of dynamic mismatches can be much more difficult. The use of standard regression techniques to estimate these *plant-model mismatches* can be extremely difficult due to the inherent non-linearity and dynamic nature of these mismatches.

In the past, methods have been developed to obtain optimal operation using nominal models with some degrees of uncertainties in model parameters (Walsh et al., 1995). In most cases, the model parameters are related to time invariant variables like chemical reaction rate constants, relative volatility, plate efficiencies, etc. The parameters are updated to match the final time constraints (i.e. amount of distillate, product composition, etc., as obtained by the actual process). No attempt has been made to obtain optimal control policies for dynamic processes with due consideration to the dynamic mismatches (between the model and the actual process) of the state variables. Neural network technique is one of the methods employed successfully in the past to model complex steady state processes (Savkovic-Stevanovic, 1994; Woinaroschy et al., 1994). Use of neural network techniques capturing process dynamics is also evident in the literature (Bhat and McAvoy, 1990; Morris et al., 1994). Instead of developing a complex and detailed dynamic process model to minimise the *plant-model mismatches*, we propose in this work, a hybrid scheme where a simple model is coupled with neural network techniques to develop the full process model. In this work an optimal control framework is also developed to obtain optimal operation of dynamic processes described by a hybrid model.

2. The Model and the Actual Process

Dynamic processes are often represented by a set of DAEs of the form:

$$f(t, x'(t), x(t), u(t), v) = 0, \quad [t_0, t_F] \tag{1}$$

where t is the time, $x(t)\varepsilon R^n$ is the set of all state variables, x'(t) denotes the derivatives of x(t) with respect to time, $u(t)\varepsilon R^m$ is a vector of control variables, and $v\varepsilon R^p$ is a vector of time invariant parameters (design variables). The time interval of interest is $[t_0, t_F]$ and the function $f{:}RxR^nxR^nxR^mxR^p \rightarrow R^n$ is assumed to be continuously differentiable with respect to all its arguments (Morison, 1984).

In many chemical processes, especially inherently dynamic processes, it is not always possible to model the actual process. Therefore, the states predicted by using the model (Eq. 1) will be different than that of the actual process and will result in *plant-model mismatches*. The implementation of the optimal operating policies obtained using the model will not result in a true optimal operation. Regardless of the nature of the mismatches, a true process can be described (Agarwal, 1996) as:

$$f(t, \underline{x}'(t), \underline{x}(t), u(t), \underline{v}, e_x(t)) = 0, \qquad [t_0, t_F] \tag{2}$$

where $\underline{x}(t)$ is the true set of all state variables, $\underline{x}'(t)$ denotes the derivatives $\underline{x}(t)$ with respect to time; $\underline{v}(t)$ is the true set of time independent design variables; $e_x(t)$ is the set of *plant-model mismatches* for the state variables x; and the control vector u, and the function f are identical to those used in the model (Eq. 1).

The error $e_x(t)$ is in general time dependent and describes the entire deviation due to *plant-model mismatches*. Structural incompleteness in the model, reformulation of the model equations as needed by a particular solution algorithm, discrepancy between v and $\underline{v}$, inaccurate initial estimate of $x(t_0)$ of the model, inaccuracies in the measurement of u, unmeasured disturbances, simplified assumption in the estimation of thermo-physical properties of the process, etc., can result in these mismatches (Agarwal, 1996). The error $e_x(t)$ takes into account of all these sources of mismatches.

At any time t, the true estimation of the state variables requires instantaneous values of the unknown mismatches $e_x(t)$. To find the optimal control policies in terms of any decision variables (say z) of a dynamic process using the model will require accurate estimation of $e_x(t)$ for each iteration on z during repetitive solution of the optimisation problem. Although estimation of *plant-model mismatches* for a fixed operating condition (i.e. for one set of z variables) can be obtained easily, the prediction of mismatches over a wide range of operating conditions can be very difficult.

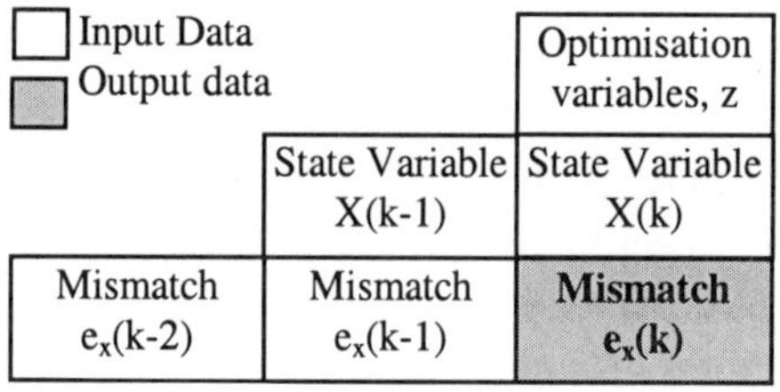

Figure 1. General Input/Output Map of Neural Network

3. Hybrid Modelling of Dynamic Process

In this work we model the actual process (Eq. 2) by combining a simple dynamic model (of type Eq. 1) and the *plant-model mismatches* ($e_x(t)$) model.

3.1. Modelling of Dynamic Plant-model Mismatches

As the mismatches of the state variables of a dynamic system (i.e. instant distillate and reboiler compositions in batch distillation) are dynamic in behaviour, they have to be treated as such and not as static processes. To develop them from first principles would be very difficult due to their non-linear dynamic behaviour and it would also be difficult to quantify them in terms of the original state variables.

However, neural networks have been known to be able to approximate non-linear continuous functions with a high degree of accuracy (Cybenko, 1989; Hussain et al., 1995). In this work neural network techniques are used to model these *plant-model mismatches*. This method would also be suitable and appropriate in dealing with the estimation of these mismatches on-line, due to its fast implementation time. Although black box in nature, it has the ability to approximate any function mapping of system inputs to outputs, from known input-output data. The method of training the neural network to perform systems identification i.e. prediction of the mismatches at discrete-time intervals is called forward modelling. In this procedure, the neural network is fed with various input data to predict the *plant-model mismatch* (for each state variable) at the present discrete time. The general input-output map for the neural network training can be seen in Fig. 1. The data are fed in a moving window scheme. In this scheme, all the data are moved forward at one discrete-time interval until all of them are fed into the network. The whole batch of data is fed into the network repeatedly until the required error criterion is achieved. The error between the actual mismatch (obtained from the simulation results) and that predicted by the network is used as the error signal to train the network (see Fig. 2). This is the classical supervised learning problem, where the system provides target values directly to the output co-ordinate system of the learning network.

In this work the prediction of mismatch profiles starts from discrete point 3. Time t=0 represents discrete point 1 where the mismatch is assumed to be zero for all state variables. At discrete point 2, mismatches are initialised with given values (obtained by judging the trend in all the data set used for the training of the neural network).

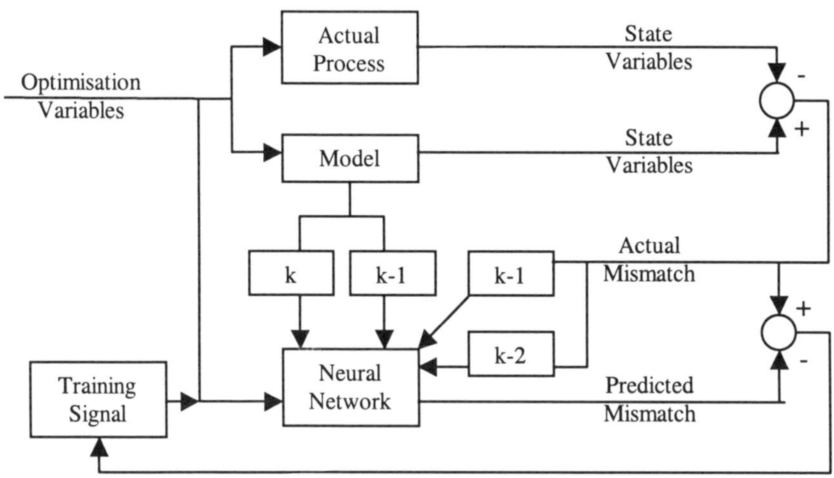

Figure 2. Forward Modelling of State Variable Mismatch by Neural Network

4. Optimal Control Formulation and Solution Using Model

In the past many authors considered Eq. 1 as the true representative of the actual dynamic process and developed optimal control (often known as dynamic optimisation) algorithms for such processes. For a given initial conditions $x(t_0)$ and v, the optimal operation of a dynamic process can be obtained by controlling u(t) optimally, while maximising (or minimising) an objective function of the form:

$$J = F(t_F, x'(t_F), x(t_F), u(t_F), v) \quad (3)$$

subject to bounds on u(t) and interior point or terminal constraints. Finite dimensional representation of the control vectors (using control vector parameterisation technique) has been considered in the past by many authors to transform the optimal control problem (DAE optimisation problem) to non-linear programming problem (Vassiliadis et al., 1994; Mujtaba and Macchietto, 1996, 1998) of the form:

$$\underset{z}{\text{Min (or Max)}} \quad J(z) \quad (4)$$

subject to: Equality constraints (eq. 1)
Inequality constraints
(bounds on control, etc.)

where, z is the parameterised control vector to be optimised.

Figure 3 illustrates a typical computation sequence for the solution of optimisation problem presented by Eq. 4. The calculation sequence starts with an initial estimate of the vector z. For each iteration, (of the OPTIMISER) DAE optimisation requires full integration of the model equations from $t = [0, t_F]$ to evaluate the objective function J and the constraints (h and g) which are then passed to the OPTIMISER. The OPTIMISER then takes a step in z and the process is repeated until convergence is achieved within an acceptable accuracy.

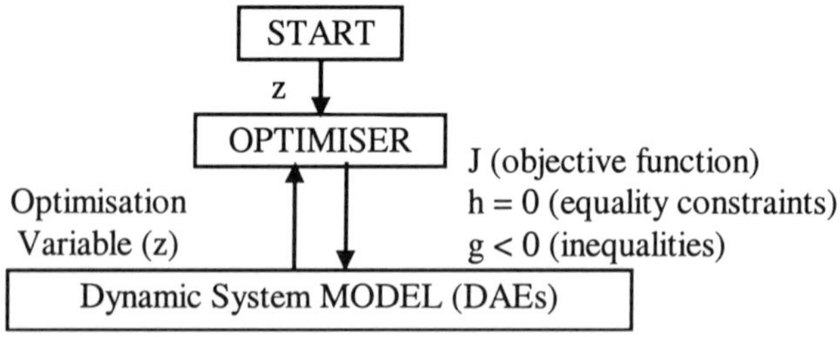

Figure 3. Computational Sequence of Dynamic Optimisation Problem

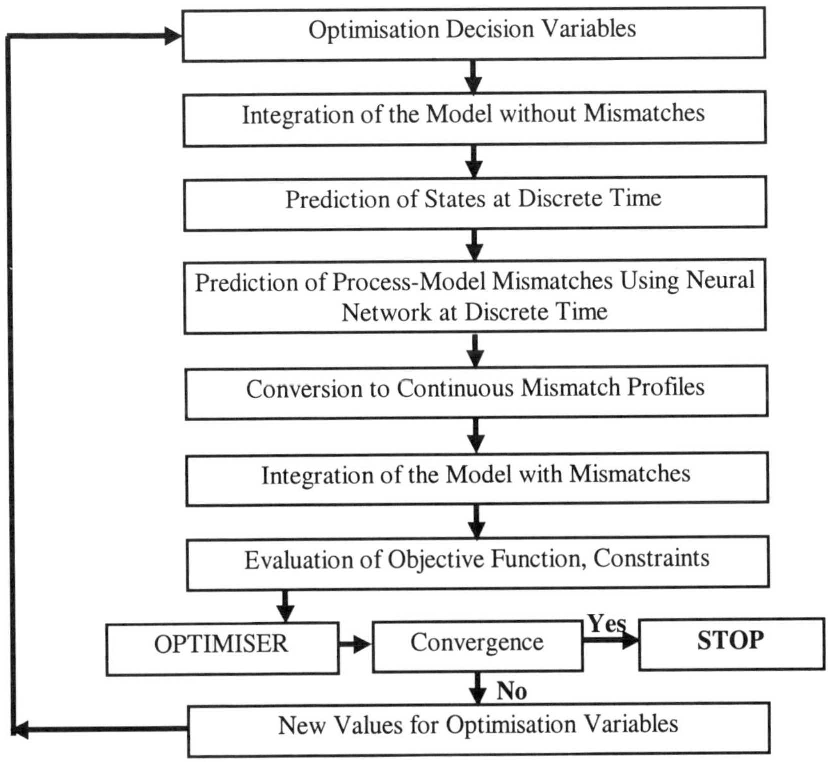

Figure 4. General Optimisation Framework For Dynamic Processes with Plant-model Mismatch

5. Dynamic Optimisation Framework Using Hybrid Model

5.1. General Strategy

Figure 4 illustrates a general optimisation framework (developed by Mujtaba and Hussain, 1998) to obtain optimal operation policies for dynamic processes with *plant-model mismatches*.

Dynamic sets of *plant-model mismatches* data is generated for a wide range of the optimisation variables (z). These data are then used to train the neural network. The trained network predicts the *plant-model mismatches* for any set of values of z at discrete-time intervals. During solution of the dynamic optimisation problem, the model has to be integrated many times, each time using a different set of z. The estimated *plant-model mismatch* profiles at discrete-time intervals are then added to the simple dynamic model during the optimisation process. To achieve this, the *discrete plant-model mismatches* are converted to continuous function of time using linear interpolation technique so that they can easily be added to the model (to make the hybrid model) within the optimisation routine. One of the important features of the framework is that it allows the use of discrete process data in a continuous model to predict discrete and/or continuous mismatch profiles.

5.1.1. Generation of Discrete Mismatch Profiles

The development and training of the neural network estimators for mismatches requires both the state variables (predicted by the model) and the mismatches at discrete points for a wide range of each optimisation variables. The number of sets of state variable and mismatch data for each type of state variable depends on the non-linearity and complexity of the system concerned.

The state variable profiles of the model are assumed to be continuous and are obtained by integration of the DAEs over the entire length of the time. Also efficient integration methods (as available in the literature) are based on variable step size methods and not on fixed step size method where the step sizes are dynamically adjusted depending on the accuracy of the integration required. In this work, therefore, the discrete values of the state variables are obtained using linear interpolation technique. For example, if the values of a state variable predicted by the model are $x_{d,k}$ and $x_{d,k+1}$ at t_k and t_{k+1}, then at any discrete t_i, which lies within $[t_k, t_{k+1}]$, the state variable value ($x_{d,i}$) is calculated using the following expression:

$$x_{d,i} = \frac{x_{d,k+1} - x_{d,k}}{t_{k+1} - t_k}(t_i - t_k) + x_{d,k} \qquad (5)$$

Usually discrete points are of equal length ($\Delta = t_{i+1} - t_i$), which usually represents the sampling time of the actual process. Now, if the state variable of the actual process at discrete time t_i, is given by $\underline{x}_{d,i}$, the discrete mismatch at t_i will therefore be $e_{xd,i} = \underline{x}_{d,i} - x_{d,i}$.

5.1.2. Continuous Mismatch Profiles During Optimisation sequence

The mismatch estimator of the neural networks estimates mismatches only at fixed discrete points. Therefore, to use the optimisation framework presented in Fig. 4 requires estimation of mismatches at variable discrete points (these points should coincide with those by the DAE integrator). This is again achieved by interpolation techniques. For example, if the values of a mismatch predicted by the estimator are $e_{xd,i}$ and $e_{xd,i+1}$ at discrete points t_i and t_{i+1} (fixed $\Delta = t_{i+1} - t_i$) then at any variable discrete point (by the integrator) t_k, which lies within $[t_i, t_{i+1}]$, the mismatch value ($e_{xd,k}$) is calculated using the following expression:

$$e_{xd,k} = \frac{e_{xd,i+1} - e_{xd,i}}{t_{i+1} - t_i}(t_i - t_k) + e_{xd,i} \tag{6}$$

6. Hybrid Model Development for Pilot Batch Distillation Column

Mujtaba and Hussain (1998) implemented the general optimisation framework based on the hybrid scheme for a binary batch distillation process. It was shown that the optimal control policy using a detailed process model was very close to that obtained using the hybrid model.

In this work, instead of using a rigorous model (as in the methodology described above), an actual pilot plant batch distillation column is used. The differences in predictions between the actual plant and the simple model (Mujtaba, 1997) are defined as the dynamic plant-model mismatches. The mismatches are modelled using neural network techniques as described in earlier sections and are incorporated in the simple model to develop the hybrid model that represents the predictions of the actual column.

The pilot-plant consists of an Aldershaw 35mm column consisting of a 5L reboiler, 40 plates and weir column surround by a pressure jacket, and a total condenser (Fig. 5). The column is charged initially with the mixture to be separated and there is one outlet for the product to be collected and two sampling points at which the temperature sensors are placed. In this work we considered methanol-water system with an initial charge of 900 ml of methanol and 2100 ml of water giving a total of 85.04 gmol of the mixture with <0.25, 0.75> mole fractions for methanol and water respectively.

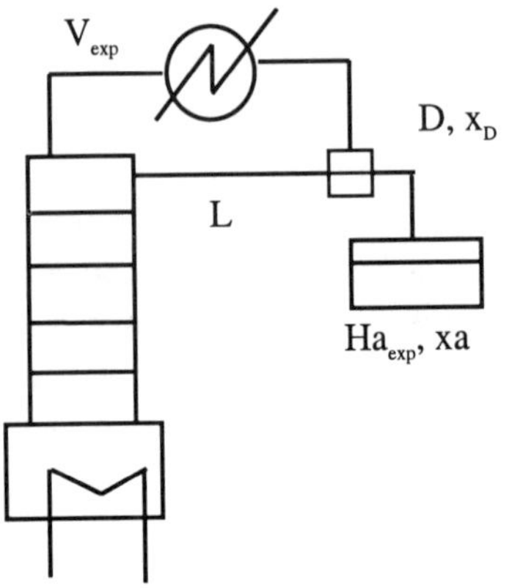

V_{exp} = *Experimental Vapour Flowrate*, mol/min
D = *Distillate rate*, mol/min
L = *Reflux rate* = V_{exp}, mol/min
Ha_{exp} = *Experimental Accumulated Distillate Hold-up*, mol
xa = *Accumulated Distillate Composition*, mole fraction
x_D = *Instant Distillate Composition*, mole fraction

Figure 5. Schematic of Batch Distillation Column

6.1. Relation Between Experimental Reflux Ratio (R_{exp}) and Model Reflux Ratio (R_{model})

Many industrial users of batch distillation (Chen, 1998) find it difficult to implement the optimum reflux ratio profiles, obtained using rigorous mathematical methods, in their pilot plants. This is due to the fact that most models for batch distillation available in the literature treat the reflux ratio as a continuous variable (either constant or variable) while most pilot plants use an on-off type (switch between total reflux and total distillate operation) reflux ratio controller. In this work we have developed a relationship between the continuous reflux ratio used in a model and the discrete reflux ratio used in the pilot plant. This allows easy comparison between the model and the plant on a common basis.

The reflux in the column is produced by a simple switching mechanism that is controlled by a solenoid in a cyclic pattern (on-off). The valve is open for a fixed period of time (to withdraw distillate) and is closed for a fixed period of time (to return the reflux to the column). In this column the valve is always open for 2 seconds and then closed for 2x(R_{exp}) second, where R_{exp} is the reflux setting. Therefore, for a total batch operating time t_{diff}, the total opening time for the valve can be given by,

$$t_{open} = \frac{t_{open} t_{diff}}{t_{open} + t_{close}} = \frac{2}{2(1 + R_{exp})} t_{diff} \tag{7}$$

If V_{exp} is the vapour rate in the condenser, then the distillate rate is:

$$D = V_{exp} \text{ (when the valve is open)} \tag{8}$$

and the reflux rate is:

$$L = V_{exp} \text{ (when the valve is closed)} \tag{9}$$

Therefore, the total amount of distillate collected (Ha_{exp}) over a period of t_{diff} (assuming V_{exp} is constant over that period) can be given by,

$$Ha_{exp} = V_{exp} t_{open} = V_{exp} \frac{1}{(1 + R_{exp})} t_{diff} \tag{10}$$

However, most of the simple models (e.g. Diwekar, 1995; Mujtaba 1997) relate the amount of distillate collected (Ha_{model}) with the vapour boil-up rate in the column (V_{model}), the continuous reflux ratio (R_{model}) and the total operating time (t_{diff}) by,

$$Ha_{model} = V_{model} \cdot (1 - R_{model}) t_{diff} \tag{11}$$

where R_{model} is defined as an internal reflux ratio.

If V_{exp} is the same as V_{model} then Eq. 10 and Eq. 11 give the desired relationship between R_{exp} and R_{model} as,

$$R_{model} = 1 - \frac{1}{(1 + R_{exp})} = \frac{R_{exp}}{(1 + R_{exp})} \tag{12}$$

However, in the pilot plant it is not possible to maintain a constant V_{exp} throughout the operation. Rather the heat input to the column is fixed. This results in a dynamic profile for V_{exp} over the operating time t_{diff}, as will be discussed next.

6.2. *Estimation of V_{exp}*

For a given R_{exp}, the distillate rate (D) (the amount of distillate over a small interval of time) can be estimated and Eq. 8 gives the corresponding V_{exp}. It is observed that the value of V_{exp} decreases with time (Greaves, 1999). This is due to gradual depletion of the light component from the column leaving behind the heavy component in the column. Since the heat of vaporisation of the heavy component is higher than that of the light component, a fixed heat duty gradually reduces the rate of vapour being produced. It is also observed that at any given time between [0, t_{diff}] the value of V_{exp} is higher for higher R_{exp}. This is due to the fact that the rate of depletion of the lighter component from the column is lower at higher reflux ratio and therefore a fixed heat duty gives a higher vapour rate.

Hence, in this work, for a given R_{exp}, the vapour rate profile is averaged to obtain an average V_{exp} to be used as V_{model} in the model. Figure 6 shows the average V_{exp} vs. R_{exp} curve and Eq. 13 gives the corresponding relationship.

$$V_{exp} = \left[a\left(\frac{1}{1+R_{exp}}\right)^2 + b\left(\frac{1}{1+R_{exp}}\right) + c \right]^{-1} \tag{13}$$

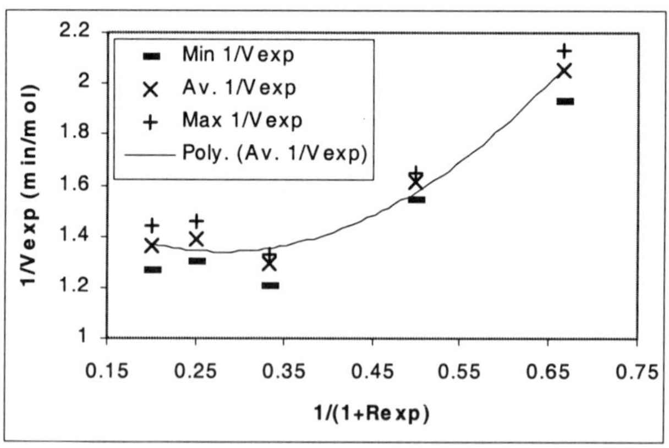

Figure 6. Vapour Load vs. Reflux Ratio

6.3. Plant-Model Simulation

We carried 5 experiments in total using the pilot-plant for different R_{exp}. The accumulated distillate composition and distillate hold-up profiles are shown in Fig. 7 and Fig. 8 respectively.

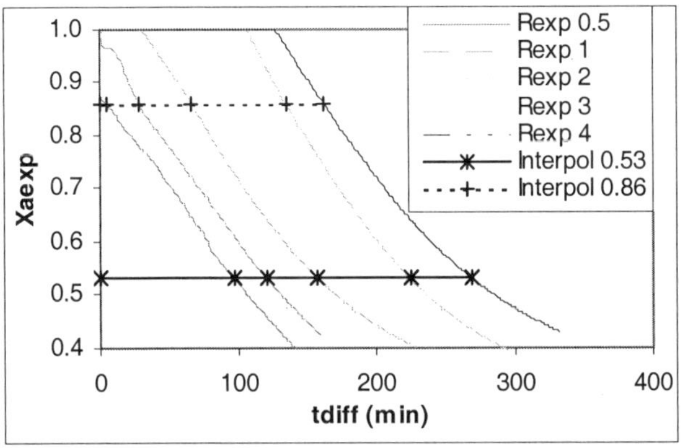

Figure 7. Accumulate Distillate Composition vs. Batch Time for Ha* = 15 mol

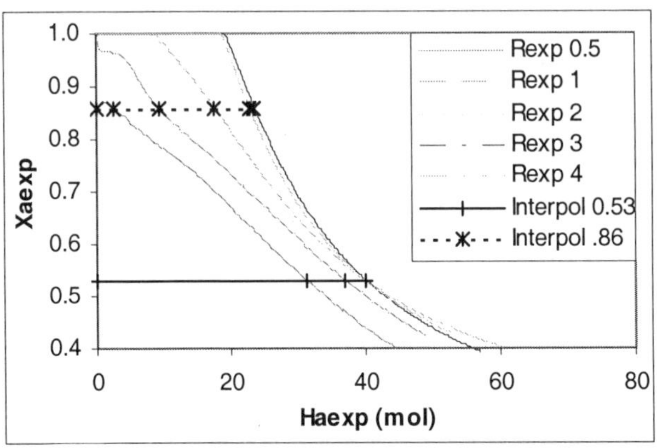

Figure 8. Accumulate Distillate Composition vs. Amount of Accumulate

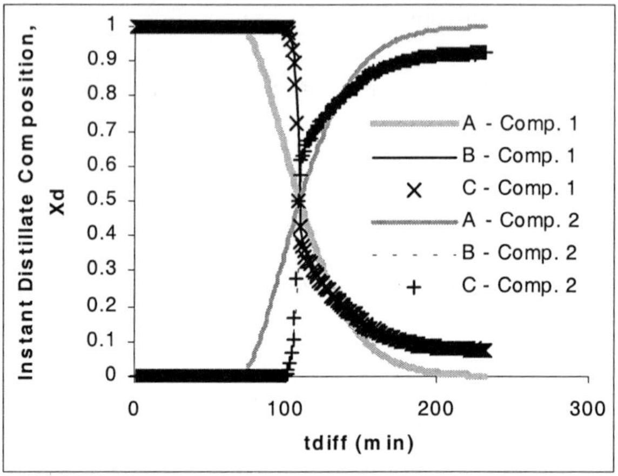

Figure 9. Experimental, Simulation Results. and Dynamic Plant-model Mismatch Model (R_{exp}= 2).

In this work, the simple model of Mujtaba (1997) is used to simulate the plant. For each R_{exp}, R_{model} and V_{model} (=V_{exp}) are calculated using Eq. 12 and Eq. 13. The simulated and experimental instant distillate composition profiles are shown in Fig. 9 for R_{exp} = 2 (corresponding R_{model} = 0.666). Curves A and B show the model and pilot plant predictions respectively. Figure 9 clearly shows that there are large *plant-model mismatches* in the composition profiles although for a given batch time of t_{diff} = 220 min the amount of distillate achieved by the experiment was the same as that obtained by the simulation. These *plant-model mismatches* can be attributed to factors such as: use of constant V_{model} instead of a dynamic one; constant relative volatility parameter used in the model and uncertainties associated with it; actual efficiency of the plates.

6.4. Hybrid Model

The four experiments done previously with R_{exp} (= 0.5, 1, 3, 4) were used to train the neural network and the experiment with R_{exp} = 2 was used to validate the system. Dynamic models of *plant-model mismatches* for three state variables (i.e. X) of the system are considered here. They are the instant distillate composition (x_D), accumulated distillate composition (xa) and the amount of distillate (Ha). The inputs and outputs of the network are as in Fig. 1. A multilayered feed forward network which is trained with the back propagation method using a momentum term as well as an adaptive learning rate to speed up the rate of convergence is used in this work. The error between the actual mismatch (obtained from simulation and experiments) and that predicted by the network is used as the error signal to train the network as described earlier.

Figure 9 also shows the instant distillate composition profile for $R_{exp} = 2$ (which is used to validate the network) using the simple model coupled with the dynamic model for the *plant-model mismatches* (curve C). The predicted profile (curve C) shows very good agreement with the experimental profile (curve B). Similar agreements have been obtained for the accumulated distillate amount and composition profiles (Greaves, 1999).

7. Optimal Control of Batch Distillation

A batch distillation column operates at some optimal reflux ratio until a certain objective is achieved (e.g. *maximum distillate, minimum time* or *maximum profit*). A dynamic optimisation problem (optimal control) can be formulated and solved using rigorous model (simple, detailed or hybrid) based mathematical techniques (Logsdon and Biegler, 1993; Mujtaba and Macchietto, 1993; Mujtaba and Hussain, 1998) to generate optimal reflux ratio profile that will achieve the objective. The formulation for *minimum batch time* will be used in this study, as described next.

7.1. Problem Formulation for Minimum Batch Time

The dynamic optimisation problem with an objective to minimise the batch time can be described as:

Given:	the column configuration, the feed mixture, vapour boil-up rate, a *separation task*
Determine:	optimal reflux ratio
So as to minimise:	the operation time
Subject to:	equality and inequality constraints (e.g. model equations, bounds, etc.)

Mathematically the problems can be written as:

$$\min_{R(t)} \quad t_{diff}$$

subject to:

$$\text{Model Equations} \quad \text{(equality constraints)}$$
$$x_a \geq x_a^* \quad \text{(inequality constraints)}$$
$$H_a \geq H_a^* \quad \text{(inequality constraints)}$$
$$R_{model}^{l} < R_{model} < R_{model}^{u} \quad \text{(inequality constraints)}$$

where H_a, x_a are the amount of distillate and its composition at the end of the operation time t_{diff} and H_a^*, x_a^* are the given amount of distillate and its purity (*separation task*). R^l_{model}, R^u_{model} are the lower and upper bounds of R_{model} within which it is optimised.

Solution of the above optimisation problem using rigorous mathematical methods have received considerable attention in the past and therefore it is not intended here to duplicate such effort. However, it is worth mentioning here that these techniques require the repetitive solution of the model equations (to evaluate the objective function and the constraints and their gradients with respect to the optimisation variables) and therefore computationally can be very expensive.

However, in this work, we present two simple algorithms which is computationally less expensive to obtain the minimum batch time for a given *separation task*. These algorithms are the results of the application of some of the unique properties of batch distillation process in the general optimisation framework discussed earlier.

For a particular mixture with a given column (fixed number of plates, heat duty, etc.) and operating policy (reflux ratio, column pressure, etc.), the residue or distillate composition will follow well defined distillation maps (Bernot et al., 1991) which is also evident in Fig. 7 of this work. Therefore, for a given reflux ratio (e.g. $R_{exp} = 3$), each point on the distillate curve in Fig. 8 corresponds to a series of (Ha_{exp}, xa_{exp}) values and for each set of (Ha_{exp}, xa_{exp}) (the *separation task*) the minimum batch time, $t_{diff,min}$, can be read from Fig. 7. For example, for a given $Ha_{exp} = 20$ mol and $xa_{exp} = 0.960$, the minimum batch time is 138 minutes when $R_{exp} = 4$. However, this minimum batch time may not be in all cases the true minimum batch time for a given *separation task*. This is due to the fact that the R_{exp} may not be the optimum one for the given task (as can be seen later). The algorithms we proposed for finding the minimum batch time is as follows:

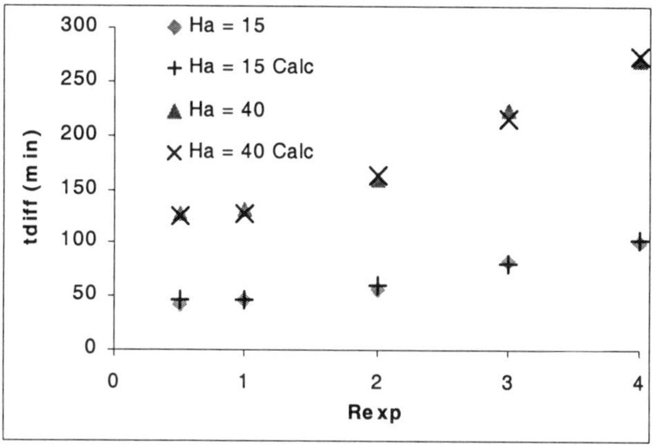

Figure 10. Batch Time vs. Reflux Ratio (Eq. 10)

8. Algorithms for Finding Minimum Batch Time

8.1. Algorithm -1: Experiment Based

For a given *separation task* (Ha^*, xa^*) Eq. 10 can be (replacing V_{exp} from with Eq. 13) rearranged as:

$$t_{diff} = \frac{Ha^*(1+R_{exp})}{f(R_{exp})} = Ha^* g(R_{exp}) \qquad (14)$$

where f and g are non-linear functions of R_{exp}.

For a given Ha^*, Eq. 14 shows that the batch time (and so does the distillate composition, xa) increases nonlinearly with the reflux ratio. Figure 10 shows these values for Ha^* = 15 mol and 40 mol respectively along with the corresponding experimental values. Although each point on any of these curves gives the minimum time for the corresponding (Ha^*, xa), only one point which is the true minimum batch time will correspond to the desired *separation task* (Ha^*, xa^*) and the optimum R_{exp}.

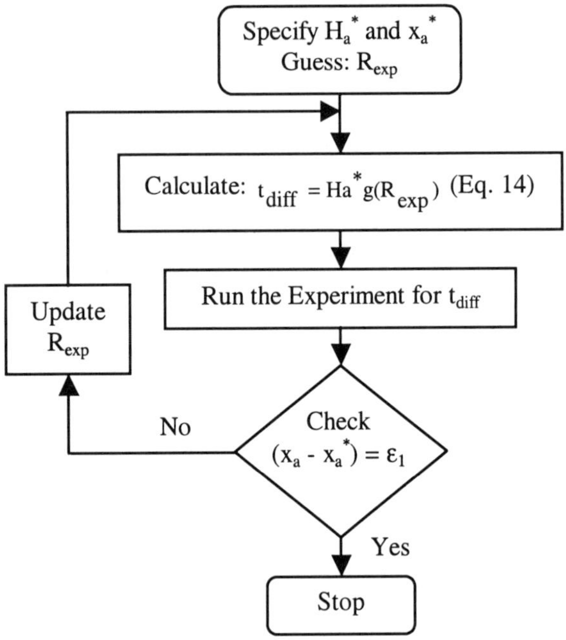

Figure 11. Experiment Based Algorithm -1 for Calculating Minimum Batch Time. ε_1 is a small number.

In this work we propose the algorithm shown in Fig. 11 for calculating the optimum reflux ratio and the minimum time for a given *separation task*. It is recommended to start with a low value of R_{exp} and gradually increase it and stop at where xa ~ xa^*. This approach will require a few iterations to achieve the minimum batch time. Calculations with a large initial value of R_{exp} do not guarantee the optimum reflux ratio and the true minimum batch time at the first point where xa ~ xa^* and therefore may require more iterations. This is explained with reference to two given *separation tasks* as summarised in Table 1 and Table 2 respectively.

The optimum reflux ratio and the minimum time for separation task 1 are 3 and 80.62 min (Table 1). The separation task 2 could be achieved using 3 different reflux ratio (Table 2) but however, $R_{exp} = 2$ gives the true minimum batch time which is about 40% lower than the batch time required to achieve the same separation with $R_{exp} = 4$.

Table 1. Separation Task 1: Ha*=15, xa*= 0.999

R_{exp}	$1/V_{exp}$	t_{diff}	xa_{exp}
0.5	2.07	46.47	0.731
1	1.58	47.27	0.792
2	1.36	60.93	0.906
3	**1.34**	**80.62**	**0.999**

Table 2. Separation Task 2: Ha*=40, xa*= 0.53

R_{exp}	$1/V_{exp}$	t_{diff}	xa_{exp}
0.5	2.07	123.92	0.439
1	1.58	126.06	0.499
2	**1.36**	**162.49**	**0.529**
3	1.34	214.98	0.531
4	1.32	273.91	0.532

The main advantage of the algorithm-1 is that for a given reflux ratio it will estimate the duration of the batch. However, the major disadvantage of this approach is that a time consuming experiment is to be carried out for each new value of the R_{exp} until the given *separation task* can be achieved in minimum time.

8.2. Algorithm -2: Model Based

To reduce the time consuming and expensive experiments of algorithm-1 by a considerable amount, we propose a second algorithm based on simple model and neural network techniques as shown in Fig. 12.

The algorithm-2 has been tested for the *separation tasks* used in algorithm-1 and they are in very good agreement. For a given *separation task*, while the algorithm-1 requires approximately 3 to 4 set of experiments for a total period of 18-22 hours, the algorithm-2 requires only about half an hour of computation time and about 4-5 hours of experiment to achieve the given *separation task*.

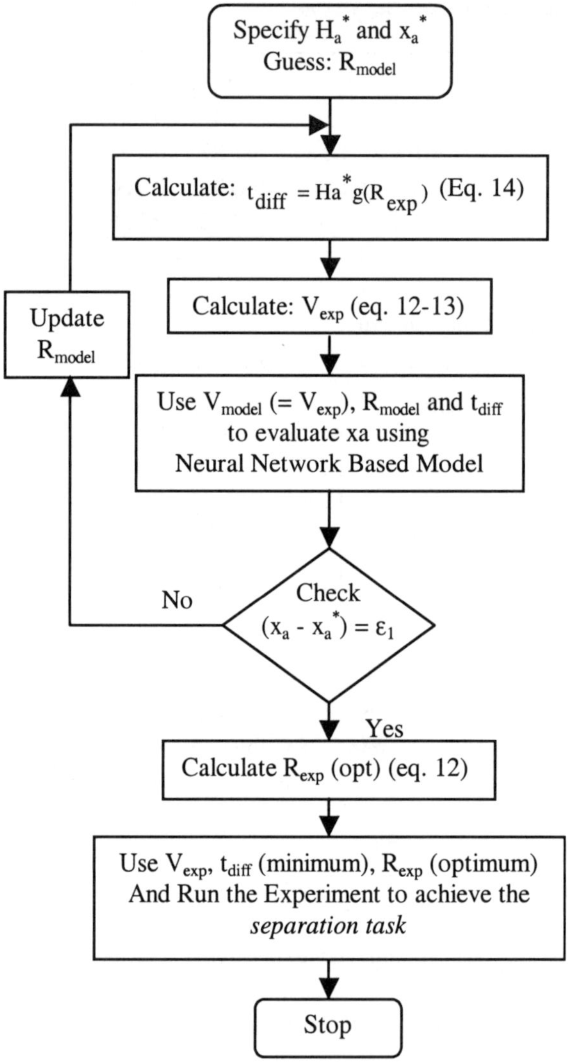

Figure 12. Model Based Algorithm -2 for Calculating Minimum Batch Time

9. Conclusions

In this work, we have discussed a general hybrid model based optimisation framework to obtain optimal control policies of dynamic processes. The hybrid scheme consisting of a simple process model and neural network technique is considered to accurately model a process and to capture the dynamic *plant-model mismatches*.

The general optimisation framework is then implemented in a pilot batch distillation column. The hybrid model was developed based on a simple model and a series of experiments in the column. A correlation between the reflux ratio used in the model (treated as a continuous variable) and that used in the pilot plant (treated as a discrete variable) is developed between the model and the pilot plant. Taking advantage of some of the inherent properties of batch distillation process two useful optimisation algorithms have been developed to obtain the optimum reflux ratio to minimise batch time for a given *separation task*.

The algorithm-1 is based on a series of experiments that can substantially reduce the trial and error method of optimising operating conditions as widely practised in industries. The algorithm-2 (new algorithm) is a simpler version of the general optimisation framework. The new algorithm is relatively simple and does not need sophisticated numerical methods for the solution of the optimisation problem as was required for the general optimisation framework. However, it is important to mention that the new algorithm is only specific to the batch distillation process.

We believe that the technique developed in this work will also be suitable for real on-line applications where *plant-model mismatches* inherently exists at all times.

However, for highly non-linear profiles of state variables, switching from continuous to discrete or from discrete to continuous using linear interpolation technique may not be efficient and non-linear interpolation technique may need to be employed.

Notation

D = Distillate flow rate, mol/min
ε_1 = Finite small positive number
H_a = Accumulated distillate hold-up, mol
L = Reflux rate, mol/min

R = Reflux ratio
t_{diff} = Total operation time, min
V = Vapour flow rate, mol/min
x_a = Accumulated distillate composition, mole fraction
x_d = Instant distillate composition, mole fraction

Subscripts

$_{exp}$ = experiment
$_{model}$ = model

References

Agarwal, M., *Batch Processing Systems Engineering: Fundamentals and Applications for Chemical Engineering*, G.V. Reklaitis et al. eds., Series F: Computer and Systems Sciences, Springer Verlag, Berlin, **143** (1996), 295.

Barolo, M, Guarise, G.B., Rienzi, S.A., Trotta, A., *IEC Res.* **33** (1994), 3160.

Bernot, C., Doherty, M.F. and Malone, M.F.,. *Chem. Engng. Sci.* **45** (1991) 1207.

Bhat, N. and McAvoy, T.J., *comput. chem. engng.* **14** (1990), 573.

Bosley, J.R. Jr. and Edgar, T.F., *Proceedings of 5th International Seminar on Process Systems Engineering*, Kyongju, Korea, 30 May - 3 June, **1** (1994), 477.

Bosley, J.R. Jr. and Edgar, T.F., *Proceedings of 5th International Seminar on Process Systems Engineering*, Kyongju, Korea, 30 May - 3 June, **1** (1994), 477.

Chen, C.L., 1998, *private communications*, E Tech., London.

Cuthrell, J.E. and Biegler, L.T., *comput. chem. engng.* **13** (1989), 49.

Cybenko, G., *Math. Cont. Sig. Syst.* **2** (1989), 303.

Diwekar, U.M., *Batch distillation: Simulation, optimal design and control* (Taylor and Francis, Washington, DC, 1995).

Farhat, S., Czernicki, M., Pibouleau L. and Domenech, S., *AIChE J.* **36** (1990), 1349.

Greaves, M.A., *Study of Batch Distillation*, Internal Report, (University of Bradford, 1999)

Henry, R.M., Mujtaba, I.M., Kamel, F.N. and Sabri, Y., *The Chem. Engr.*, November (1997), 32.

Hussain, M.A., Allwright, J.C. and Kershenbaum, L.S., *Proceedings of IChemE - Advances in Process Control 4*, (1995), York, 27-28 September, 195.

Logsdon, J.S. and Biegler, L.T., *IEC Res.* **32** (1993), 700.

Logsdon, J.S., Diwekar, U.M. and Biegler, L.T., *Trans IChemE.* **68** (1990) Part A:434.
Luus, R., *J.Proc. Cont.* **4** (1994), 218.
Macchietto, S. and Mujtaba, I.M., *Batch Processing Systems Engineering: Fundamentals and Applications for Chemical Engineering*, G.V. Reklaitis et al. eds., Series F: Computer and Systems Sciences, Springer Verlag, Berlin, Vol. **143** (1996), 174.
Morison, K.R., *PhD thesis*, (Imperial College, London, 1984).
Morris, A.J., Montague, G.A. and Willis, M.J., *Trans. IChemE.* **72** (1994) Part A, 3.
Mujtaba, I.M., *Trans. IChemE.* **75** (1997), Part A, p609.
Mujtaba, I.M., Hussain, M.A., *comput. chem. engng.* **22** (1998), S621.
Mujtaba, I.M. and Macchietto, S., *comput. chem. engng.* **17** (1993), 1191.
Mujtaba, I.M. and Macchietto, S., *J. Proc. Cont.* **6** (1996), 27.
Mujtaba, I.M. and Macchietto, S., *Chem. Eng. Sci.* **53** (1998), 2519.
Savkovic-Stevanovic, J., *comput. chem. engng.* **18** (1994), 1149.
Sorensen, E. and Skogested, S., *Chem Eng Sci*, **51** (1996), 4949.
Vassiliadis, V.S., Sargent, R.W.H, and Pantelides, C.C., *IEC Res.* **33** (1994), 2123.
Walsh, S., Mujtaba, I.M. and Macchietto, S., *Acta chimica Slovenica.* **42** (1995), 69.
Woinaroschy, A., Isopescu, R. and Filipescu, L., *Chem. Eng. Technol.* **17** (1994), 269.

Acknowledgements

The University of Bradford Studentship to M.A. Greaves and the UK royal Society support to M.A. Hussain are gratefully acknowledged.

8. HIERARCHICAL NEURAL FUZZY MODELS AS A TOOL FOR PROCESS IDENTIFICATION: A BIOPROCESS APPLICATION

L. A. C. MELEIRO, R. MACIEL FILHO
Laboratory of Optimization, Design and Advanced Control (LOPCA)
DPQ/FEQ, State University of Campinas - UNICAMP, CP 6066, CEP 13081-970
Campinas - SP, Brazil.

R. J. G. B. CAMPELLO, W. C. AMARAL
Laboratory of Computer Engineering and Industrial Automation (LCA)
DCA/FEEC, State University of Campinas - UNICAMP, CP 6101, CEP13083-970
Campinas - SP, Brazil.

Hierarchical structures have been introduced in the literature to deal with the dimensionality problem, which is the main drawback to the application of neural networks and fuzzy models to the modeling and control of large-scale systems. In the present work, hierarchical neural fuzzy models are reviewed focusing on an industrial application. The models considered here consist of a set of Radial Basis Function (RBF) networks formulated as simplified fuzzy systems and connected in a cascade fashion. These models are applied to the modeling of a Multi-Input/Multi-Output (MIMO) complex biotechnological process for ethyl alcohol (ethanol) production and show to adequately describe the dynamics of this process, even for long-range horizon predictions.

1. Introduction

The capacity for memory storage and processing of the latest computational devices has allowed the development of more complex and efficient mathematical models of static and dynamic nonlinear systems. Two important classes of nonlinear models are the feedforward architectures of Neural Networks (Haykin, 1999) and Fuzzy Systems (Yager and Filev, 1994), especially because these models are universal approximators, i.e., they can approximate to arbitrary accuracy any continuous mapping defined on a compact (closed and bounded) domain (Kosko, 1992; 1997). However, due to their generic structures, both the neural and fuzzy models usually require the estimation of a large number of parameters. Generally, the number of parameters and data needed to provide a desired accuracy increases exponentially with the dimension of the input space of the mapping to be approximated. This is the well-known problem called "Curse of Dimensionality" (Kosko, 1997; Haykin, 1999).

In order to get around the curse of dimensionality problem in fuzzy control, Raju *et al.* (1991) proposed a hierarchical structure of fuzzy systems in which a set of subsystems connected in a cascade architecture is used instead of a single fuzzy system. In this hierarchical structure the number of fuzzy rules increases linearly (instead of exponentially) with the dimension of the input space, thus allowing the application of fuzzy control to large-scale systems (Jamshidi, 1997). Recently, Wang (1998; Chen and Wang, 2000) applied this hierarchical structure to fuzzy modeling. In his approach, Wang implemented the hierarchical subsystems using special kinds of Takagi-Sugeno models, constructed step by step, and showed that the resulting model is a universal approximator. Although the result that hierarchical models can be constructed as universal approximators (despite their reduced structure) is surprising, Wang's approach is too restrictive for most real world applications. The main reason is that the analytical construction of the model assumes that the input-output mapping to be approximated is a black-box, i.e., its analytic formula is unknown but the output value related to any input in the domain is available. This assumption is too strict, especially in dynamic system identification. In this context, a numerical approach to the estimation of hierarchical models from a finite data set became desirable. Wang (1999) proposed the use of backpropagation techniques to estimate the hierarchical models in this way. However, the formulation provided is restricted to models with just three input variables.

A similar approach with a generic formulation suited for models with any number of input variables was proposed in (Campello and Amaral, 1999; 2000). A review of this approach focusing on an industrial application is considered in the present work, in which a special kind of fuzzy system is used as the subsystems in the hierarchical models. This fuzzy system, called Simplified Relational Structure (Oliveira and Lemos, 1997), is under certain conditions completely equivalent to a radial basis function neural network (Broomhead and Lowe, 1988). Its formulation is, however, easier to manipulate since it deals separately with each input variable. The backpropagation equations for the numerical optimization of the hierarchical models are derived from this formulation. The set of equations to compute the gradient vector of the cost function to be minimized is written in a recursive manner. This means that the local gradient with respect to the parameters in a given subsystem is derived from the local gradient (previously computed) related to the subsequent subsystem. From the gradient information, the conjugate gradient algorithm of Fletcher and Reeves (Bazaraa and Shetty, 1979) can be used to carry out the optimization of the models. This algorithm is well suited to large-scale problems since it does not demand the computation of the Hessian matrix or its inverse, thus having small storage requirements. Also, it ensures the convergence of the optimization procedure (with a second-order rate).

The industrial application considered here is concerned with a biotechnological process. Biotechnology has become increasingly important in the activities of contemporary society as a "clean" and safe technology when compared to traditional chemical processes. Moreover, it provides extremely useful and valuable products in several industrial areas (pharmaceutical, foods, fuels, etc.). Biotechnological processes are characterized by their complex dynamics, such as inverse response, dead time and strong nonlinearities, especially because the main driving force of these processes is microorganisms (cells) that are very sensitive to any environmental variations in the fermentation broth (e.g., temperature, substrate concentration, pH, among others). For these reasons modeling, simulation, and control of those systems are problems that have not yet been totally resolved. Therefore, they are still a relevant and timely research theme (Meleiro and Maciel Filho, 2000).

The underlying problem here refers to an important class of biotechnological industrial processes. The case study is a typical large-scale industrial plant to produce ethanol from sugar cane syrup. The process operational conditions are those typically found in the Brazilian industrial distilleries. A hierarchical neural fuzzy model of this process is estimated and validated using data for those typical conditions and has been shown to adequately describe the process dynamics. The model has also presented a good performance for long-range horizon predictions, having a great potential for use in advanced control strategies.

2. RBF Networks as Simplified Fuzzy Relational Structures

Consider a generic Multi-Input/Single-Output (MISO) system given by $y = F(x_1, \cdots, x_n)$, where F is a nonlinear operator which maps the inputs $x_i \ (i = 1, \cdots, n)$ into the output y. This system can be modeled using a simplified relational structure (Oliveira and Lemos, 1997), given by the following equation:

$$\hat{y} = \Psi^T \Omega \tag{1}$$

where $\hat{y}$ is the model output, $\Omega\,(m\times 1)$ is the parameter vector and $\Psi\,(m\times 1)$ is the fuzzy input vector. The vector Ψ is given by the Kronecker Product ($\otimes$) of the individual fuzzy inputs, i.e.,

$$\Psi = X_1 \otimes X_2 \otimes \cdots \otimes X_n \tag{2}$$

The inputs $X_i\ (i = 1, \cdots, n)$ are derived from the nonfuzzy inputs x_i as follows:

$$X_i = \left[\mathcal{X}_{i_1}(x_i)\quad \mathcal{X}_{i_2}(x_i)\quad \cdots\quad \mathcal{X}_{i_{c_i}}(x_i)\right]^T \tag{3}$$

where $\mathcal{X}_{i_j}(\cdot)$ is the j-th fuzzy set of the i-th input variable (with c_i fuzzy sets).

It is possible to demonstrate (see the Appendix) that the fuzzy model given by the equations presented above is completely equivalent to an RBF neural network with Gaussian activation functions whenever Gaussian fuzzy sets are used. The analogy between these models is illustrated in Figs. 1 and 2 for a two-dimensional input space. Figure 1 shows Gaussian fuzzy sets defined on the domains of the input variables x_1 and x_2 of a simplified fuzzy structure. Figure 2 shows the activation functions of the equivalent RBF network. It can be noted that the fuzzy sets are the projections of the multivariate activation functions on the unidimensional spaces of the input variables.

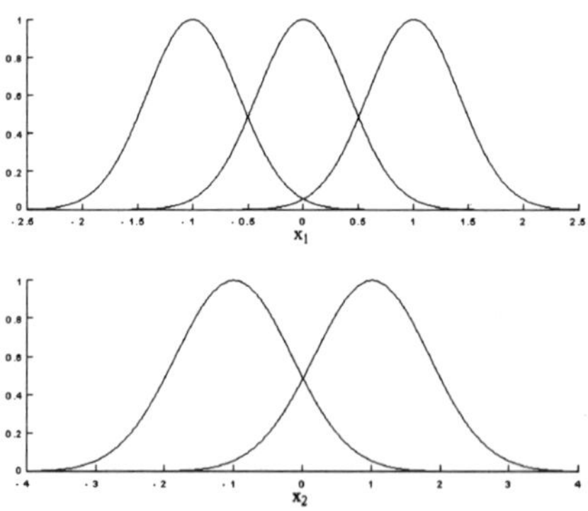

Figure 1. Fuzzy sets of a fuzzy model with two inputs: $\mathcal{X}_{1_i}(x_1), i = 1,2,3$ (above); $\mathcal{X}_{2_i}(x_2), i = 1,2$ (below).

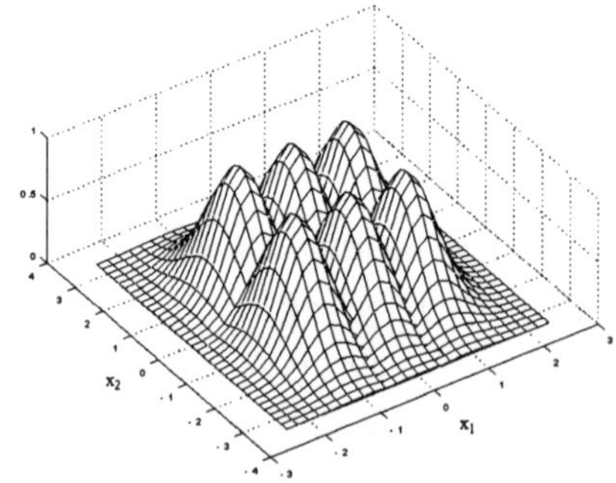

Figure 2. Activation functions (six neurons) of the equivalent RBF neural network.

2.1. Model Structure

The model given by Eqs. (1), (2), and (3) follows the conventional structure of fuzzy models (FM) and feedforward neural networks (NN) shown in Fig. 3-a. The main problem of this structure is discussed in the sequel.

Consider, for simplicity and without loss of generality, that $c_i = c$ for $i = 1, \cdots, n$ in Eq. 3. Then, it can be seen from Eq. 2 that the number of elements of both vectors Ψ and Ω in (1) is given by $m = c^n$. This is the number of fuzzy rules associated with the model or, alternatively, the number of neurons in the equivalent RBF network. This is also the number of parameters to be estimated (synaptic weights in the RBF networks) if the fuzzy sets/activation functions are kept constant. On the other hand, if the centers and widths of the fuzzy sets/activation functions can be varied, then the number of parameters to be estimated becomes $p = c^n + 2nc$. Note that the approximation capacity of the model depends directly on *c*.

The exponential relationship between the number of inputs, *n*, and the number of fuzzy rules/neurons, *m*, is shown in Fig. 3-b for typical values of *c*. This figure illustrates the dimensionality problem in nonhierarchical models, i.e., the increase in the number of fuzzy rules/neurons needed to cover the input space with given "density" as an exponential function of the number of inputs.

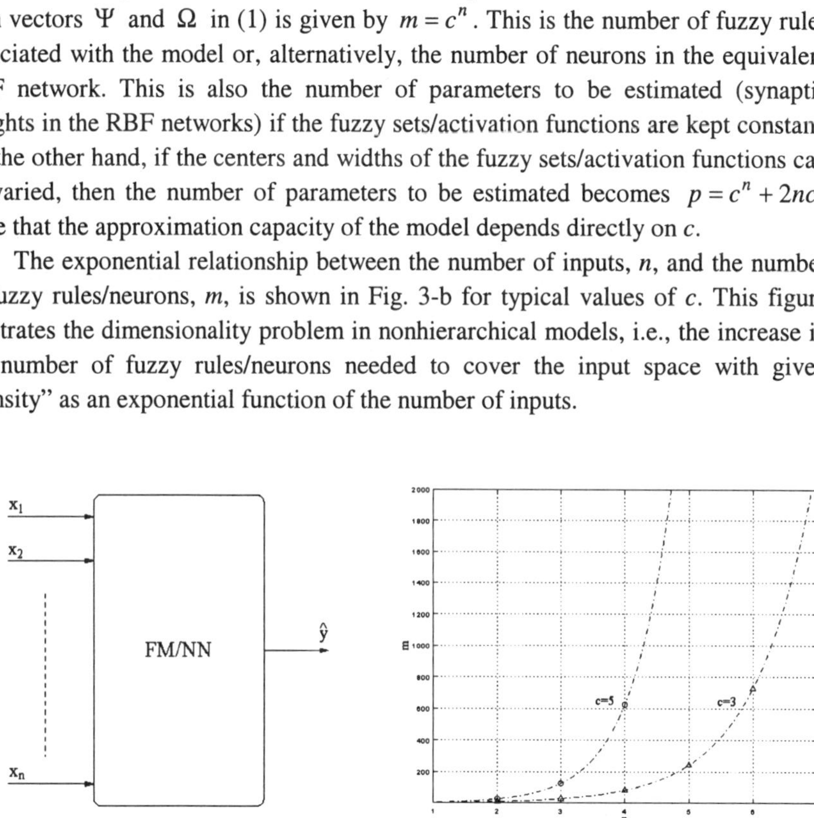

(a) (b)

Figure 3. (a) Nonhierarchical model. (b) Relationship between the number of inputs, *n*, and fuzzy rules/neurons, *m*.

3. Hierarchical Models

As outlined in Section 1, an alternative to get around the dimensionality problem of the conventional (nonhierarchical) models is the hierarchical structure shown in Fig. 4-a, where n-1 submodels (processing blocks) with two-dimensional input spaces are connected in a cascade architecture. Since the processing blocks have two inputs each, the number of fuzzy rules/neurons in each block is c^2. Consequently, the total number of fuzzy rules/neurons in the model is $m = (n-1)c^2 \; (n \geq 2)$. This is also the number of parameters to be estimated if the fuzzy sets/activation functions are kept constant. Otherwise, the number of free design parameters becomes $p = (n-1)c^2 + 2nc + 2(n-2)c = (n-1)c^2 + 4(n-1)c$.

The relationship between the number of inputs, n, and the number of fuzzy rules/neurons, m, is displayed in Fig. 4-b for typical values of c. This figure shows that the rate of growth in the number of fuzzy rules/neurons as a function of the number of inputs is constant (for $n \geq 2$). This is a significant advantage in comparison with the behavior of the nonhierarchical structure shown in Fig. 3-b.

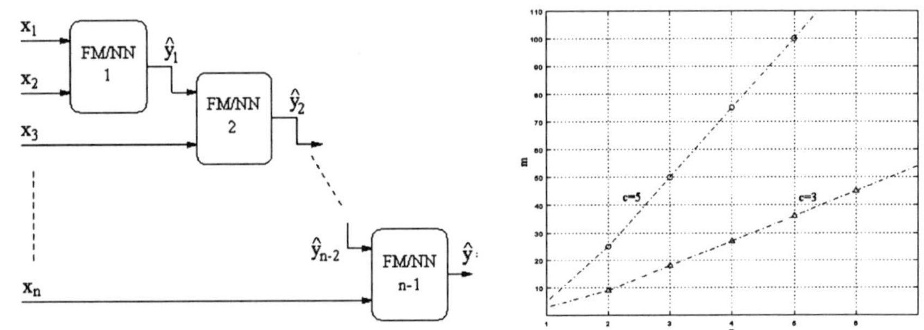

(a) (b)

Figure 4. (a) Hierarchical model. (b) Relationship between the number of inputs, n, and fuzzy rules/neurons, m.

3.1. *Formulation*

Based on the formulation presented in Section 2, the equations which describe the model shown in Fig. 4-a are

$$\hat{y}_i = \Psi_i^T \Omega_i = \sum_{l=1}^{c^2} \Psi_{i_l} \Omega_{i_l}, \qquad i = 1, \cdots, n-1 \tag{4}$$

$$\Psi_i = \begin{bmatrix} \Psi_{i_1} & \cdots & \Psi_{i_{c^2}} \end{bmatrix}^T = \begin{cases} X_{i+1} \otimes \hat{Y}_{i-1}, & i = 2, \cdots, n\text{-}1 \\ X_{i+1} \otimes X_i, & i = 1 \end{cases} \tag{5}$$

$$X_j = \left[X_{j1}(x_j) \;\; X_{j2}(x_j) \;\; \cdots \;\; X_{jc}(x_j) \right]^T, \qquad j = 1, \cdots, n \tag{6}$$

$$\hat{Y}_h = \left[Y_{h1}(\hat{y}_h) \;\; Y_{h2}(\hat{y}_h) \;\; \cdots \;\; Y_{h_c}(\hat{y}_h) \right]^T, \qquad h = 1, \cdots, n-2 \tag{7}$$

where $X_{(.)}$ and $Y_{(.)}$ are the Gaussian fuzzy sets associated with the inputs $x_{(.)}$ and hidden outputs $\hat{y}_{(.)}$, respectively:

$$X_{jl}(x_j) = \exp\left(-\frac{(x_j - \theta_{jl})^2}{\sigma_{jl}^2} \right) \tag{8}$$

$$Y_{hl}(\hat{y}_h) = \exp\left(-\frac{(\hat{y}_h - \phi_{hl})^2}{\varphi_{hl}^2} \right) \tag{9}$$

where $\theta_{jl}\,(\phi_{hl})$ and $\sigma_{jl}\,(\varphi_{hl})$ are the center and the width of the l-th fuzzy set associated with the j-th input x_j (h-th hidden output $\hat{y}_h$), respectively.

3.2. Optimization Problem

Consider a set of N input/output data pairs, i.e. $\{x_1(k), \;\cdots,\; x_n(k),\; y(k)\}_{k=1}^N$, measured from a system to be modeled. Then, a hierarchical model of the system can be estimated by solving the following optimization problem:

$$\min_{\Gamma} \; J = \frac{1}{2} \sum_{k=1}^{N} (y(k) - \hat{y}(k))^2 \tag{10}$$

where Γ denotes the set of all free design parameters of the model. Problem (10) can be solved using unconstrained optimization techniques (Bazaraa and Shetty, 1979), such as the conjugate gradient algorithm of Fletcher and Reeves. These techniques require the computation of the gradient vector of the cost function J with respect to the set of parameters Γ. The set Γ related to the hierarchical models considered here is constituted by the parameter vectors $\Omega_{(\cdot)}$ in Eq. 4 as well as the centers ($\theta_{(\cdot)}$ and $\phi_{(\cdot)}$) and widths ($\sigma_{(\cdot)}$ and $\varphi_{(\cdot)}$) of the fuzzy sets in Eqs. 8-9. Taking the derivatives of J in Eq. 10 with respect to these parameters by applying the chain rule through Eqs. 4-7 yields the following[3]:

$$\frac{\partial J}{\partial \Omega_{hi}} = -\sum_{k=1}^{N} e(k)\Psi_{h_i}(k)\lambda_h(k), \qquad \begin{array}{l} i = 1,\cdots,c^2 \\ h = 1,\cdots,n-1 \end{array} \tag{11}$$

$$\frac{\partial J}{\partial \phi_{hi}} = -\sum_{k=1}^{N}\left[e(k)\lambda_{h+1}(k)\left(\sum_{l=0}^{c-1}\Omega_{(h+1)(lc+i)}X_{(h+2)(l+1)}(k)\right)\frac{\partial \mathcal{Y}_{h_i}(\hat{y}_h(k))}{\partial \phi_{hi}}\right], \qquad \begin{array}{l} i = 1,\cdots,c \\ h = 1,\cdots,n-2 \end{array} \tag{12}$$

$$\frac{\partial J}{\partial \theta_{hi}} = -\sum_{k=1}^{N}\left[e(k)\lambda_{h-1}(k)\left(\sum_{j=1}^{c}\Omega_{(h-1)(ic-c+j)}\hat{Y}_{(h-2)_j}(k)\right)\frac{\partial \mathcal{X}_{h_i}(x_h(k))}{\partial \theta_{hi}}\right], \qquad \begin{array}{l} i = 1,\cdots,c \\ h = 3,\cdots,n \end{array} \tag{13}$$

$$\frac{\partial J}{\partial \theta_{2i}} = -\sum_{k=1}^{N}\left[e(k)\lambda_{1}(k)\left(\sum_{j=1}^{c}\Omega_{1(ic-c+j)}X_{1_j}(k)\right)\frac{\partial \mathcal{X}_{2_i}(x_2(k))}{\partial \theta_{2i}}\right], \qquad i = 1,\cdots,c \tag{14}$$

$$\frac{\partial J}{\partial \theta_{1i}} = -\sum_{k=1}^{N}\left[e(k)\lambda_{1}(k)\left(\sum_{l=0}^{c-1}\Omega_{1(lc+i)}X_{2(l+1)}(k)\right)\frac{\partial \mathcal{X}_{1_i}(x_1(k))}{\partial \theta_{1i}}\right], \qquad i = 1,\cdots,c \tag{15}$$

where $X_{h_i}(k)$, $\hat{Y}_{h_i}(k)$ and $e(k)$ are defined as $\mathcal{X}_{h_i}(x_h(k))$, $\mathcal{Y}_{h_i}(\hat{y}_h(k))$, and $y(k) - \hat{y}(k)$, respectively, and $\lambda_{(\cdot)}(k)$ is written recursively as

$$\lambda_q(k) = \lambda_{q+1}(k)\sum_{j=1}^{c}\left[\left(\sum_{l=0}^{c-1}\Omega_{(q+1)(lc+j)}X_{(q+2)(l+1)}(k)\right)\frac{\partial \mathcal{Y}_{q_j}(\hat{y}_q(k))}{\partial \hat{y}_q(k)}\right], \quad q = 1,\cdots,n-2 \tag{16}$$

---

[3] The procedure is the same as that used to compute the local gradient equations in Multilayer Perceptron (MLP) Networks (Haykin, 1999), *mutatis mutandis*.

with $\lambda_{n-1}(k) = 1$. The derivatives with respect to the widths of the fuzzy sets can be obtained from Eqs. (12), (13), (14), and (15), substituting $\sigma_{(\cdot)}$ and $\varphi_{(\cdot)}$ for $\theta_{(\cdot)}$ and $\phi_{(\cdot)}$, respectively. Whenever Gaussian fuzzy sets are used, the implicit derivatives in the equations presented above are rewritten from Eqs. (8) and (9) as

$$\frac{\partial Y_{qj}(\hat{y}_q(k))}{\partial \hat{y}_q(k)} = \frac{-2}{\varphi_{qj}^2}\left(\hat{y}_q(k) - \phi_{qj}\right) Y_{qj}(\hat{y}_q(k))$$

$$\frac{\partial Y_{h_i}(\hat{y}_h(k))}{\partial \phi_{h_i}} = \frac{2}{\varphi_{h_i}^2}\left(\hat{y}_h(k) - \phi_{h_i}\right) Y_{h_i}(\hat{y}_h(k)); \quad \frac{\partial Y_{h_i}(\hat{y}_h(k))}{\partial \varphi_{h_i}} = \frac{2}{\varphi_{h_i}^3}\left(\hat{y}_h(k) - \phi_{h_i}\right)^2 Y_{h_i}(\hat{y}_h(k))$$

$$\frac{\partial X_{h_i}(x_h(k))}{\partial \theta_{h_i}} = \frac{2}{\sigma_{h_i}^2}\left(x_h(k) - \theta_{h_i}\right) X_{h_i}(x_h(k)); \quad \frac{\partial X_{h_i}(x_h(k))}{\partial \sigma_{h_i}} = \frac{2}{\sigma_{h_i}^3}\left(x_h(k) - \theta_{h_i}\right)^2 X_{h_i}(x_h(k))$$

3.3. Parameter Initialization

First, it is assumed that the input and output variables are normalized within a certain interval, such as [-1,1]. This procedure is usually adopted to avoid numerical problems during the training phase of neural networks (Haykin, 1999). Under this assumption, the fuzzy sets can be initialized empirically by a homogeneous distribution (within the normalization interval) of the sets associated with a given input or hidden output, which means equally spaced centers and standard deviations (widths) equal to the distance between two consecutive centers. The fuzzy sets of the input variables could optionally be initialized using fuzzy clustering techniques (Bezdek, 1981).

The parameter vectors $\Omega_{(\cdot)}$ should be initialized randomly with zero mean and absolute values small enough so that the initial values of the hidden outputs belong (at least approximately) to the normalization interval.

4. Industrial Process for Ethanol Production

4.1. Introduction

The Brazilian alcoholic fermentation processes arose from the production of the sugar cane liquor (aguardiente). Later, these processes were applied to the production of ethanol from molasses obtained from the sugar production plants. In the 1980s, the Federal Government's "Pro-Alcohol" program created an incentive for the use of ethanol as an alternative fuel in automobiles. As a consequence, there was an increase in research focusing on improving the productivity and yield of these processes. Current research concentrates on the optimization of continuous operation of the processes.

As stated previously, an industrial plant for the production of ethanol is considered in the present work. Because of difficulties encountered when working directly with the plant in operating mode, especially because of the high costs involved in interruptions in its operation for tests, it has been chosen to work with a simulator whose kinetic parameters have already been validated in the real plant. This simulator was developed by Andrietta (1994); Andrietta and Maugeri, 1994), who modeled the set of biochemical reactions of the process (by means of a set of nonlinear ordinary differential equations called kinetic model) and optimized an operational region - which was further implemented in the plant - in such a way so as to achieve satisfactory productivity values without affecting its operational and economic feasibility. Within this pre-optimized operational region, controllers should be applied to act on the manipulated variables of the process to optimize its real time yield, even in the presence of disturbances.

Controllers of particular interest to this type of process, which usually has slow dynamics, transport delay, and restrictive operational conditions, are the so-called Predictive Controllers (Model-Based Predictive Controllers - MBPC or MPC) (Clarke, 1994). These controllers demand a model of the process to predict its response to excitations and/or measurable disturbances. Such a model should be feasible for implementation in a computer as well as mathematically suitable for the formulation of the control law. Due to the complexity of the ethanol production process (which will be discussed in Section 4.2), the identification of an accurate nonlinear model for further use in designing effective predictive controllers for this process becomes necessary. This is the main objective of this work.

4.2. Plant Description

The fermentative process for ethanol production is illustrated in Fig. 5. The system is a typical large-scale industrial process composed of four tank reactors (fermenters) arranged in series and operated with cell recycling to produce ethanol from sugar cane syrup. The process is fed with a mixture composed of sugars (Total Reducing Sugars - TRS) as well as sources of nitrogen and mineral salts, called feed medium.

The feed medium is converted into ethanol by a fermentation process carried out using the yeast *Saccharomyces cerevisae*.

Since the behavior of the microorganisms is very sensitive to their environmental conditions, some of them are purged and the remaining cells are submitted to an acid treatment and dilution before being recycled into the first reactor. The recycling procedure is important because the generation of new microorganism colonies is an expensive and time-consuming process. A set of centrifuges split the fermented medium, which is formed of a mixture of water, CO_2, sugars, microorganisms (30-45g/l of cells), and alcohol, into two phases. The heavy phase contains most of the cells (160-200g/l) while the light phase contains at most 3g/l of cells and is 9-12% alcohol. The light phase is then sent to the distillation unit, where the alcohol is extracted.

Each reactor has an external system of heat exchangers with independent control loops (PI controllers) whose objective is to maintain the temperature of the reactants (fermentation broth) constant at an ideal level for the fermentation process. The set point for the temperature was optimized by (Andrietta, 1994; Andrietta and Maugeri, 1994) to maximize the efficiency of the reactions (conversion) of the industrial plant.

In the simulator of the plant some simplifications related to apparatus that are not represented in Fig. 5 are also considered. One of them is an independent internal control loop to regulate the liquid volumes of the tanks, which are represented in the simulator by the condition of equal flow rates in all the tanks. Another simplification refers to the flow control valves of the feed medium and recycling. The dynamics of these devices can be neglected without loss of generality since they are much faster than the other dynamics of the process. In addition, the hypothesis of perfect stirred tanks (Andrietta, 1994; Andrietta and Maugeri, 1994) is adopted, i.e., it is assumed that the reactions occur homogeneously inside the tanks. This is a good approximation with respect to the kinetic model that was validated by Andrietta, but it influences the dynamic representation of the process by the elimination of the transport delays existing in real situations. Therefore, this simplification is going to be reconsidered in future work.

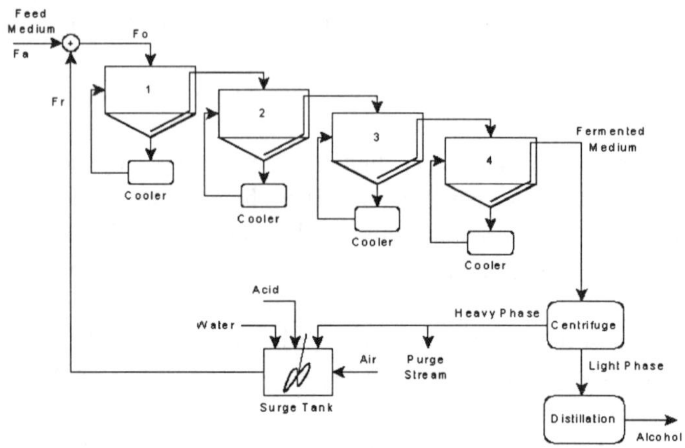

Figure 5. Schematic illustration of the industrial plant for ethanol production.

As mentioned previously, the industrial process for ethanol production is a highly nonlinear process. The main nonlinearities arise from the behavior of the microorganisms. Increasing the feed medium flow rate, for example, the TRS concentration inside the tanks also increases. Under this condition, ethanol production from the biological conversion of the sugar tends to increase. However, an excessive amount of sugar, which exceeds the microorganisms' processing capacity, will not be converted into ethanol (substrate inhibition phenomenon). This excess of sugar will appear in the final product, thus characterizing a drop in the conversion efficiency as well as a waste of raw material and energy. Another problem caused by the substrate inhibition effect is a decrease in the microorganism reproduction, which is reflected directly in the alcohol production. This inhibitory effect can also be caused by an excess of alcohol in the fermentative broth, which in turn can cause the death of cells. Besides, low levels of substrate can also cause the death of cells. All of these factors influence the dynamics and the efficiency of the fermentation process. More details on this process can be found in (Andrietta, 1994), and a set of trials illustrating its inverse responses and/or strongly nonlinear behavior is presented in (Dechechi, 1998).

Considering these characteristics, the fundamental objective of the study of the fermentative process for ethanol production is to generate models and controllers in such a way so as to maximize its efficiency, i.e., to maximize ethanol concentration and minimize TRS concentration in the outlet of the fourth tank, while maintaining the stability of the microorganism colony.

4.3. Input, Output, and Disturbances Variables

Considering the pre-optimized operational conditions of the plant discussed in Section 4.1, the input, output, and disturbance variables of the process are

- *Feed Medium Flow Rate* (F_a [m^3/h]): This is the main manipulated input variable. The universe of discourse of this variable is the interval [50, 150]. This interval is conservative in terms of the economic and operational viability of the plant. It represents the upper and lower bounds for the substrate feed flow of the microorganisms and comprises the limitations related to valve operation and tank volumes as well.

- *Recycle Rate* (t_r [dimensionless]): This variable relates the feed medium flow rate, F_a, with the cell recycle flow rate (F_r [m^3/h]) and, accordingly, with the real inlet feed flow rate in the first tank (F_0 [m^3/h]), as shown in Fig. 5. This relationship is given by

$$F_0 = F_a + F_r = \frac{F_a}{(1-t_r)} \tag{17}$$

Thus, a recycle rate of 0.3 implies $F_a = 0.7F_0$ and $F_r = 0.3F_0$. This is the pre-optimized industrial operation value for the plant. This nominal value can eventually be changed by a plant worker (operator) to fix problems in the microorganism colony.

The recycle rate can be considered either as a measurable disturbance or as an input manipulated by an automatic controller. In the first case, where the disturbances are accomplished manually as a function of the operator experience, a variation of ±10% around the nominal value is allowed, resulting in the interval [0.27,0.33]. In the second case, where the automatic manipulation is based on an optimum criterion and upon a reliable model of the plant, the interval referred to can be expanded to approximately ±90% of the nominal value, i.e., [0.05,0.55].

- *TRS Concentration in the Feed Medium* (S_0 [g/l]): The nominal value of this variable under real operational conditions is 180g/l. However, since it depends on the sugar cane used, it is important to take into account possible disturbances of at least ±5% around this value. In this case, this variable becomes a measurable disturbance (input) belonging to the interval [170, 190].

The output variables of interest (according to the control objectives discussed in Section 4.2) are

- *Outlet Ethanol Concentration in the Fourth Tank* (P_4 [g/l]).
- *Outlet TRS Concentration in the Fourth Tank* (S_4 [g/l]).
- *Outlet Cells Concentration in the Fourth Tank* (M_4 [g/l]).

where the outlet product of the fourth tank is the fermented medium (see Fig. 5).

5. Hierarchical Neural Fuzzy Modeling of the Ethanol Production Process

5.1. Data Generation and Sampling

The strategy of dealing with a validated simulator of the actual process makes it possible to generate identification data as desired. Thus, a representative data set, which contains the input and output signals of the process related to 5000 hours of its simulated operation, was generated. In these data the manipulated input F_a is a sequence of steps, each of which with a period of 10h (long enough so that the process can nearly reach the steady state) and amplitude uniformly distributed within the operational interval [50,150]. The inputs t_r and S_0 are also sequences of steps. However, they were given Gaussian distributions for the amplitudes and periods of 25h and 50h, respectively, so that they can express the underlying statistical characteristics of the measurable disturbances, i.e., their greatest probability of taking the respective nominal values or their neighboring values most of the time. To accomplish this, the Gaussian probability distribution functions were centered on the nominal values of t_r and S_0, i.e., S_0 = 180g/l and t_r = 0.3. The standard deviations were set as 1/6 of the respective operational intervals (S_0 = [170,190] and

t_r = [0.27,0.33]) in such a way that the amplitudes of the randomly generated steps belong to these intervals with a probability of approximately 99%.

The data was sampled using the traditional procedure in which the sampling period T is derived in such a way that (Åström and Wittenmark, 1997)

$$N_T = \frac{T_S}{T} \approx 4 \text{ to } 10 \tag{18}$$

where T_s is the rise time of the process and N_T is the number of data samples during this time. In the case of nonlinear multivariate processes, however, the rise time depends on the input values and may be different for each output. In these cases, the faster dynamics should be considered. The rise time of the industrial process considered in the present work was estimated roughly between 2h and 3h by means of simulation experiments. Hence, Eq. 18 yields T ≈ [12min,45min]. In addition, the sampling period must be a multiple of 15min, which is the typical interval between samples of the chromatograph (the device used for measuring the TRS concentrations involved in the process). For the sake of the considerations mentioned above, a value of T = 30min was selected. This value is large enough to avoid numerical problems and gives rise to a set of 10000 discrete-time data which will be used in the sequel, one half intended for the estimation of a hierarchical model of the process and the other half intended for the validation of this model.

5.2. Structure Selection

Structure here refers to three distinct concepts: *i*) the regressors, so-called regression vectors, of the input and output variables of the dynamic hierarchical model; *ii*) the hierarchical order of the variables; and *iii*) the internal structural characteristics of the model itself.

The model regressors were selected experimentally. Roughly speaking, a set of Single-Input/Single-Output (SISO) modeling procedures were carried out for every input-output pair of the process. Independent data sets were used during this modeling phase; each of these consisted of 1000 input-output data pairs where the respective input was set as described for the main data set in Section 5.1 and the others were kept constant at their nominal values. Due to the simplicity of the SISO representation as well as the number of experiments needed to evaluate different regressors for each of the 9 (3 × 3) input-output pairs, a simple nonhierarchical RBF neural network (as described in Section 2) was used, with c = 5 and trained rapidly through the Recursive Least Squares (RLS) algorithm (Ljung, 1999). The regressors were selected according to the performances of the resulting RBF models in one-step-ahead prediction and synthetic data (recursive or open-loop simulation) as well. Priority was given to those smaller regressors, especially with respect to the output variables because they are subject to prediction errors (over greater-than-one horizons). The regressors selected from the above-mentioned methodology were $[F_a(\text{k-1})\ F_a(\text{k-2})]$, $[S_0(\text{k-1})]$, $[t_r(\text{k-1})]$, $[P_4(\text{k-1})]$, $[S_4(\text{k-1})]$, and $[M_4(\text{k-1})]$.

The results presented by the SISO models (that take into consideration only one output, simply neglecting the effect of the others) showed that it would be possible to derive a model with multiple inputs and outputs (MIMO) by means of three completely independent MISO models. This is a very desirable property since the independence of the models implies the independence of their prediction errors, which in turn can result in a better overall modeling performance over long-range prediction horizons. The internal structure of these three MISO hierarchical models was defined as in Section 3.1 with $c = 5$ (which is a common value, at least in the context of fuzzy logic (Pedrycz, 1995)). Parameter initialization was performed as described in Section 3.3.

The hierarchical order of the variables was selected using a priori knowledge of the basic behavior of the process. First, the variable F_a was placed on the first hierarchical level (first processing block, which has the most complex mapping from the respective inputs into the model output) since it is the main input variable of the process, besides having the largest regressor. Contrariwise, the feedback of the output variable of each model was placed on the last (lowest) level based on the insights into their relevance obtained during the regressor selection phase. Finally, to select the hierarchical order of the disturbance variables, a set of dynamic response experiments (Dechechi, 1998) performed in the same ethanol production plant focused in the present work were considered. The experiments referred to above point out that the nonlinearities associated with the disturbances in S_0 are stronger than those associated with the disturbances in t_r. Hence, the hierarchical order of the former was set greater than that of the latter.

5.3. Model Estimation and Validation

To estimate the MIMO hierarchical model of the ethanol production process, each of the three independent MISO models was trained (optimized) using the 5000 estimation data throughout 1000 epochs, when the optimization procedure could no longer improve the models and no significant overfitting had as yet occurred. The evolution of the model accuracy throughout the optimization procedure is shown in Fig. 6.

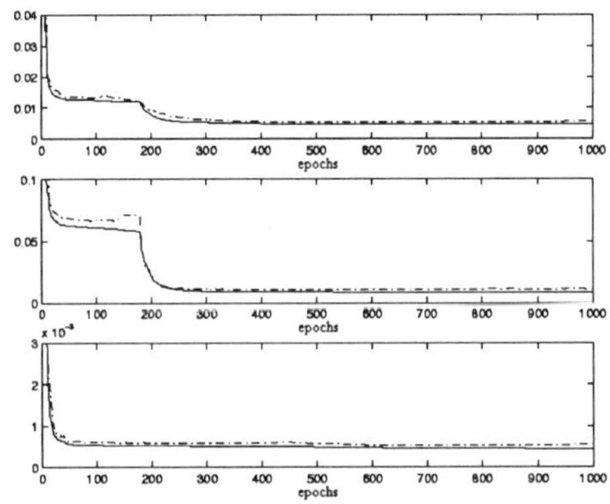

Figure 6. Evolution of the Mean Squared Errors (MSEs) between the outputs of the process and the model: Estimation data (solid line) and validation data (dashed line); Outputs P_4, S_4 and M_4 (Top to bottom).

These errors, however, are not comparable with each other since the universes of discourse of the output variables are far too different. To allow a quantitative comparison of the modeling performances for P_4, S_4 and M_4, the following Normalized Mean Squared Error (NMSE) is used:

$$NMSE\left(\{y(k), \hat{y}(k)\}_{k=1}^{N}\right) = \frac{1}{N}\sum_{k=1}^{N}\left(\frac{y(k) - \hat{y}(k)}{\bar{y}(k)}\right)^2 \quad (19)$$

where $y(k)$ denotes one of the outputs of the process, $\hat{y}(k)$ represents the respective prediction of the model and

$$\bar{y}(k) = y(k) - \frac{1}{N}\sum_{k=1}^{N} y(k) \quad (20)$$

is a normalization term.

The simulation performance of the model using the 5000 validation data is shown in Table 1, where columns from left to right present the output variables and the respective NMSEs with respect to one-step-ahead prediction and synthetic data, respectively. The simulation curves related to the first 500 samples are illustrated in Figs. 7, 8, and 9.

Table 1. Simulation results for the MIMO hierarchical neural fuzzy model of the ethanol production process.

Output	NMSE 1 step pred.	NMSE s. data
P_4	4.13	223.20
S_4	1.23	11.41
M_4	0.50	93.31

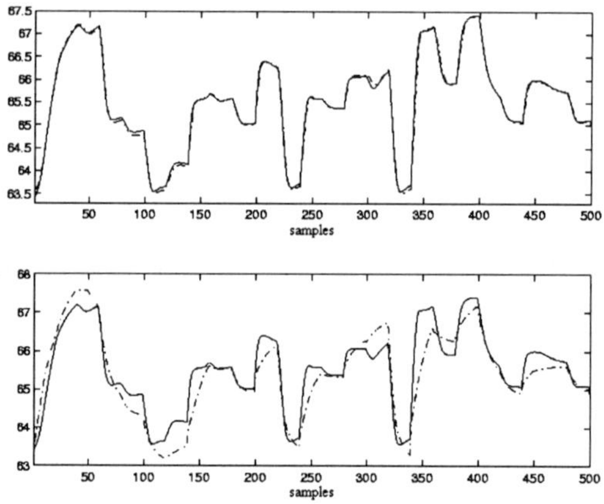

Figure 7. Ethanol concentration in the outlet of the fourth tank (P_4 [g/l]) (solid line) and the respective model output (dashed line) for the validation data: One-step-ahead prediction (above) and synthetic data (below).

The figures illustrate a good performance of the model for both one-step-ahead prediction and synthetic data, specially considering the long prediction horizons involved in the simulations. It can be noted that the largest modeling errors are related to the output P_4. However, in the context of model-based control applications, such as those based on adaptive and predictive controllers, this model can be adequate even with respect to P_4 since such controllers can generally be tuned using short prediction horizons, especially in the case of stable processes.

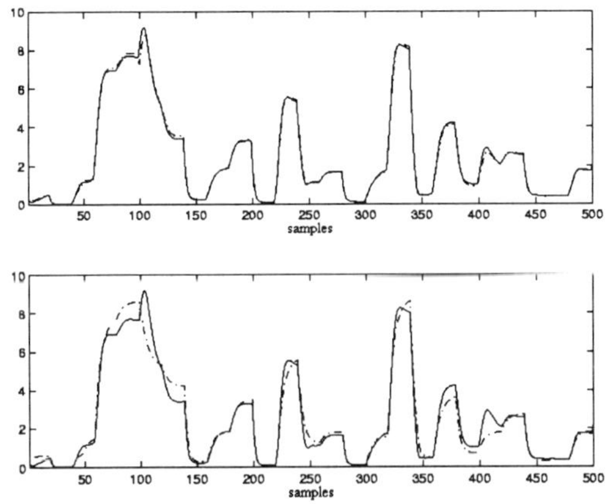

Figure 8. TRS concentration in the outlet of the fourth tank (S_4 [g/l]) (solid line) and the respective model output (dashed line) for the validation data: One-step-ahead prediction (above) and synthetic data (below).

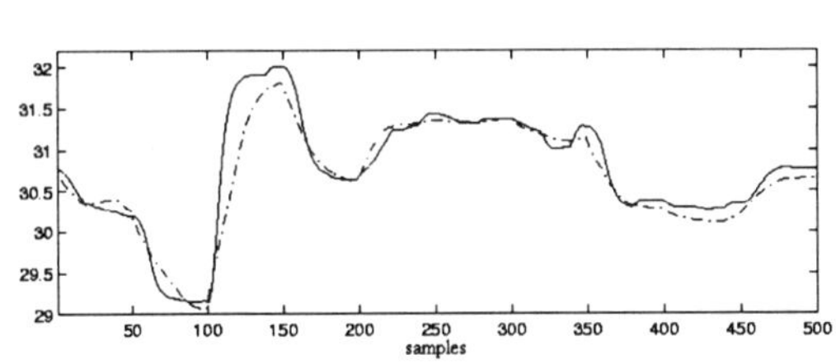

Figure 9. Microorganism concentration in the outlet of the fourth tank (M_4 [g/l]) (solid line) and the respective model output (dashed line) for the validation data: One-step-ahead prediction (above) and synthetic data (below).

Another important point is that, in most of the control applications to this kind of process undertaken up to now, the main controlled variable has been the output S_4, whose NMSE in synthetic data is the smallest.

6. Conclusions and Perspectives

Hierarchical neural and fuzzy systems have been shown to be effective tools for large-scale modeling and control problems. In the present work a hierarchical neural fuzzy model was used for the identification of a complex biotechnological process for ethanol production. In fact, ethanol is a powerful "clean" and renewable source of fuel whose use in automobiles has been encouraged by the Brazilian Government since the 1980s due to its considerable economic impact.

It has been possible to derive a MIMO hierarchical model of the process by means of three completely independent MISO models. Simulations have shown that the resulting model can adequately represent the system with a reduced number of free design parameters. Since the model has shown a good performance in long-range horizon predictions, it has great potential for use in advanced control strategies.

In future work the authors intend to extend the hierarchical model of the ethanol production process by utilizing the orthonormal basis function approach presented in Oliveira *et al.* (1999) in such a way so as to avoid the feedback of prediction errors as well as the regressor estimation task. The authors also intend to design controllers for this process based on the models developed.

References

Andrietta S. R., *Modeling, Simulation and Control of Industrial-Scale Process for Continuous Alcoholic Fermentation*, Ph.D. thesis (FEA/UNICAMP – Campinas-SP-Brazil, 1994, in Portuguese).

Andrietta S. R. and Maugeri F., *Advances in Bioprocess Engineering*, (Kluwer Academic Publishers, 1994), 47-52.

Åström K. J. and Wittenmark B., *Computer Controlled Systems* (Prentice Hall, 1997), 3rd Edition.

Bazaraa M. S. and Shetty C. M., *Nonlinear Programming and Algorithms*, (John Wiley & Sons, 1979).

Bezdek J. C., *Pattern Recognition with Fuzzy Objective Function Algorithm*, (Plenum Press, 1981).
Broomhead D. S. and Lowe D., *Complex Systems.* **2** (1988), 321-355.
Campello R. J. G. B. and Amaral W. C., in *Proc. IV Brazilian Symposium on Intelligent Automation (in Portuguese)*, São Paulo-Brazil (1999), 449-454.
Campello R. J. G. B. and Amaral W. C., in *IEEE-INNS-ENNS International Joint Conference on Neural Networks (to be published)*, Como-Italy (2000).
Chen W. and Wang L.-X., *Information Sciences.* **123** (2000), 241-248.
Clarke D. W., ed., *Advances in Model Predictive Control*, (Oxford University Press, 1994).
Dechechi E. C., *Modern Adaptive Predictive Control "Multivariate DMC"*, Ph.D. thesis (1998), FEQ/UNICAMP - Campinas-SP-Brazil (in Portuguese).
Haykin S., *Neural Networks: A Comprehensive Foundation*, (Prentice Hall, 1999), 2nd Edition.
Jamshidi M., in *Proc. 7th IFSA World Congress*, (Prague-Czech Republic, 1997), 324-329.
Kosko B., *Neural Networks and Fuzzy Systems: A Dynamical System Approach to Machine Intelligence*, (Prentice Hall, 1992).
Kosko B., *Fuzzy Engineering*, (Prentice Hall, 1997).
Ljung L., *System Identification: Theory for the User*, (Prentice Hall, 1999), 2nd Edition.
Meleiro L. A. C. and Maciel Filho R., in *Computers and Chemical Engineering*, **24** (2000), 925-930.
Oliveira G. H. C. et al., in *Proc. 8th IEEE Internat. Conf. On Fuzzy Systems*, (Seoul-Korea, 1999), 957-962.
Oliveira J. V. and Lemos J. M., in *Proc. 7th IFSA World Congress*, (Prague-Czech Republic, 1997), 330-335.
Pedrycz W., *Fuzzy Sets Engineering*, (CRC Press, 1995).
Raju G. U., Zhou J. and Kisner R. A., in *Int. J. Control*, (1991), 1201-1216.
Wang L.-X., in *Fuzzy Sets and Systems.* **93** (1998), 223-230.
Wang L.-X., in *IEEE Trans. Fuzzy Systems.* **7** (1999), 617-624.
Yager R. R. and Filev D. P., *Essentials of Fuzzy Modeling and Control*, (John Wiley & Sons, 1994).

Acknowledgements

The first, second and fourth authors acknowledge the funding received from CNPq, the Brazilian National Research Council. The third author acknowledges the assistance of FAPESP, the Research Foundation of the State of São Paulo, in the form of fellowship 99/03902-6.

Appendix

Equation 1 can be rewritten explicitly as

$$\hat{y} = \sum_{l_1}^{c_1} \cdots \sum_{l_n}^{c_n} \Psi_{l_1,\cdots,l_n} \Omega_{l_1,\cdots,l_n} \tag{21}$$

where $\Psi_{l_1,\cdots,l_n}$ and $\Omega_{l_1,\cdots,l_n}$ denote elements of Ψ and Ω, respectively, in such a way that $\Psi = \begin{bmatrix} \Psi_{1,\cdots,1} & \cdots & \Psi_{c_1,\cdots,c_n} \end{bmatrix}^T$ and $\Omega = \begin{bmatrix} \Omega_{1,\cdots,1} & \cdots & \Omega_{c_1,\cdots,c_n} \end{bmatrix}^T$. Each element $\Psi_{l_1,\cdots,l_n}$ can be written from Eqs. (2) and (3) as

$$\Psi_{l_1,\cdots,l_n} = X_{1l_1}(x_1)\ X_{2l_2}(x_2) \cdot \ \cdots \ \cdot X_{nl_n}(x_n) \tag{22}$$

Whenever Gaussian fuzzy sets are used, Eq. 22 can be rewritten from Eq. 8 as

$$\Psi_{l_1,\cdots,l_n} = \exp\left(-\frac{\left(x_1 - \theta_{1l_1}\right)^2}{\sigma^2_{1l_1}} \right) \cdot \ \cdots \ \cdot \exp\left(-\frac{\left(x_n - \theta_{nl_n}\right)^2}{\sigma^2_{nl_n}} \right) \tag{23}$$

Since the product of n unidimensional Gaussian functions is an n-dimensional Gaussian function, Eq. 23 results in

$$\Psi_{l_1,\cdots,l_n} = \exp\left(-\left(\mathbf{x} - \Theta_{l_1,\cdots,l_n}\right)^T \Lambda^{-1} \left(\mathbf{x} - \Theta_{l_1,\cdots,l_n}\right) \right) \tag{24}$$

where

$$\mathbf{x} = \begin{bmatrix} x_1 & x_2 & \cdots & x_n \end{bmatrix}^T \tag{25}$$

$$\Theta_{l_1,\cdots,l_n} = \begin{bmatrix} \theta_{1l_1} & \theta_{2l_2} & \cdots & \theta_{nl_n} \end{bmatrix}^T \tag{26}$$

$$\Lambda = \begin{bmatrix} \sigma_{1_{l_1}}^2 & 0 & \cdots & 0 \\ 0 & \ddots & & \vdots \\ \vdots & & \ddots & 0 \\ 0 & \cdots & 0 & \sigma_{n_{l_n}}^2 \end{bmatrix} \tag{27}$$

From Eqs. 21 and 24 it can be seen that the model output is given by a weighted sum of multivariate Gaussian functions, which is precisely the architecture of an RBF neural network (Haykin, 1999).

PART III ESTIMATION AND CONTROL

9. ADAPTIVE INVERSE MODEL CONTROL OF A CONTINUOUS FERMENTATION PROCESS USING NEURAL NETWORKS

M. A. HUSSAIN
Chemical Engineering Department, University Malaya,
50603 Kuala Lumpur, Malaysia

Many difficult techniques involving non-linear models have been proposed and applied for the control of non-linear processes in the past. However, most of these techniques are complex and difficult to obtain and implement. They are also restricted to limited ranges of operations. In recent years, neural networks have emerged as an attractive method that is easily implemented in various model-based control techniques. One such technique is the internal model control method, which incorporates approximations of both system model and its inverse in the control algorithm. In this article, the application of this neural network based IMC strategy on a highly non-linear system such as the continuous fermentation process will be shown. The control strategy regulates the biomass concentration by manipulating the dilution rate within the reactor. Acceptable performance was achieved for the set point regulation under various internal and external disturbances but with some offsets in the output. An adaptive scheme using the modified sliding window approach was further applied, which eliminated these offsets completely from the system under the same disturbances. The comparison between the conventional and adaptive IMC technique will be further highlighted in this article.

1. Introduction

Neural networks, being versatile in nature, can also be easily incorporated in various model-based control techniques. One such technique is the inverse-model-based control strategy. The ease and speed of applying this method relative to other possible methods (such as the predictive schemes) for many applications is clearly evident. This method relies heavily on the availability of the inverse of the system's model, which acts as the controller in this scheme, which may be difficult to obtain analytically for most nonlinear systems. Since neural networks have the potential to model any system, the use of neural network for modeling these inverses and hence utilized them in the inverse-model-based-control strategies is highly promising. One method of such nature is the nonlinear internal model control (IMC) technique, basically an extension of the linear IMC method. In this scheme, both the forward and inverse models are used directly as elements within the feedback loop. Details of these models will be shown later. However, this methodology normally does not eliminate offsets in its output due to the difficulty in getting an exact neural network model by offline training. For this reason, an adaptive scheme for online identification needs to be further applied to this strategy.

Two strategies concerning adaptive neural network-based control that have been studied are based on single-input single-output system, as can be seen in Chen and Khalil (1992), and Polycarpou and Mears (1998). Chen and Khalil (1992) designed an adaptive control by using feedback linearization which is derived in terms of some unknown nonlinear functions. This function is modeled by multi-layered neural networks and the weights of neural network are updated and used to generate the control. Polycarpou and Mears (1998) designed a stable adaptive neural controller for an uncertain nonlinear dynamical system with unknown nonlinearities.

Furthermore, Van Breusegem *et al.* (1991) designed an on-line adaptation scheme for neural network model in fermentation processes to integrate progressively dynamic changes by modifying the set of weights. The adaptive algorithm employed by Van Breusegem *et al.* (1991) is the sliding window learning scheme. This scheme is used to refresh progressively the knowledge integrated in the neural model. This procedure is inspired by the recursive parameter estimation techniques which are widely used in identification and control. The adaptation scheme restricts the memory of the neural network by adding the effects of new data and by removing progressively the influence of "obsolete" data.

The work presented by Mills *et al.* (1994) proposes and demonstrates a useful method for implementing neural network models for adaptive control. The performance of adaptation has been sufficiently raised to allow practical adaptive control to be considered. Then the new adaptive method is combined with multistep non-linear predictive control techniques to form an adaptive neural controller. The performance of this controller is evaluated using two simulated realistic processes: level control of a conical tank and multivariable control of an industrial evaporator. The results indicate that the techniques have good practical potential for adaptive control of non-linear processes. The adaptive neural predictive control (ANPC) is an amalgamation of multistep non-linear predictive control technique, history-stack learning technique and offset compensation method.

The work proposed in this article make use of the sliding window approach to update the neural network forward and inverse models in the internal model control strategy online. This updating involves changes in the target values for both these forward and inverse models. Simulation for a disturbance rejection case study in a fermentation process is shown in later sections to demonstrate the utility of this method.

2. Continuous Fermentation Process

The process that has been used to study the application of this neural network based strategy is the fermentation system as shown in Fig. 1. The process consists of a constant volume reactor in which a single, rate-limiting substrate promotes biomass growth and product formation (Agrawal et al., 1989).

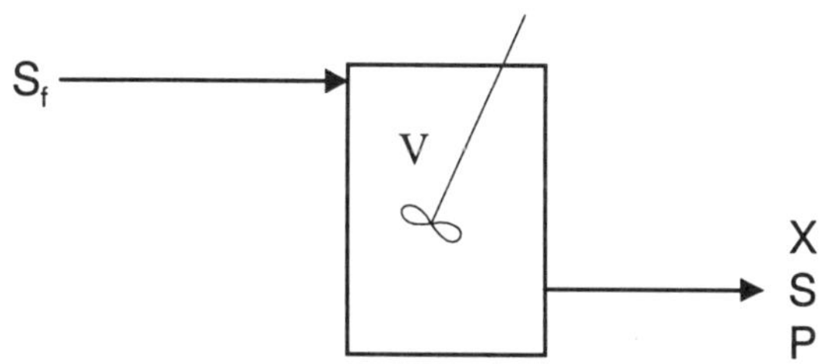

Figure 1. Continuous Fermentation System

By assuming constant yields, a process model with three non-linear ordinary differential equations can be obtained as follows,

$$\dot{X} = -DX + \mu(S, P)X \tag{1}$$

$$\dot{S} = D(S_f - S) - \frac{1}{Y_{X/S}}\mu(S, P)X \tag{2}$$

$$\dot{P} = -DP + [\alpha\mu(S, P) + \beta]X \tag{3}$$

where *X*, *S* and *P* are the biomass, substrate and product concentrations respectively; *D* is the dilution rate; S_f is the feed substrate concentration; and $Y_{X/S}$, α and β are yield parameters. The specific growth rate μ is modeled as,

$$\mu(S, P) = \frac{\mu_m\left(1 - \frac{P}{P_m}\right)S}{K_m + S + \frac{S^2}{K_i}} \tag{4}$$

where μ_m is the maximum specific growth rate; and P_m, K_m, and K_i are constant parameters. The nominal operating conditions are shown in Table 1.

The control objective in this continuous fermentation system is to maximise the steady-state biomass production. This can be a difficult task since parameters such as the maximum specific growth rate μ_m and cell-mass yield $Y_{X/S}$ may exhibit significant time-varying behavior. It can however be shown that near optimal steady-state performance can be achieved by manipulation the dilution rate D and regulating the biomass concentration X at a constant value. Hence, in this simulation, the input is D, output is X and the states are that of X, S and P respectively.

i.e. $u = D, \ y = X$

Note that, All figures and the text referring to Y is that of the biomass concentration, X

Table 1. Nominal operating conditions for the fermenter model

Variable	Value	Variable	Value
$Y_{X/S}$	0.4 g/g	α	2.2 g/g
β	0.2 h^{-1}	μ_m	0.48 h^{-1}
P_m	50 g/L	K_m	1.2 g/L
K_I	22 g/L	S_f	20 g/L
D	0.202 h^{-1}	X	6.0 g/L
S	5.0 g/L	P	19.14 g/L

3. Internal Model Control (IMC) Strategy

In this scheme, both the forward and inverse models are used directly as elements within the feedback loop. The network inverse model is utilized in the control strategy by simply cascading it with the controlled system or plant. In this case the neural network, acting as the controller, has to learn to supply at its output, the appropriate control parameters, u for the desired targets, ysp at its input. In addition, the neural network forward model is placed in parallel with the plant, to cater for plant or model mismatches and the error between the plant output and the neural net forward model is subtracted from the set point before being feedback into the inverse model (See Fig. 2.). A Filter, F can be introduced prior to the controller in this approach to incorporate robustness in the feedback system (especially where it is difficult to get exact inverse models) (Hussain, 1999). In order to reduce the overshoot, a first order filter is added in the control loop. The filtering action can be represented by:-

$$Xsp(k) = \alpha_f \times Xsp(k-1) + (1-\alpha) \times e(k)$$

where $e(k) = Xsp(k) - Xdiff(k-1)$ and α_f = the filtering parameter

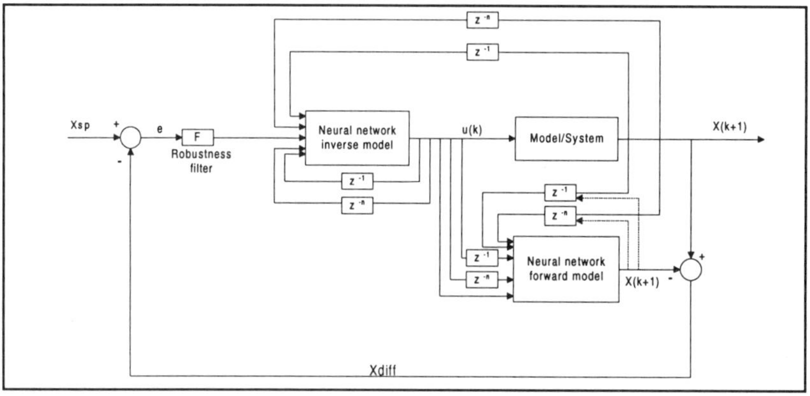

Figure 2. Internal Model Control Strategy

In order to implement this strategy the important components needed are the forward neural network and inverse neural network models respectively. The method of obtaining these models accurately will be detailed out next.

3.1. Forward Modeling

The procedure of training a neural network to represent the dynamics of the system is referred to as forward modeling. Two training data sets are used in the training, which are switched from one to the other during training to improve the identification process. The training database for the network was developed by changing the dilution rate, D in a random stepwise fashion from its steady state value (0.202 h^{-1}) where each step lasts for 100 hours. The changes introduced in D as well as the resulting changes in X, S and P can be seen in Figs. 3 and 4 for the first and second training data set respectively. To validate the training, a ramp input signal is introduced into the system as Fig. 5. The sampling time is determined by studying the open loop response of the system. A nonlinear system will behave differently when the manipulated variable is increased and decreased from the steady state value. The smallest response time of the open loop response is observed and the sampling time is taken as 10% of this observed time. Sampling time is taken at 0.5 hour and total sampling time is 500 hours.

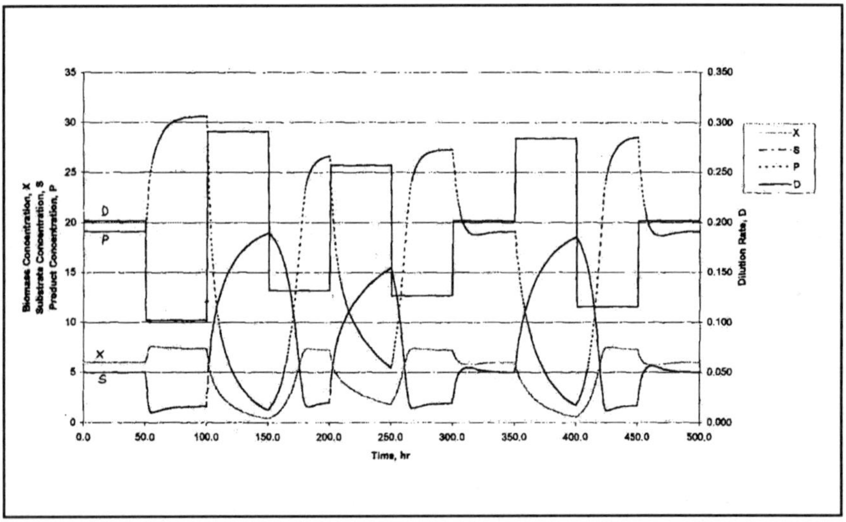

Figure 3. First Training Data

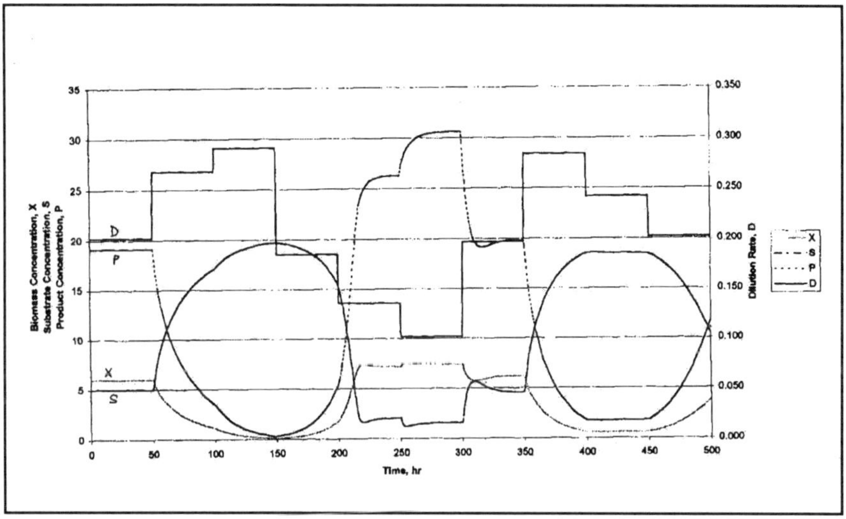

Figure 4. Second Training Data

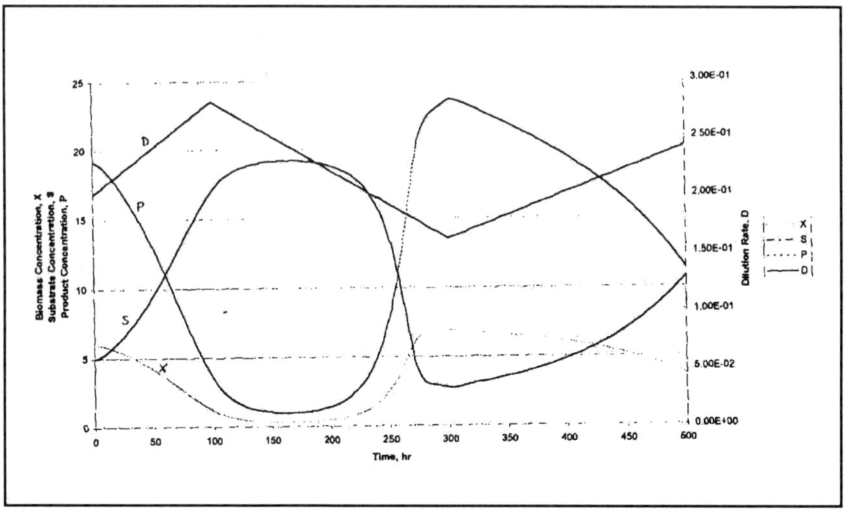

Figure 5. Validation Data

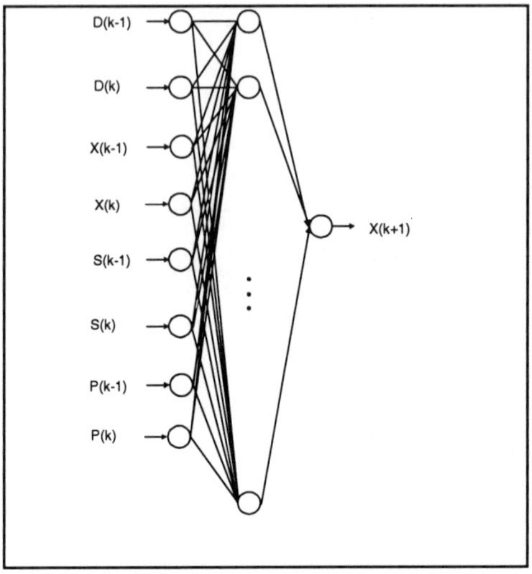

Figure 6. Forward Model Architecture

The architecture of the forward neural network can be seen in Fig. 6. The inputs to the network consists of present and past values of X, S, P and D values. The desired network output are the future X values. Those input and output values are fed into the network in the moving window approach. From these training exercise, the neural network architecture produced is an 8 nodes input layer, 10 nodes of hidden layer and a 1-output layer system. The activation function utilized is the sigmoidal function in both the hidden and output layer. The average sum of squared error (ASS) for the training and validation are listed in Table 2. The result for validation can be seen in Fig. 7. Both training and validation have shown satisfactory results, and the final forward model obtained represents the model to be utilised in the IMC method.

Table 2. ASS error for the Training and Validation for Forward Model

	Training Set 1	Training Set 2	Validation Set
Sum Square Error	7.06×10^{-3}	5.85×10^{-3}	4.31×10^{-3}
Num of Epochs	100	100	-
Num of data, n_p	999	999	999
ASS	7.07×10^{-6}	5.86×10^{-6}	4.31×10^{-6}

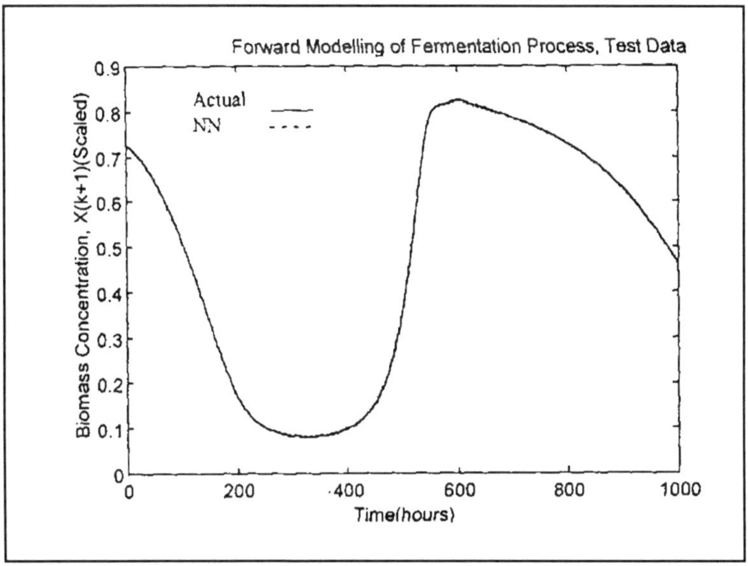

Figure 7. Forward Validation data

3.2. Inverse Modeling

Inverse modeling is referred to the training of the neural network in predicting the input to the plant given past data of the inputs and outputs together with the desired output. The model plays an important role in designing control system. Similar to the forward modeling methodology two training data sets were used, which were switch from one to the other during training to improve the identification process.

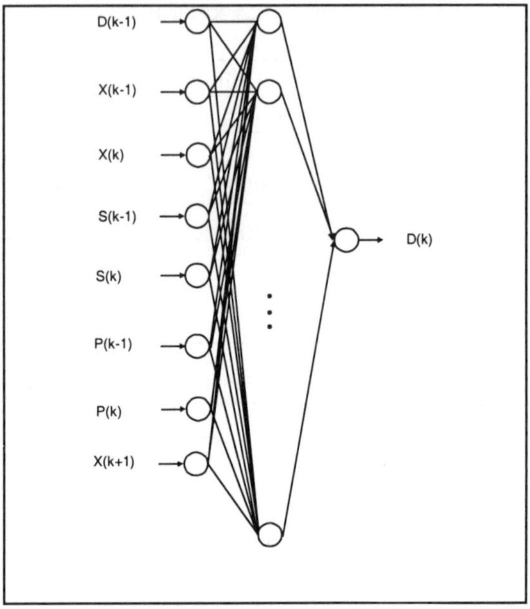

Figure 8. Inverse Neural Network Architecture

The data sets generated from the training and validating the inverse model were similar to the forward model. During training the network is fed with the required future value, X(k+1), together with the present and past input and outputs to predict the current input or control action, D(k) as seen in Fig. 8. From these the training, the network architecture and activation functions that were chosen are similar to the forward model.

The result for validating the model can be seen in the Fig. 9. The ASS error values of the training and validation are listed in Table 3. Both training and validation again have shown satisfactory results, resulting in the use of this inverse neural network model for the IMC simulation later.

In order to compare the conventional IMC with the adaptive approach in removing offsets under disturbances, their simulation results for the closed loop control system will be shown after the description of the adaptive approach below.

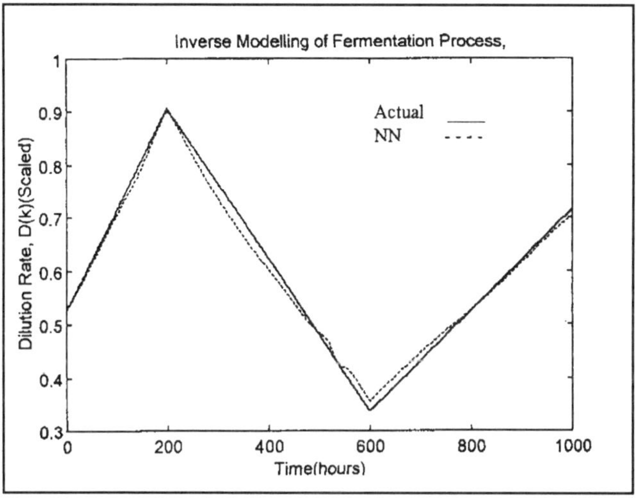

Figure 9. Inverse Validation Result

Table 3 ASS error for the Training and Validation for Inverse Model

	Training Set 1	Training Set 2	Validation Set
Sum Square Error	9.95×10^{-1}	4.42×10^{-1}	2.19×10^{-1}
Num of Epochs	1000	1000	-
Num of data, n_p	999	999	999
ASS	9.96×10^{-4}	4.42×10^{-4}	2.19×10^{-4}

4. Adaptive Internal Model-Based Control Strategy

Adaptive control technique can be added into the internal model control (IMC) strategy to handle the plant/model mismatch or disturbance. Offsets are normally observed when a disturbance or plant mismatch occurs in the IMC system. Therefore an improved adaptive scheme using the "Modified Sliding Window Learning Method" is used to increase the ability of IMC controller to handle the plant/model mismatch or disturbance in this work, as described below.

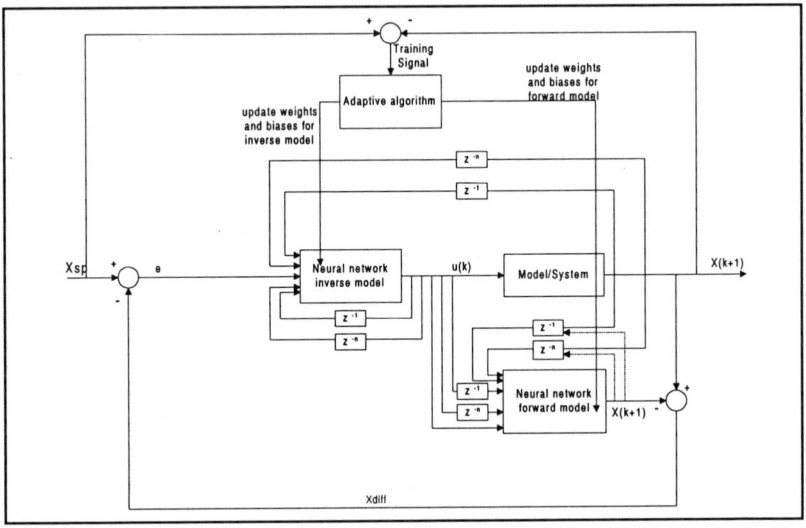

Figure 10. Adaptive IMC Structure

4.1. Modified Sliding Window Learning Method

In this adaptive scheme, the sliding window learning method (Van Breusegem et al., 1991) had been modified and adopted in the IMC control strategy for the biomass fermentation system. The online adaptive strategy can be seen in Fig. 10. The modified sliding window adaptive algorithm is used to refresh progressively the knowledge integrated in the neural network in the inverse and forward model.

The sliding window learning procedure is as follows:

1. Obtain an initial inverse and forward neural model from the previous experimental or simulated data and construct it within the IMC control strategy.
2. Choose the length L of the learning window

For each sampling instant k greater than L or equal to L,

3. Form a new learning data set with the last L successive pairs of input and output data vectors corresponding to the data from sampling instant (k-L+1) to k for inverse model and forward model

4. Teach the network with the newly formed learning data set to update the weights of the current neural model
5. Repeat steps 3 for next sampling instant

The learning procedure consists of two successive steps i.e. construction of an updated learning data set and updating of the neural model with the learning algorithm. In this study, the Levenberg Marquardt method had been used as the learning algorithm. The teaching in step 4 goes on for a maximum number of iterations or until the desired sum square error is achieved.

These steps had been proposed in order to enhance the learning speed for the inverse and forward model. Besides the speed, this modificd method is expected to give better disturbance rejection performance compared to IMC controller alone. When a disturbance or plant mismatch occurred in the system, the training data set for the inverse or forward modelling would no longer be valid because the plant/model output would change at the same time. This would require a new training data set. With this modified sliding window learning method, the targeted output for the last L successive pairs of data (i.e., biomass concentration for forward model, dilution rate for inverse model) is changed immediately when the difference between the desired output and plant output (biomass concentration) exceeded the maximum allowable error. The target values for forward and inverse modelling also are changed to provide an entirely new set of training data for the identification step. With this new learning data set, the network is taught to update the weights and biases of both models continuously.

4.2. Updating Target Values For Online Forward Modeling

Referring to Fig. 10, when there is no plant/model mismatch and disturbance in the systems, the neural network have been considered trained properly to represent the forward and inverse of the plant/model and hence the plant output and neural network forward model output would equal the setpoint. Thus, when there is a mismatch or disturbance in the system, the objective will be to update the weights and biases in the neural network forward model so that the output from this forward model is equal to the setpoint again.

With the modified sliding window learning scheme, when the difference between the set point and plant output exceeded the maximum allowable error, the last 19 successive pairs of input data vectors for the neural network forward model are collected. The target output from the forward model for these 19 successive pairs of input is set equal to the setpoint. With this, a new learning data set is formed. The network is taught with the newly formed learning data set to update the weights and biases of the forward neural network model. The steps are then repeated for the next sampling instant. In this way, we can obtain a new neural network forward model that represents the plant with mismatch or disturbance in a fast, adaptable approach.

4.3. Updating Target Values For Online Inverse Modeling

When the difference between the set point and plant output exceeded the maximum allowable error, the last 11 successive pairs of input data vectors for the neural network inverse model are collected. The target output, i.e., the control action (dilution rate) for these 11 successive pairs of input were changed with the formula proposed below.

$$T(k) = T(k-1) + C \times Z$$

where

$T(k)$ = Column matrix of target value for the chosen successive pair of inputs, i.e., from time t = (k-L+1) to t = k
k = present time
$T(k-1)$ = Column matrix of target value from time t = (k-L) to t =k-1
C = Constant factor
$\tilde{Z}$ = Matrix with size L × 1, where each element is equal to B

The value of B depends on the system, whether it is reverse acting or direct acting. In this system,

$$B = \begin{cases} 1 & xdiff > 0 \\ 0 & xdiff = 0 \\ -1 & xdiff < 0 \end{cases}$$

where *xdiff* is the difference between the plant output (in this case the plant output is biomass concentration) and forward modeling output for an sampling instant as given by the formula below:

$$xdiff(k) = X(k) - Xfor(k)$$

and X(k) = biomass concentration
$Xfor$(k) = biomass concentration predicted by NN forward modelling

Constant C is an adjustable adaptive value and depends on the plant or system. In this study, the value C of 0.008 gives satisfactory results, thus it is used throughout all the internal and external parameter changes.

With this, new learning data set is formed. The network is taught with the newly formed learning data set to update the weight and biases of the inverse model. The steps are then repeated for next sampling instant. In this way, we can obtain an updated neural network inverse model that would produce an appropriate control action obtained in an adaptable approach.

5. Simulation Results (Conventional and Adaptive IMC Strategies).

In this section, we study the effect of disturbance for the conventional and adaptive IMC strategy. Two types of disturbances have been studied, that is the internal and external disturbance. Internal disturbance in this study includes changes in two yield parameters, i.e. α and β while the external disturbance is the feed substrate concentration, Sf.

Disturbance is injected into the plant between time t = 200 hours to t = 400 hours. The prior-to-disturbance control action is implemented from time t = 200 hours to t = 300 hours in order to see the full action of the IMC controller when implemented after time t = 300 hours.

Figures 11 to 13 (on the right) shows the effect of implementing the modified sliding window learning method (MSWL) adaptive strategy in the IMC controller under these disturbances. The result with the conventional IMC method (without adaptive control) is put next to it to compare the results when implementing adaptive and no-adaptive strategies. In the adaptive scheme, the maximum iteration for the inverse and forward modeling training is 10 cycles.

With internal disturbance, i.e. changes in α and β (Figs. 11-12), large offsets were observed when these disturbance were introduced into the system at time t = 200 hours. However, when the conventional IMC controller was implemented at t = 300 hours, these disturbances were rejected but with some offset still remaining in the output. However, when the MSWL adaptive strategy was applied at t = 300 hours, these offsets were eliminated completely. Hence the MSWL approach has been able to successfully adapt the changing models online since the relationship between the set point and the inverse and forward models had been learnt during online training. This is done by the method outlined previously, the forward model is trained with desired set point; the inverse model is trained with the target which takes into account the trend and sign of the error between plant and the neural network forward model output.

For the external disturbance, that is the change in feed substrate concentration, S_f, from nominal value to 24 g/l, the MSWL also shows good performance in rejecting the disturbance by totally removing the offset. For S_f = 24 g/l, the conventional IMC controller alone cannot handle this disturbance, but produced oscillations and unstable behavior of the plant. The results using the standard sliding window approach of Van Breusegem show similar unstable results as the IMC and hence not shown here. Details of the results can be seen in L. S. Hang and H. P. Yee (1999).

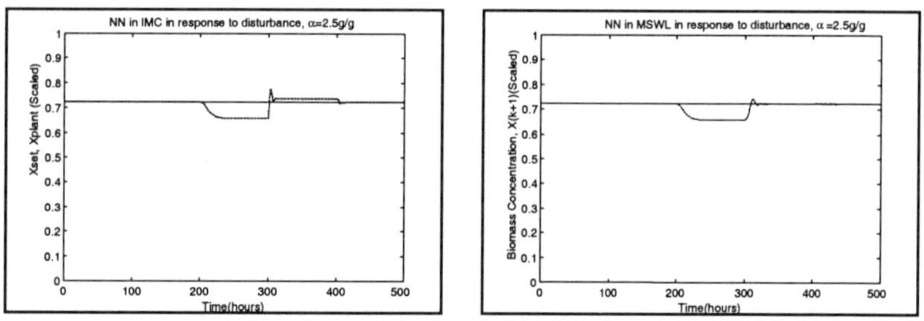

Figure 11. System Response in disturbance, α=2.5g/g

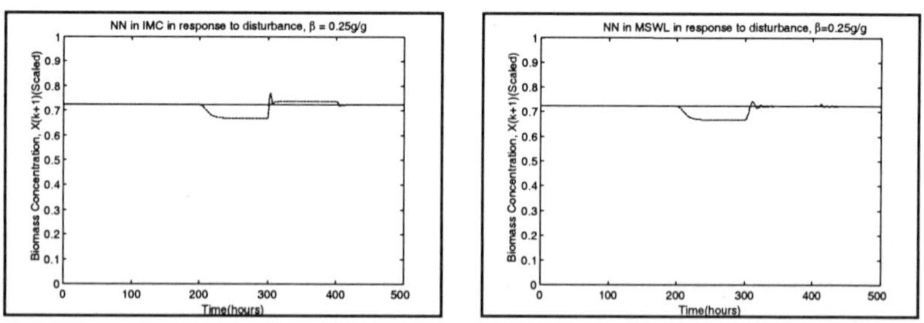

Figure 12. System Response in disturbance, β=0.25g/g

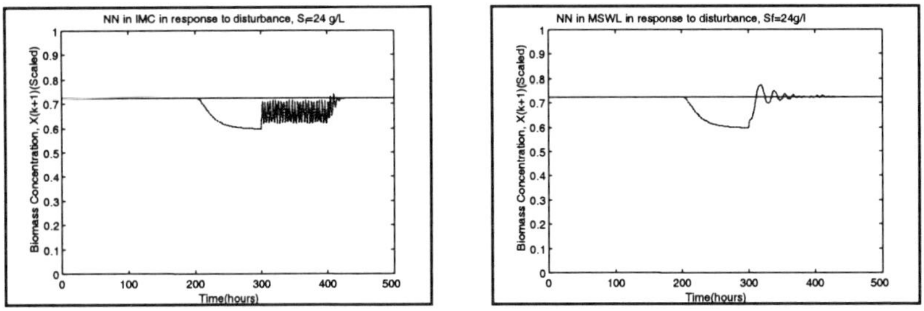

Figure 13. System Response in disturbance, S_f=24 g/L

6. Discussion

In this paper, we have successfully shown an adaptive technique for readapting the neural network forward and inverse model used in the internal model control strategy. The adaptive technique basically adapts the modified sliding windows learning scheme (MSWL) together with the changing of target values for the forward and inverse model online. In conclusion, the MSWL method has shown satisfactory performance in the IMC technique for handling the disturbances by eliminating the offsets and stabilising the system, which could not be done by the IMC control method alone and the standard sliding window approach, particularly to the continuous fermentation process. The applications of this technique to other system are also ongoing at present

References

Agrawal, D., Koshy, G. and Ramseier, M., *Biotech. Bioeng*. **33** (1989), 115.
Chen F.C. and Khalil H.K., *Int. J. Control*. **55** (1992), 1299-1317.
Hussain M. A., *Artificial Intelligence in Engineering*. **13** (1999), 55-68.
Liew S. H. and Ho P. Y., *Application of Neural Networks in Chemical Engineering Processes*, Research Report, (Chemical Engineering Department, University of Malaya, 1999).
Mills P.M., Zomaya A.Y. and Tade M.O., *Int. J. Control*. **60** (1994), 1163-1192.
Polycarpou M.M. and Mears M.J., *Int. J. Control*. **70** (1998), 363-384.
Van Breusegem V., Thibault J., Cheruy A., *Can. J. Chem. Eng*. **69** (1991), 481-487.

Acknowledgments

Acknowledgments are to be made to my students Liew Siew Hang, Ho Pei Yee and Ahmad Khairi Abdul Wahab for their assistance in producing the results in this work as well as the University Of Malaya for the facilities made available in this project.

10. SET POINT TRACKING IN BATCH REACTORS: USE OF PID AND GENERIC MODEL CONTROL WITH NEURAL NETWORK TECHNIQUES

N. AZIZ, I. M. MUJTABA
Computational Process Engineering Group,
Department of Chemical Engineering, University of Bradford, West Yorkshire BD7 1DP, UK.

M. A. HUSSAIN
Department of Chemical Engineering
University of Malaya, 59100 Kuala Lumpur, Malaysia

In batch reactors, the optimal reactor temperature profiles which maximise the conversion to the desired product is obtained by solving dynamic optimisation problems off-line. The control Vector Parameterisation (CVP) technique is used to pose the dynamic optimisation problems as Non-linear Programming Problems which are solved using the Successive Quadratic Programming (SQP) based optimisation technique. Two different types of controllers are used here to track the optimal batch reactor temperature profiles (set points). They are Generic Model Control (GMC) and dual mode (DM) control with Proportional-Integral-Derivative (PID). The Neural Network technique is used as the on-line estimator to estimate the amount of heat released by the chemical reaction within the GMC strategy. The GMC controller coupled with the Neural Network based heat-release estimator is found to be more effective, robust and stable compared to PID controllers in tracking the optimal reactor temperature profiles of various reaction schemes. Two different exothermic reaction schemes are used in order to illustrate the idea.

1. Introduction

The necessity of rapid change from one process to another with minor modifications and with a relatively small amount of production material with high added value has made batch operations very popular, especially in the fine chemical industry (Zaldivar and Hernandez, 1992). Since the reactor is the heart of any batch process, it has become an essential unit operation in almost all batch-processing industries. It has inherent kinetic advantages over continuous reactors for some reactions (primarily those with slow rate constants). The control of a batch reactor consists of charging the reactor, controlling the reactor temperature to meet some processing criterion, and shutting down and emptying the reactor. For an exothermic reaction, heat may be required at the beginning to obtain the desired reaction temperature, and then cooling is used to maintain the proper reaction temperature. The control of batch reactors is more difficult to achieve than that of continuous processes due to the inherent unsteady-state dynamic nature. Consequently, modelling of such reactors results in a system of Differential Algebraic Equations (DAEs).

The aim of the fine chemical industries is to produce high quality and high purity product in small quantities while controlling polluting waste materials and losses of raw materials. Therefore optimisation of batch operating conditions such as temperature, operating time, etc. is important to obtain maximum yield of the desired product in a minimum time or at minimum cost, as well as to reach the specific final conditions of the products (including waste products) in terms of quality and quantity. As far as overall profitability is concerned, it is very important to operate batch reactors efficiently and economically. Every small improvement in the process may result in considerable reduction in the production costs. Because of the necessity to answer to strict constraints and objectives, the optimisation problems encountered in the fine chemical industries can be very complex.

In the past, many researchers have studied the dynamic optimisation (optimal control) of batch reactors. They determined the optimum reactor temperature for different reaction schemes which maximises the yield, productivity, profit etc. (Logsdon and Biegler, 1993; Luus, 1994; Vassiliadis et al., 1994; Garcia et al., 1995; Carrasco and Banga, 1997; Aziz and Mujtaba, 1998). However, all these researchers considered only the off-line optimisation problems. None of them have implemented these results on-line. Designing controllers to implement the optimal control profiles or tracking the dynamic set points on-line are an important area of research for inherently dynamic batch processes.

Cott and Macchietto (1989) have used the Generic Model Control (GMC) algorithm proposed by Lee and Sullivan (1988) as the controller in order to track the reactor temperature set point (T_{rsp}). To estimate the heat-release on-line they used a three-term difference equation and exponential filters as the estimator. Later, Kershenbaum and Kittisupakorn (1994) considered the same reaction scheme of Cott and Macchietto and also used the GMC algorithm for the controller. However the extended Kalman filter was used as the on-line heat-release estimator.

In this work, we also consider the GMC controller but use Neural Network techniques for on-line estimation of the heat-release. We demonstrate the idea using two case studies with two reaction schemes. The first case study deals with consecutive exothermic batch reactions (Aziz and Mujtaba, 2000). The second case study uses the reaction scheme considered by Cott and Macchietto (1989). In both cases an off-line dynamic optimisation problem is solved with fixed batch time to find the optimum temperature profile that will maximise the conversion of the desired product. The Control Vector Parameterisation (CVP) technique (Aziz and Mujtaba, 1998, 2000) has been used to pose the dynamic optimisation (optimal control) problem as a Non-linear Programming Problem (NLP) which is solved using the Successive Quadratic Programming (SQP) based optimisation technique (Chen, 1988). The optimum temperature profile thus obtained (off-line) is used as the set point to be tracked (on-line) by the GMC controller.

2. Dynamic optimisation of batch reactors

The dynamic optimisation problem to maximise the conversion to the desired product (*maximum conversion problem*) can be described as:

Given	the fixed volume of the reactor and the batch time.
Optimise	the reactor temperature profile
So as to maximise	the conversion of the desired product
Subject to	constraint bounds on the reactor temperature, reactor model, etc.

Mathematically the optimisation problem (OP) can be written as:

$$\max_{T(t)} \quad X$$

$$\text{s.t.} \quad f(t, \underline{x}'(t), \underline{x}(t), \underline{u}(t), \underline{v}) = 0 \text{ (model)}$$

$$t_f = t_f^*$$

$$T_L \leq T \leq T_U$$

where X is the conversion to the desired product and T is the reactor temperature. T_L and T_U are the lower and upper bounds of the reactor temperature; t_f^* is the fixed batch time. The details of the solution method for this optimisation problem can be found in Aziz and Mujtaba (1998, 2000).

3. Generic Model Control (GMC) Strategy

Generic Model Control (GMC), a model-based control strategy developed by Lee and Sullivan (1988), is one of several advanced process control algorithms developed recently. The GMC uses non-linear models of a process to determine the control action. The desired response can be obtained by incorporating two tuning parameters. The main advantage of the GMC is that the non-linear process models do not need to be linearised because it directly inserts non-linear process models into the controller itself. In addition, the GMC algorithm is relatively easy to implement.

The GMC control algorithm can be written as:

$$\frac{dx}{dt} = K_1(x_{sp} - x) + K_2\int_0^t (x_{sp} - x)dt \tag{1}$$

where x is the current value and x_{sp} is the desired value of the control variable.

The first expression in the algorithm (K_1 $(x_{sp} - x)$) is to bring the process back to steady-state due to change in dx/dt. In order to make the process have a zero offset, the second expression ($K_2 \int (x_{sp} - x)\, dt$) is introduced. Details of this GMC method can be seen in Lee and Sullivan (1988).

The batch reactor system of interest in our control strategy is shown in Fig. 1. For temperature control of the batch reactor, a process model relating the reactor temperature, T_r, to the manipulated variable i.e. the jacket temperature, T_j, is required. Assuming that the amount of heat retained in the walls of the reactor is small in comparison to the heat transferred in the rest of the system, an energy balance around the reactor contents gives the following model:

$$\frac{dT_r}{dt} = \frac{Q_r + UA(T_j - T_r)}{\rho_r C_p V_r} \tag{2}$$

Replacing T_r for x and T_{rsp} for x_{sp} in Eq. 1, combining Eq. 1 and 2 and finally solving for the manipulated variable, T_j, the control formulation under the GMC is given by:

$$T_j = T_r + \frac{V_r C_p \rho_r}{UA}\left[K_1(T_{rsp} - T_r) + K_2\int_0^t (T_{rsp} - T_r)dt\right] - \frac{Q_r}{UA} \tag{3}$$

where T_j gives the jacket temperature trajectory required so that the reactor temperature, T_r, follows the desired trajectory incorporating the values of GMC tuning parameters, K_1 and K_2.

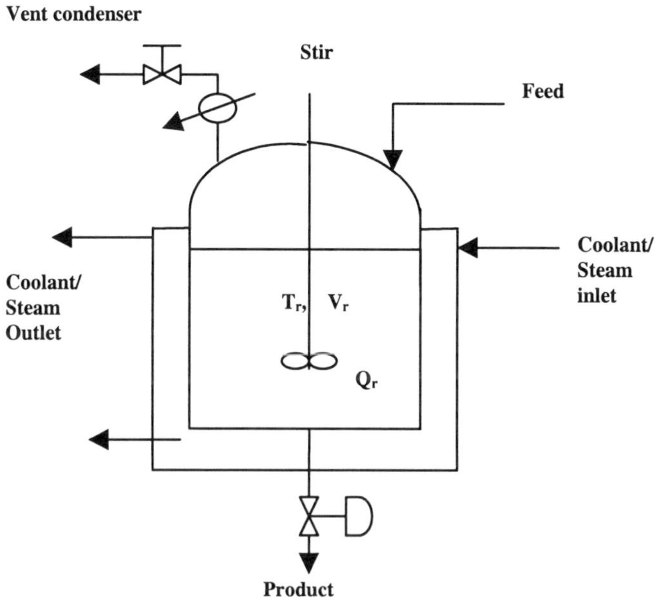

Figure 1. Schematic diagram of a jacketed batch reactor

The discrete form of Eq. 3 for the kth time interval is implemented for the on-line control and is given by:

$$T_j(k) = T_r(k) + \frac{V_r C_p \rho_r}{UA}\left[K_1(T_{rsp} - T_r(k)) + K_2 \sum_0^k (T_{rsp} - T_r(k))\Delta t\right] - \frac{Q_r}{UA} \quad (4)$$

where Δt is the sampling time.

However, Eq. 4 gives the actual jacket temperature, $T_j(k)$ which is not the jacket temperature set point, $T_{jsp}(k)$, needed to control the reactor temperature at its set point T_{rsp}. It is reasonable to assume that the dynamics of the jacket temperature control are approximately first order (Liptak, 1986) with time constant τ_j and hence, the T_{jsp} can be further calculated using the following equation:

$$T_{jsp}(k) = T_j(k-1) + \frac{\tau_j}{\Delta t}\left[T_j(k) - T_j(k-1)\right] \quad (5)$$

4. On-Line Estimation of the Heat-Release using Neural Network Method (GMC Strategy)

The success of the GMC controller as formulated in Eq. 4 is largely dependent on the ability to measure, estimate, or predict the heat-release, Q_r at any given time. As Neural Networks have been proven to be an accurate and fast on-line dynamic estimator, they may be used to carry out the task in this work (Hussain, 1999). A multilayered feedforward network is used which is trained using the back-propagation method. The back-propagation method has been chosen since it is the most well-known and widely-used algorithm associated with the training of a feed-forward Neural Network. The Neural network systems identification steps can be seen in Fig. 2.

The multi-layered feed-forward Neural Network consists of a set of nodes which are arranged in layers. The nodes in each layer are connected to all the nodes in the layer above/next to it and all the signals propagate in a forward direction through the network layers. There are no self-connections, lateral connections or back connections. In each node (in hidden and output layer), a constant bias is added. The outputs of nodes in one layer are transmitted to nodes in another layer through connections which incorporate weighting factors that amplify or attenuate such outputs. The net input to each node (except for input layer nodes) is the sum of the weighted outputs of the nodes in the prior layer. Each node is activated in accordance with the input to the node, the activation function and the bias/threshold of the node. There are various types activation functions available but in this work, the log-sigmoid function has been used in both the hidden and output layer. The architecture of the multi-layered feed-forward Neural Network can been seen in Fig. 3.

The numbers of hidden layers and nodes may vary in different applications and depend on user specification. No specific technique is available to decide the number and it is totally based on experience and carried out through trial-and-error procedure. In this work, the 3 layer Neural Network with one hidden (middle) layer consisting of 18 or 20 nodes (depending on the case study) is used.

Since the process being studied is a dynamic system, it is necessary to feed the Neural Network with past historical data. Here the input layers consists of the present and past values of T_r ($T_r(k-2)$, $T_r(k-1)$, $T_r(k)$), T_j ($T_j(k-1)$, $T_j(k)$) and the past values of Q_r ($Q_r(k-1)$) and the output layer estimates the value of the heat-release, Q_r at time interval k.

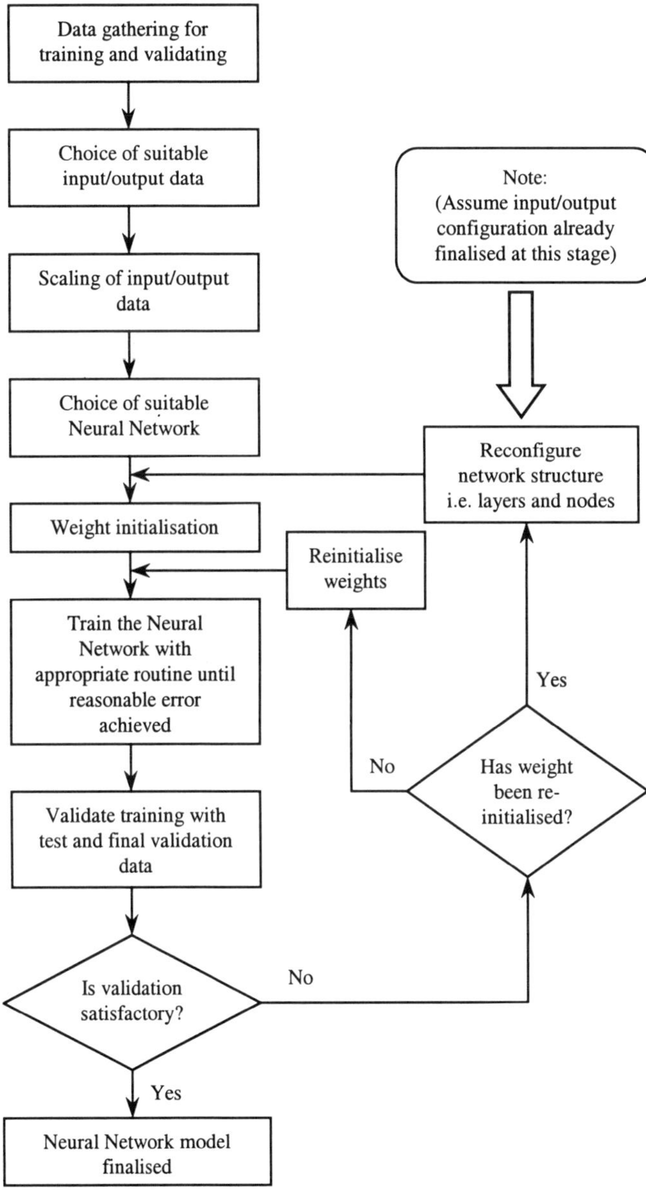

Figure 2. Neural Network systems identification—Basic steps

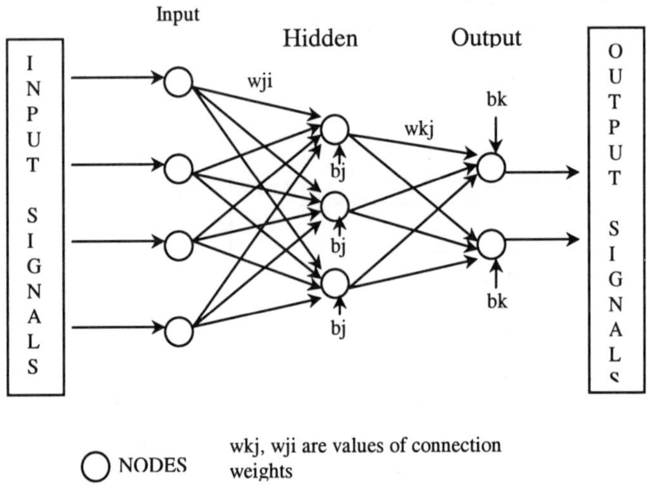

Figure 3. Multi-layered Feed-forward Neural Network Topology

T_r (k-2)	T_r (k-1)	T_r (k)
□ Input ■ Output	T_j (k-1)	T_j (k)
	Qr (k-1)	**Q_r (k)**

Figure 4. Input/Output Map of Neural Network

With 6 inputs, the Neural Network is trained through forward modelling methodology to obtain the value of the output i.e. present value of Q_r. All the data are moved forward at one discrete-time interval until all of them are fed into the network in a moving window scheme. All data are fed into the Neural Network repeatedly until the training error criterion is achieved. In this work, training error is set at 1×10^{-8}. After this step, the designed Neural Network with its weights, biases and chosen functions is validated/tested with a new set of data before being used in the GMC strategy. The input-output map for the Neural Network training can be seen in Fig. 4. Here the Neural Network is placed in parallel with the estimator for

Q_r and the error between them and Neural Network output (i.e. prediction error) is used as the training signal for the Neural Network (see Fig. 5). The estimated Q_r is then used in Eq. 4 to estimate the value of T_j and is then applied in the GMC strategy as illustrated in Fig. 6 and for two case studies as will be discussed below.

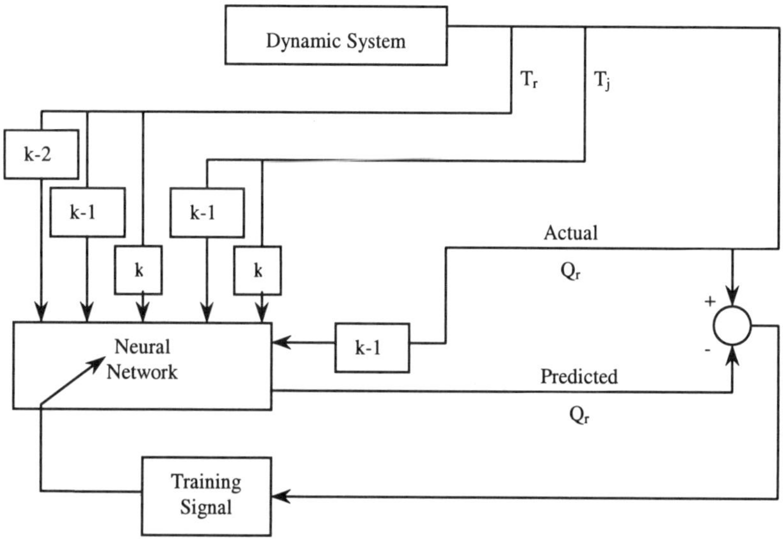

Figure 5. Forward modelling of heat-release by Neural Network

5. Dual-Mode Control (DM) Strategy

Dual-mode control (DM) is the most commonly-used strategy in batch reactors that have initial heat-up (i.e. for exothermic reaction). This is an on-off control type strategy. First, maximum heating (on) is applied until the reactor temperature is within a specified degree of the set point and then maximum cooling (off) is applied when the temperature has reached its final desired set point. At this point, standard feedback controllers are switched on and used to maintain the temperature (constant or dynamic set points). In the standard DM strategy, the PID controller is normally used.

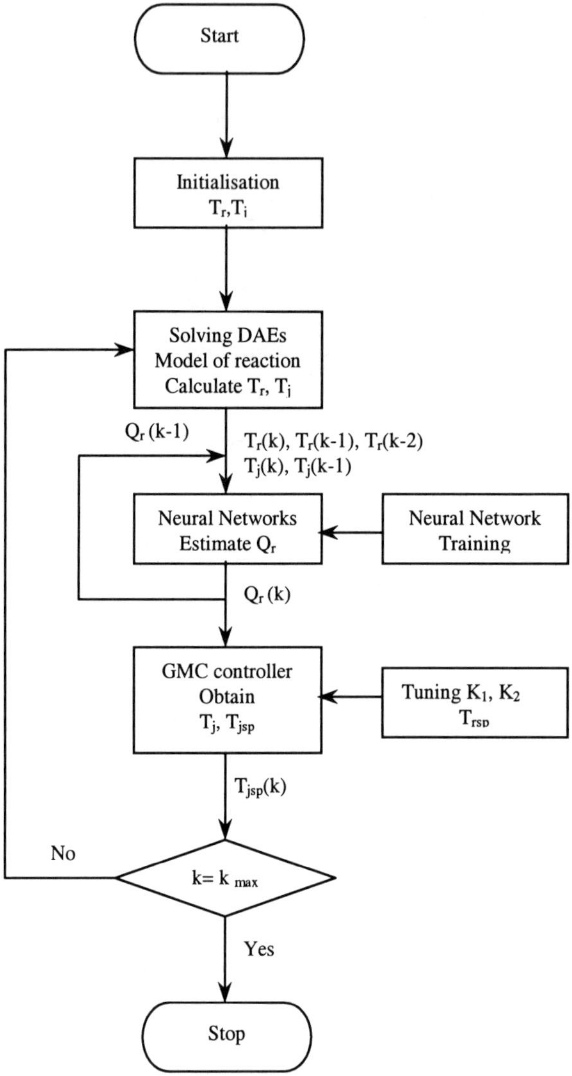

Figure 6. GMC strategy in controlling batch reactor

The DM controller consists of a sequence of control actions, each one carried out after the reactor has reached a certain condition. The sequence of operations is as follows:

1. Full heating is applied until the reactor temperature is within a certain percent (E_m) of its set point temperature.
2. Full cooling is then applied for a certain period of time (TD-1).
3. The jacket set point temperature (T_{jsp}) of controller is then set to the pre-load temperature (PL) for a certain period of time (TD-2).
4. A temperature controller (PID) is cascaded to the jacket temperature controller and its set point is set to T_{rsp}.

There are two steps applied in order to tune the DM controller. First, PID tuning parameters were tuned by performing an open-loop step response test. The Cohen and Coon method was then applied to estimate the value of PID tuning parameters (K_c, τ_I and τ_D). However the tuning parameters have been fine tuned to make the control less drastic. Second, the remaining four constants (E_m, TD-1, TD-2 and PL) were determined by running a series of simulation runs. The details of DM control and its tuning can be found in Liptak (1986).

6. Applications

6.1. Case Study 1

In this example, a consecutive exothermic batch reaction scheme (Aziz and Mujtaba, 2000) is considered. The reaction type is:

$$A \xrightarrow{k_1} B \xrightarrow{k_2} C$$

where A is a raw material, B is the desired product and C is a waste or by-product.

6.1.1. Models Equations

The conversion to B from A and conversion C from B follow a first order of reaction rate. The model equations for the batch reactor can be written as:

$$\frac{dC_A}{dt} = -k_1 C_A \tag{6}$$

$$\frac{dC_B}{dt} = k_1 C_A - k_2 C_B \tag{7}$$

$$\frac{dC_C}{dt} = k_2 C_B \tag{8}$$

$$\frac{dT_r}{dt} = \frac{(Q_r + Q_j)}{\rho\, C_p V} \tag{9}$$

$$\frac{dT_j}{dt} = \frac{(T_{jsp} - T_j)}{\tau_j} - \frac{Q_j}{V_j \rho_j C_{pj}} \tag{10}$$

$$k_1 = k_{10} \exp\left(-\frac{E_1}{RT_r} \right) \tag{11}$$

$$k_2 = k_{20} \exp\left(-\frac{E_2}{RT_r} \right) \tag{12}$$

$$Q_r = -\Delta H_1 (k_1 C_A) - \Delta H_2 (k_2 C_B) \tag{13}$$

$$Q_j = UA(T_j - T_r) \tag{14}$$

All constant parameter values are as given in Table 1.

Table 1. The constant parameter values of the model and control equation

ΔH_1= -6.50E8 J/kmol	k_{10}= 4.38E4 h^{-1}	C_P = 4200.0 J/kgK
ΔH_2= -1.20E8 J/kmol	k_{20}= 3.94E5 h^{-1}	C_{Pj} = 4200.0 J/kgK
E_1= 3.49E7 J/kmol	ρ = 800.0 kg/m^3	τ_j = 0.075 h
E_2= 4.65E7 J/kmol	R = 8314.0 J/kmol.K	ρ_j = 1000.0 kg/m^3
A = 5.25 m^2	U = 8.18E6 J/h.K.m^2	Δt = 0.01 h
V = 1.23 m^3	V_j = 0.53 m^3	

The objectives of this study are: (1) to obtain optimum reactor temperature profile to maximise the conversion to the desired product B by solving the dynamic optimisation problem presented in section 2. This does not require the full model (only requires 6 – 8, 11 – 12) to be used; (2) to track the optimum temperature profile obtained in (1) using GMC and PID controllers. This requires the full model to be used (6 – 15); (3) to compare the performance of the GMC controller with PID controller; (4) to test the robustness of both controllers.

It is assumed that the reactants are being pre-heated before they are charged into the reactor. The initial values of [C_A, C_B, C_C, T_r, T_j] are [0.975, 0.025, 0.0, 350K, 300K] respectively. The total batch time is 3.5 hours. The reactor temperature T_r (controlled variable) and T_j (manipulated variable) are bounded between 300 and 400K. The results are summarised in Table 2. It is found that the maximum conversion achieved is 0.6613 (off-line conversion achieved by solving the dynamic optimisation problem OP). The optimum temperature is 369.40, which has been used as the T_{rsp} to be tracked by a GMC (Eq. 4) and a PID controller.

Table 2. Summary of the Result for case study 1

Off-line Optimum Temperature Profile			C_B*
Temperature,K	369.40		0.6613
Switching time,h	t = 0	3.5	

Controller	C_B	CP
PID	0.6602	99.83
GMC	0.6602	99.83

GMC Tuning Parameters	DM Tuning Parameters	
$K_1 = 22.22\ h^{-1}$	$E_m = 0.5\%$	$K_c = 19.68$ h
$K_2 = 1.235\ h^{-2}$	PL = 345 K	$\tau_I = 0.07$ h
	TD-1 = 0.02 h	$\tau_D = 0.0007$ h
	TD-2 = 0.01 h	

C_B On-line conversion to B
C_B* Off-line conversion to B
CP Controller Performance (%) {(C_B/C_B*) ×100}

6.1.2. Results

It can be clearly seen that the conversion of 0.6602 achieved by the GMC coupled with Neural Network is very close to that achieved by off-line dynamic optimisation (0.6613). The GMC was also able to track the given set point very well. The response of the GMC controller is shown in Fig. 7. The performance of the GMC controller is strongly dependent on the estimation of the heat released by the reaction, Q_r. Figure 8 shows that the Neural Network was able to give a very good estimation of the heat released by the reaction and hence guarantee the good performance of the GMC controller.

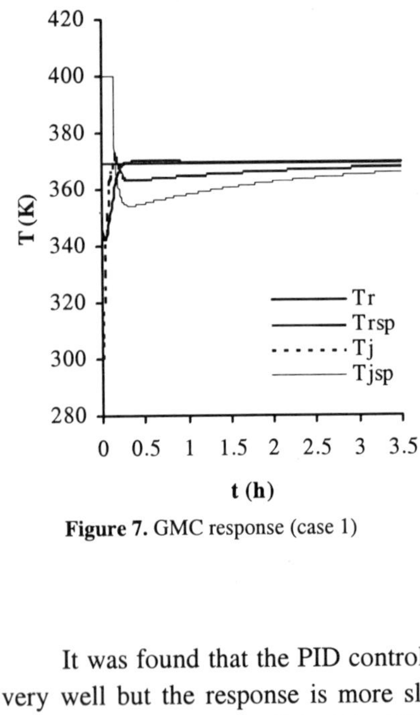

Figure 7. GMC response (case 1)

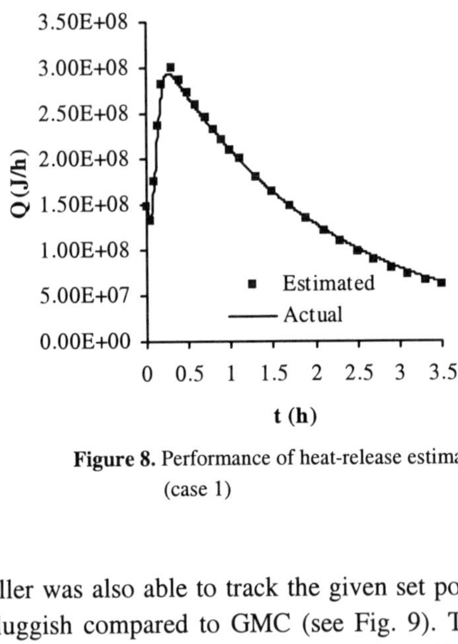

Figure 8. Performance of heat-release estimator (case 1)

It was found that the PID controller was also able to track the given set point very well but the response is more sluggish compared to GMC (see Fig. 9). The controller performance (CP) of the PID controller was the same as that of the GMC controller (99.83%). In order to study further the robustness of the controller, four tests have been carried out by changing the process parameters. The GMC and PID controllers (tuned as before) were used to control an operation where some of the conditions have changed from their true values. In the first test (TEST1), the heats of reaction were increased by 25%. The second test (TEST2) involves the increment of the rate constant by 25% from the true value. The third test (TEST3) involves 30% reduction in the weight of the initial quantities of reactants and the fourth test (TEST4) involves the reduction of heat transfer coefficient by 40% from its original value.

The results for all the tests are shown in Fig. (10-13). They show that both the GMC and PID controllers are able to accommodate these changes. The Neural Network also gives a very good estimation of heat released in every test.

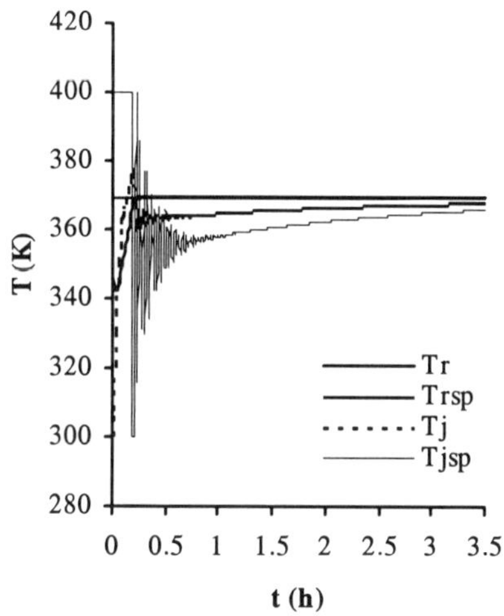

Figure 9. PID response (case 1)

6.2. Case Study 2

Here the reaction scheme is same as that used by Cott and Macchietto (1989).

The reaction scheme is: $A + B \rightarrow C; A + C \rightarrow D$
where A, B are raw materials, C is the desired product and D is the waste product.

6.2.1. Model Equations

The model equations for the batch reactor can be written as:

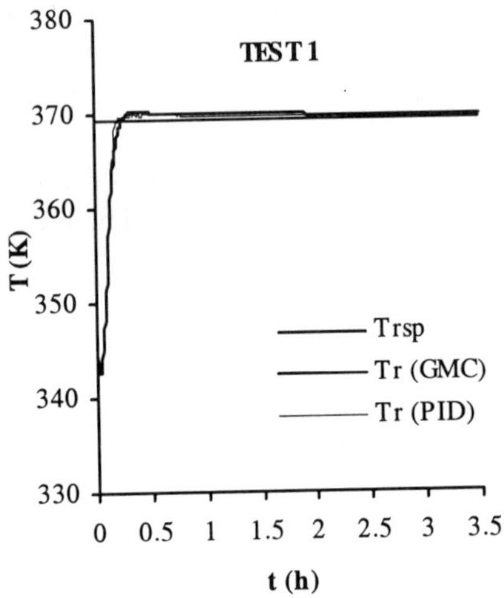

Figure 10. Controller responses for heat reaction change (case 1)

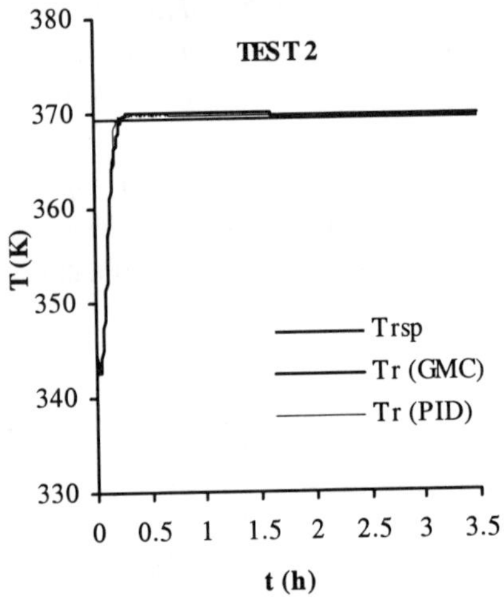

Figure 11. Controller responses for rate constant change (case 1)

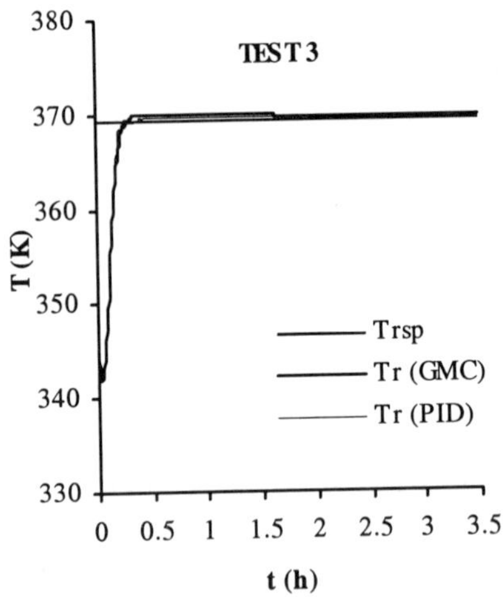

Figure 12. Controller responses for weight change (case 1)

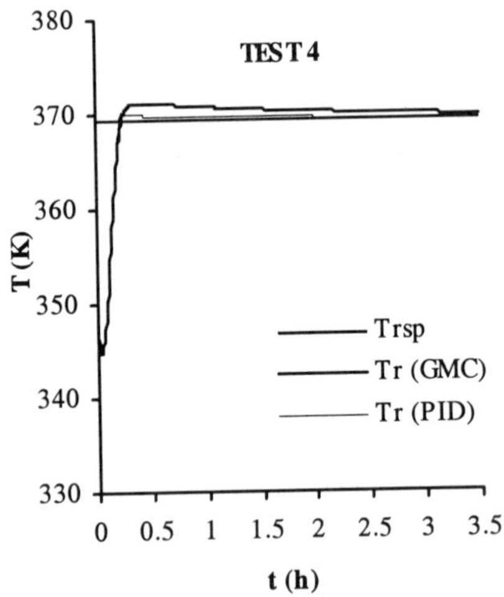

Figure 13. Controller responses for heat transfer coefficient change (case 1)

$$\frac{dM_A}{dt} = -R_1 - R_2 \tag{16}$$

$$\frac{dM_B}{dt} = -R_1 \tag{17}$$

$$\frac{dM_C}{dt} = +R_1 - R_2 \tag{18}$$

$$\frac{dM_D}{dt} = +R_2 \tag{19}$$

$$\frac{dT_r}{dt} = \frac{(Q_r + Q_j)}{M\ C_{pr}} \tag{20}$$

$$\frac{dT_j}{dt} = \frac{(T_{jsp} - T_j)}{\tau_j} - \frac{Q_j}{V_j \rho_j C_{pj}} \tag{21}$$

$$R_1 = k_1 M_A M_B \tag{22}$$

$$R_2 = k_2 M_A M_C \tag{23}$$

$$k_1 = \exp\left(k_1^1 - \frac{k_1^2}{(T_r + 273.15)} \right) \tag{24}$$

$$k_2 = \exp\left(k_2^1 - \frac{k_2^2}{(T_r + 273.15)} \right) \tag{25}$$

$$Q_r = -\Delta H_1 R_1 - \Delta H_2 R_2 \tag{26}$$

$$M_r = M_A + M_B + M_C + M_D \tag{27}$$

$$C_{pr} = \frac{C_{PA} M_A + C_{PB} M_B + C_{PC} M_C + C_{PD} M_D}{M_r} \tag{28}$$

$$Q_j = UA(T_j - T_r) \tag{29}$$

All the parameter and constant values used in the model and control equation are given in Table 3.

Here again an off-line dynamic optimisation problem (OP) is solved to find the optimum temperature profile that will maximise the product "C" and minimise the by-product "D". Two runs were carried out; RUN1 uses one control interval (time) and RUN2 uses three fixed control intervals. The batch time is 120 minutes and the initial values of [M_A, M_B, M_C, M_D, T_r, T_j] are [12.0, 12.0, 0.0, 0.0, 20.0, 20.0] respectively. The reactor temperature is used as the controlled variable and is bounded between 20 and 100°C. The manipulated variable, T_j is bounded between 20 and 120°C. The model in the dynamic optimisation problem does not require Eq. (20 – 21) and (26 – 29) to be used.

Table 3. The constant parameter values of the model and control equation

$C_{PA} = 75.3$ kJ/kmol°C	$k_1^1 = 20.9057$
$C_{PB} = 167.3$ kJ/kmol°C	$k_1^2 = 10000$
$C_{PC} = 217.6$ kJ/kmol°C	$k_2^1 = 38.9057$
$C_{PD} = 334.7$ kJ/kmol°C	$k_2^2 = 17000$
$\Delta H_1 = -41840.0$ kJ/kmol	$V_j = 0.6921\ m^3$
$\Delta H_2 = -25104.0$ kJ/kmol	$A = 6.24\ m^2$
$C_P = 1.8828$ kJ/kg°C	$\Delta t = 0.2$ min
$C_{Pj} = 1.8828$ kJ/kg°C	$\tau_j = 3.0$ min
$U = 40.84$ kJ/min.m^2.°C	$W_r = 1560.0$ kg
$\rho_j = 1000.0$ kg/m^3	

The results (optimal temperature profiles) for both runs are then used as the set point to be tracked by a GMC controller (Eq. 30) and a PID controller.

$$T_j(k) = T_r(k) + \frac{W_r C_p}{UA}\left[K_1(T_{rsp} - T_r(k)) + K_2\sum_0^k (T_{rsp} - T_r(k))\Delta t\right] - \frac{Q_r}{UA} \qquad (30)$$

The results are summarised in Table 4.

6.2.2. Results

In Table 4, it can be seen that by using three control intervals, the amount of product achieved is slightly higher than that obtained using one control interval.

Figures 14 and 15 show the response of the GMC controller in tracking the set points (T_{rsp}) and the performance of the Neural Network in estimating heat-released for RUN1 and Figs. 16-17 show those for RUN2. It can be seen that the GMC coupled with Neural Network method was able to accommodate very well both constant and dynamic set points although a little sluggishly for the latter. Again

Figs. 15 and 17 show that the Neural Network gives very good estimation for the heat released by the reaction.

Table 4 shows that for both runs the amount of desired product obtained on-line (after implementing the GMC controller) was within 4% of that obtained by off-line dynamic optimisation. This clearly shows the effectiveness of implementing the GMC controller combined with the Neural Network estimator.

Table 4. Summary of the Results for case study 2

Run	Off-line Optimum Temperature Profile					M_C*
1	Temperature,°C		92.46			6.5126
	Switching time,min	t = 0	120.0			
2	Temperature,°C		92.83	91.17	93.41	6.5171
	Switching time, min	t = 0	40.0	80.0	120.0	

Run	Controller	M_C	CP
1	PID	6.3392	97.34
	GMC	6.3270	97.15
2	PID	6.3409	97.30
	GMC	6.3309	97.14

GMC Tuning Parameters	DM Tuning Parameters	
$K_1 = 0.20\ min^{-1}$	$E_m = 5.0\%$	$K_c = 26.5381$ min
$K_2 = 1.00E\text{-}4\ min^{-2}$	PL = 46 °C	$\tau_I = 2.8658$ min
	TD-1 = 2.8 min	$\tau_D = 0.4284$ min
	TD-2 = 2.4 min	

M_C On-line Product
M_C* Off-line Product
CP Controller Performance (%) $\{(M_C/M_C^*) \times 100\}$

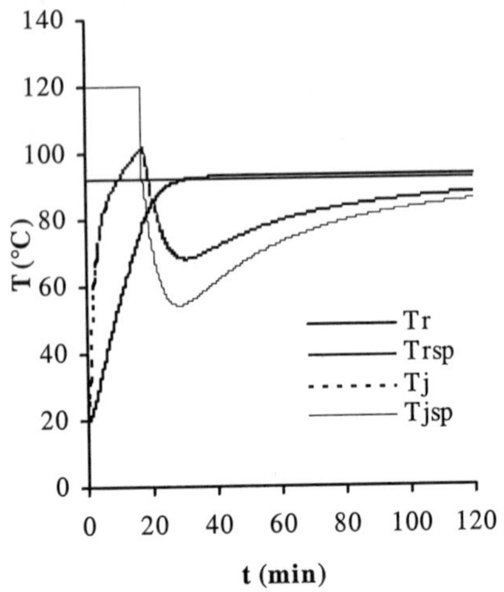

Figure 14. GMC response for RUN1 (case 2)

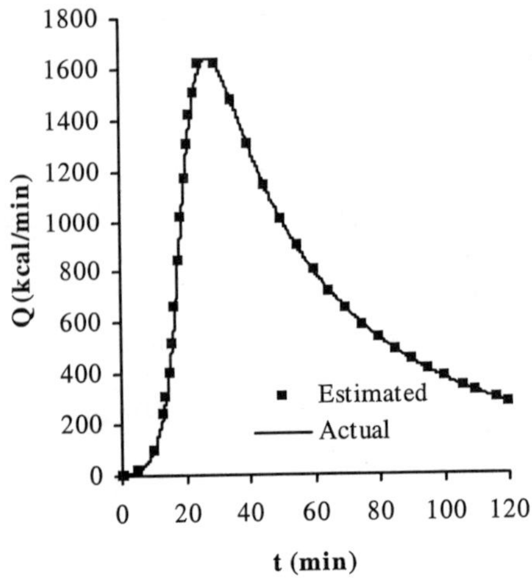

Figure 15. Performance of heat-release estimator for RUN1 (case 2)

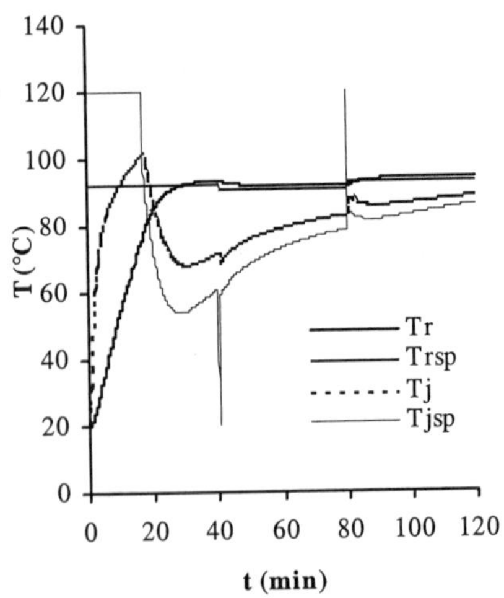

Figure 16. GMC response for RUN2 (case 2)

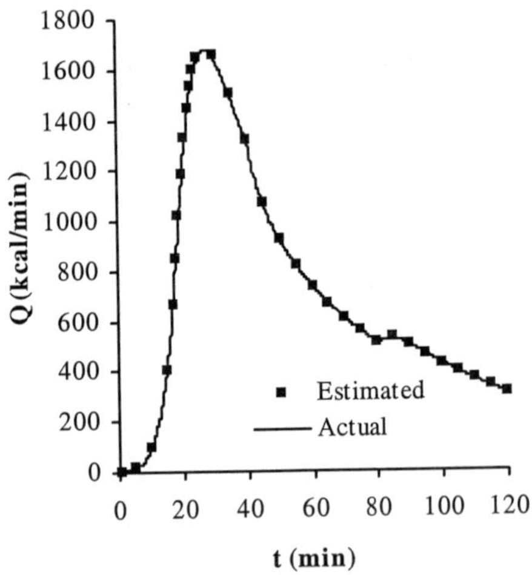

Figure 17. Performance of heat-release estimator for RUN2 (case 2)

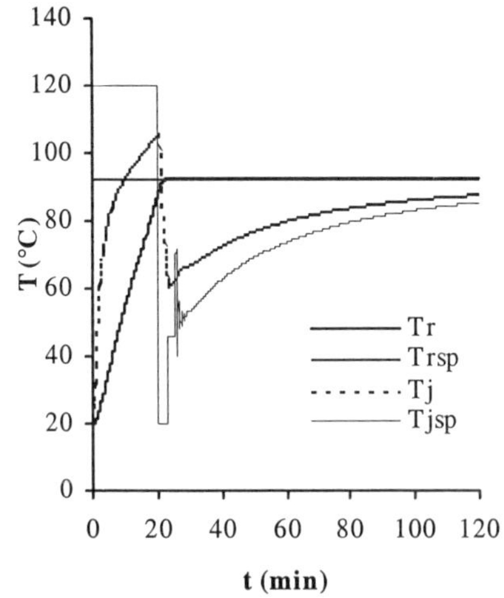

Figure 18. PID response for RUN1 (case 2)

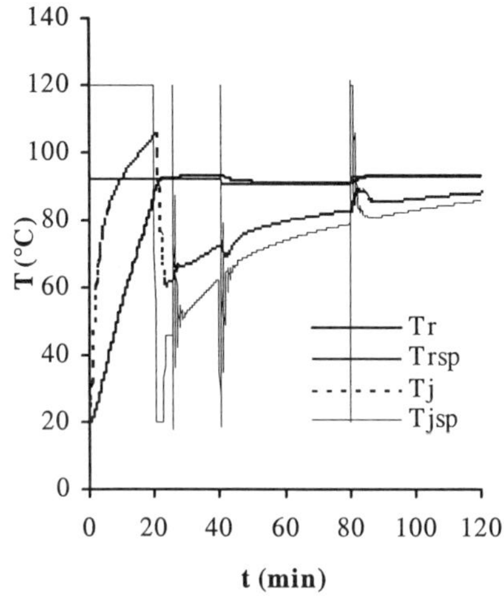

Figure 19. PID response for RUN2 (case 2)

The responses of the PID controller for RUN1 and RUN2 are shown in Figs. 18-19. Again, the PID was able to track the reactor temperature very well. Moreover, based on the amount of desired products achieved (Table 4), the controller performance (CP) using the PID is found to be slightly better than that obtained by using the GMC. This is due to the smaller rise time needed by the PID controller compared to the GMC controller. The heat-up process used by the DM controller was quicker compared to the GMC controller. However the GMC controller provides less drastic changes in the jacket temperature set point compared to the PID controller. Also it is evident that the performance of the GMC controller is more stable compared to the PID controller, the latter resulting in a more sluggish response in tracking the dynamic set points. Another advantage of the GMC is that only two parameters were needed to be tuned compared to seven parameters for the DM with PID controller. The performance of the GMC controller is strongly dependent on the estimation of the heat released (Q_r) by the reaction. The Neural Networks used in this work gives a good estimation of the heat released by the reaction (Figs. 15 and 17) and hence guarantees the good performance of the GMC controller.

Here, again the robustness of the controllers has been tested. Three tests were carried out by changing the process parameters. In all tests the controllers (tuned as before) were used to control an operation where some of the conditions have been changed from their true values. In the first test (TEST1), the heats of reaction were increased by 50%. In the second test (TEST2) the heat transfer coefficient is reduced by 40% of its original value. The third test (TEST3) involves 30% reduction in the molar (or mass) of reactants. In all tests, a constant reactor temperature set point (RUN1, Table 2) is to be tracked by both controllers.

The results for all these tests are shown in Figs. 20-22. For all tests, it can be seen that the GMC controller was able to accommodate all the changes very well compared to the PID controller. This clearly shows the robustness and the stability of the GMC method combined with the Neural Network estimator in controlling various kinds of reaction schemes while the PID controller could not handle the parameter changes in this case study.

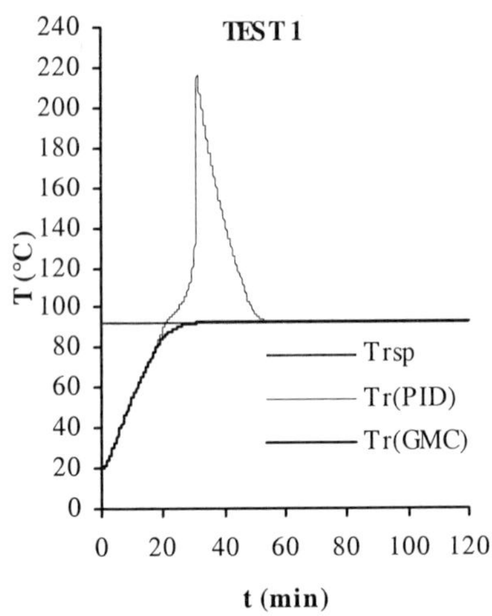

Figure 20. Controller responses for heat reaction change (RUN1-case 2)

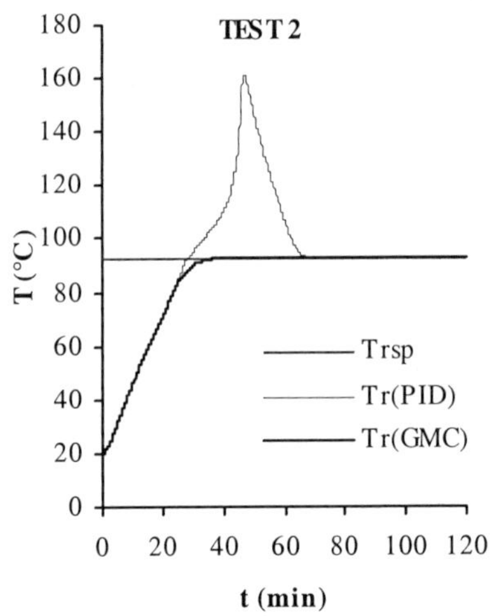

Figure 21. Controller responses for heat transfer coefficient change (RUN1-case 2)

7. Conclusions

Two different types of controllers, GMC and PID, have been used to track the optimal batch reactor temperature profiles using two different reaction schemes. The optimal profiles have been obtained by solving an off-line dynamic optimisation problem which maximise the desired product in batch reactors. Robustness tests for both controllers have been carried out by changing a number of process parameters. In the two case studies presented, the GMC controller coupled with the Neural Network based heat-release estimator has been found to be more effective, robust and stable than PID controller in tracking the optimal reactor temperature profiles of various reaction schemes.

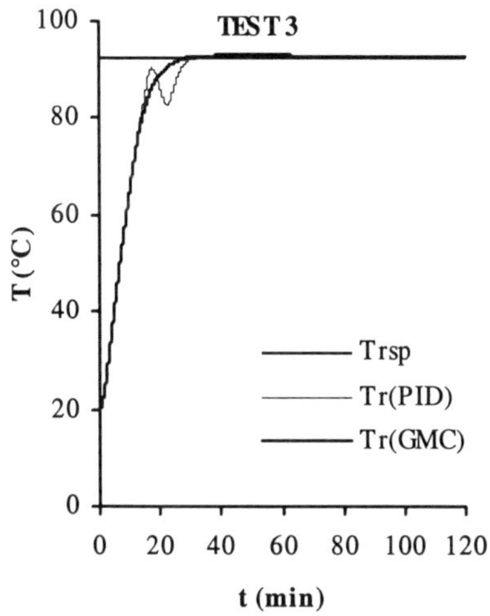

Figure 22. Controller responses for molar/weight change (RUN1-case 2)

Nomenclature

For case study 1:

V	Volume (m^3)
T	Temperature (K)
k_i	Reaction rate constant for reaction i (h^{-1})
t	time (h)
ΔH_i	Heat of reaction for reaction i (J/kmol)
Q_j	Heat input to the reactor from jacket (J/h)
Q_r	Heat released by reaction (J/h)
K_1	GMC controller constants (h^{-1})
K_2	GMC controller constants (h^{-2})
ρ	Density (kg/m^3)

U Heat transfer coefficient (J/h.°C.m^2)
A Heat transfer area (m^2)
Δt Sample interval (h)
C_p Mass heat capacity of reactant (J/kg°C)
C_i Concentration of component i (kmol/m^3)
E_i Activation energy for reaction i (J/kmol)
R Universal gas constant (J/kmol.K)

For case study 2:
V Volume (m^3)
T Temperature (°C)
k_i Reaction rate constant for reaction i (min^{-1})
t time (min)
ΔH_i Heat of reaction for reaction i (kJ/kmol)
Q_j Heat input to the reactor from jacket (kJ/min)
Q_r Heat released by reaction (kJ/min)
K_1 GMC controller constants (min^{-1})
K_2 GMC controller constants (min^{-2})
ρ Density (kg/m^3)
U Heat transfer coefficient (kJ/min.°C.m^2)
A Heat transfer area (m^2)
Δt Sample interval (min)
C_p Mass heat capacity of reactant (kJ/kg°C)
C_{Pi} Molar heat capacity of component i (kJ/kmol°C)
C_{Pr} Molar heat capacity of reactant (kJ/kmol°C)
k_1^i 'Pre-exponential' rate constant for reaction i
k_2^i Activation energy for reaction i
M_i Number of moles of component i (kmol)
R_i Rate of reaction i (kmol^2/min)
W Mass (kg)

Subscript
j Jacket
r Reactant
sp Set point

References

Aziz, N. and Mujtaba, I.M., Optimal control of batch reactors, *IChemE Advances in Process Control Conference V*, 2-3 September 1998.

Aziz, N. and Mujtaba, I.M., Dynamic optimisation of batch reactors, submitted to AIChEJ (2000).

Carrasco, E.F. and Banga, J.R., *Ind. Eng. Chem. Res.* **36** (1997), 2252-2261.

Chen, C.L., *A class of successive quadratic programming methods for flowsheet optimisation*, PhD thesis, (University of London, 1988).

Cott, B.J. and Macchietto, S, *Ind. Eng. Chem. Res.* **28** (1989), 1177-1184.

Garcia, V. *et al*, *Chem.Eng. and Biochemical Eng. J.* **59** (1995), 229-241.

Hussain, M.A., *Artificial Intelligence Eng.* **13** (1999), 55-68.

Kershenbaum, L.S. and Kittisupakorn, P., *Trans IChemE.* **72** (1994), 55-63.

Lee, P.L. and Sullivan, G.R., *comput. chem. engng.* **12** (1988), 573-580

Liptak, G., *Chem. Engng.*, May (1986), 69-81.

Logsdon, J.S. and Biegler, L.T., *comput. chem. engng.* **17** (1993), 367-372.

Luus, R., *J.Proc. Cont.* **4** (1994), 218-226.

Mujtaba, I.M. and Hussain, M.A., *comput. chem. engng.* **22** (1998), S621-S624.

Vassiliadis, V.S., Sargent, R.W. and Pantelides, C.C., *Ind. Eng. Chem. Res.* **33** (1994), 2111-2122.

Zaldivar, J.M. and Hernandez, H., *Chem.Eng.Processing.* **31** (1992), 173-180.

Acknowledgements

The Fellowship support to N.Aziz from the Universiti Sains Malaysia and the UK Royal Society support to M.A. Hussain are gratefully acknowledged.

11. INFERENTIAL ESTIMATION AND OPTIMAL CONTROL OF A BATCH POLYMERISATION REACTOR USING STACKED NEURAL NETWORKS

J. ZHANG, A. J. MORRIS
Centre for Process Analytics and Control Technology
Department of Chemical & Process Engineering
University of Newcastle, Newcastle upon Tyne NE1 7RU, U. K.

Inferential estimation and optimal control of a batch polymerisation reactor using bootstrap aggregated neural networks are presented in this contribution. In responsive agile manufacturing, the frequent change in product designs makes it less feasible to develop mechanistic model based estimation and control strategies. Techniques for developing robust empirical models from a limited data set have to be capitalised. The bootstrap aggregated neural network approach to nonlinear empirical modelling is very effective in building empirical models from a limited data set. It can also provide model prediction confidence bounds, thus, provide process operators with additional indications on how confident a particular prediction is. Robust neural network based techniques for inferential estimation of polymer quality, estimation of the amount of reactive impurities and reactor fouling during an early stage of a batch, and optimal control of batch polymerisation process are studied in this contribution. The effectiveness of these techniques is demonstrated by simulation studies.

1. Introduction

Polymer production facilities face increasing pressures for production cost reductions and more stringent quality requirements. However, product quality is a much more complex issue in polymerisation than in more conventional short chain reactions. Because the molecular architecture of the polymer is so sensitive to reactor operating conditions, upsets in feed conditions, mixing, and reaction temperature, can alter critical molecular properties such as molecular weight distributions, copolymer composition distribution, etc. Currently, the main factors limiting the development of comprehensive policies for controlling the properties of polymer products include the limited availability and the cost of on-line instrumentation, a lack of detailed understanding of the dynamics of the process and, finally, the highly sensitive and nonlinear behaviour of polymerisation processes (Kiparissides, 1996). Appropriate process control techniques and optimisation techniques provide leverage for making cost reductions and improvements in product consistency by enabling process to be operated closer to economic, plant and safety constraints.

A major problem in the control of product quality in industrial polymerisation reactors is the lack of suitable on-line polymer quality measurements. Although instruments for measuring the number average molecular weight and the weight average molecular weight are available, these instruments possess substantial measurement delays. Some of these difficult-to-measure variables can however be related to certain easily measurable variables such as temperature, solution viscosity and density of the reaction mixture. Inferential estimators, or software sensors, of these difficult-to-measure 'quality' variables can then be derived from measurements of the more easily measured process variables. The key step in inferential estimation is to establish a relationship between the difficult-to-measure quantities and the more easily measured variables. One popular approach is through the use of a first principles mechanistic model of the process and state estimation techniques such as the extended Kalman filter (Shuler and Zhang, 1985; Ellis *et al.*, 1988; Dimitrators *et al.*, 1989; Kozub and MacGregor, 1992). These approaches, however, require a deep understanding of the polymerisation process and consequently model development is usually very demanding in production processes; even for pilot plant models which involve large sets of differential, algebraic and kinetic equations (Kiparissides, 1996).

To overcome this difficulty, especially in industrial polymerisation, neural network representations based upon monitored reactor data can be developed. Neural networks have been shown to be able to approximate any continuous nonlinear functions, (Cybenko, 1989; Girosi and Poggio, 1990; Park and Sandberg, 1991) and have been widely applied to nonlinear process modelling, (Bulsari, 1995; Morris *et al.*, 1994; Zhang *et al.*, 1998; 1999). Using the learning capability of a neural network, the relationship between polymer quality variables and the on-line measured variables in the reactor can be identified from the reactor operation data.

The economic operation of polymerisation reactors requires the recovery and recycling of unreacted monomers and solvent. This inevitably introduces reactive impurities which are mainly in the form of oxygen and traces of inhibitors. Reactive impurities can rapidly consume free radicals and cease or slow down the polymerisation process. It will also make the polymerisation control strategies less effective. Most polymers are viscous and, hence, reactor fouling is an inevitable problem. Reactor fouling will reduce the heat transfer capability of a reactor and make the reactor temperature control system less effective. Severe reactor fouling can significantly deviate the reactor temperature from its normal value leading to deviations in product quality or even make the reactor inoperable.

Due to the lack of understanding of the highly nonlinear polymerisation dynamics, conventional estimation techniques, such as the Kalman filtering techniques, are usually less effective in the estimation of impurities and fouling since they rely on (reduced order) mechanistic models of polymerisation processes. In this paper, we present techniques for the estimation of reactive impurities and reactor fouling through artificial neural networks. Neural networks are used to build an inverse model of a batch polymerisation process. Given several points on the polymerisation trajectory, the neural network model is used to calculate the effective initial reaction condition. The amount of impurities and fouling are then estimated from the difference between the calculated effective initial condition and the nominal initial condition.

An issue in neural network based modelling is the network generalisation capability, i.e. how the neural network model performs when applied to unseen data. A perfect neural network model is usually very difficult, if not impossible, to develop due to the following reasons. Firstly, network training is a nonlinear optimisation problem and it can converge to a local minimum. Secondly, data collected from process instruments will inevitably contain measurement noise. A network can over-fit noise, especially when the amount of training data is limited. Recent studies have shown that an improved neural network model can be obtained by combining several non-perfect neural networks (Jordan and Jacobs, 1994; Raviv and Intrator, 1996; Sridhar *et al.*, 1996; Zhang *et al.*, 1997). The combination of multiple neural networks is known as stacked neural networks (Wolport, 1992; Sridhar *et al.*, 1996; Zhang *et al.*, 1997).

To address the problem of limited process data, bootstrap aggregated neural network models have been proposed to improve neural network model accuracy and robustness, (Wolpert, 1992; Breiman 1992). Stacked generalisation is a technique which combines different representations to improve the overall modelling capability. In the technique proposed by Zhang *et al.* (1997), process data is randomly re-sampled to form a number of different training and test data sets. Neural networks are then developed based upon each re-sampled data set. However, instead of selecting a perceived 'best' single neural network for prediction purposes, several networks are combined (aggregated) and the aggregated predictor is used as the final representation.

The chapter is structured as follows. Section 2 presents bootstrap aggregated neural network techniques for building nonlinear empirical models. Section 3 presents the batch polymerisation reactor studied. Inferential estimation of polymer quality is presented in Section 4. Section 5 presents a neural network based method for impurities and fouling estimation using neural networks. Optimal control of the polymerisation reactor is presented in Section 6. Finally, Section 7 draws some concluding remarks.

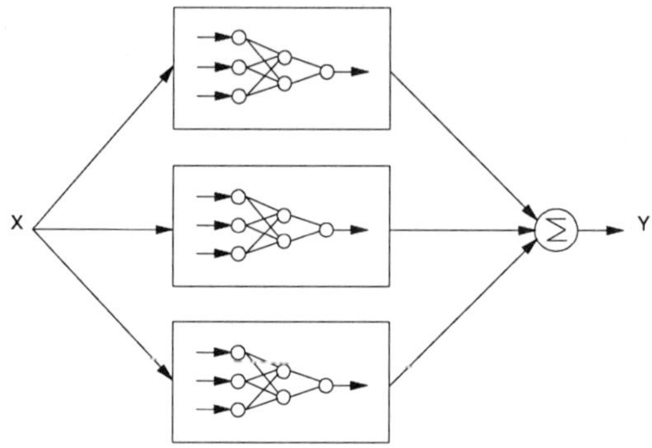

Figure 1. A stacked neural network

2. Robust Neural Networks

In recognition of the difficulty in building a perfect neural network model, several researchers have recently shown that a better neural network model can be obtained by combining several non-perfect neural network models (e.g. Sridhar *et al.*, 1996; Hashem, 1997; Zhang *et al.*, 1997). This forms a stacked neural network model.

A diagram for a stacked neural network is shown in Fig. 1, where several neural network models are developed to model the same relationship between input X and output Y and are combined together. The overall output of the stacked neural network is a weighted combination of the individual neural network outputs. This can be represented by the following equation.

$$f(X) = \sum_{i=1}^{n} w_i f_i(X) \tag{1}$$

where $f(X)$ is the stacked neural network predictor, $f_i(X)$ is the ith neural network predictor, w_i is the stacking weight for combining the ith neural network, and X is a vector of neural network inputs.

The individual neural networks can be developed on the same training data set or on bootstrap re-samples of the training data set. The experimental studies of Taniguchi and Tresp (1997) show that developing individual networks on bootstrap re-samples of the training data set gives better performance.

Stacking weights can be determined in a number of ways. A simple approach is to take equal weights for the individual networks. Another approach to obtain the weights is through multiple linear regression. However, this approach has problems due to the severe correlation among the individual predictors. Since each network is developed to model the same relationship, these networks are highly correlated. We found that obtaining stacking weights through multiple linear regression does not give good performance. This was also experienced by Breiman (1992) and he suggests to put a constraint on the stacking weights such that they are non-negative. Since the individual neural networks are highly correlated, appropriate stacking weights could be obtained through principal component regression (PCR) (Zhang *et al.*, 1997).

A problem in industrial applications of neural network models is the current lack of model prediction confidence bounds. The bootstrap re-sampling technique can be used to estimate the standard errors of model predictions (Tibshirani, 1996). Based on the estimated standard errors, confidence bounds for neural network model predictions can be calculated. Neural network prediction confidence bounds give the process operator extra information about the predictions. The process operator can accept or reject a particular prediction from a neural network model by using the associated prediction confidence bounds.

The bootstrapping method for calculating neural network prediction confidence bounds is summarised as follows:

Step 1. Generate B samples, each one of size n drawn with replacement from the n training observations $\{(x_1, y_1), (x_2, y_2), \ldots, (x_n, y_n)\}$. Denote the bth sample by $\{(x^b_1, y^b_1), (x^b_2, y^b_2), \ldots, (x^b_n, y^b_n)\}$.

Step 2. For each bootstrap sample $b = 1, 2, \ldots, B$, train a neural network model. Denote the resulting neural network weights by W^b.

Step 3. Estimate the standard error of the ith predicted value by

$$\{\frac{1}{B-1}\sum_{b=1}^{B}[y(x_i;W^b) - y(x_i;\cdot)]^2\}^{1/2}$$

where $y(x_i; .) = \sum_{b=1}^{B} y(x_i;W^b) / B$.

Step 4. Calculate the 95% confidence bounds by taking plus and minus 1.96 times the standard error of the mean of the predicted values.

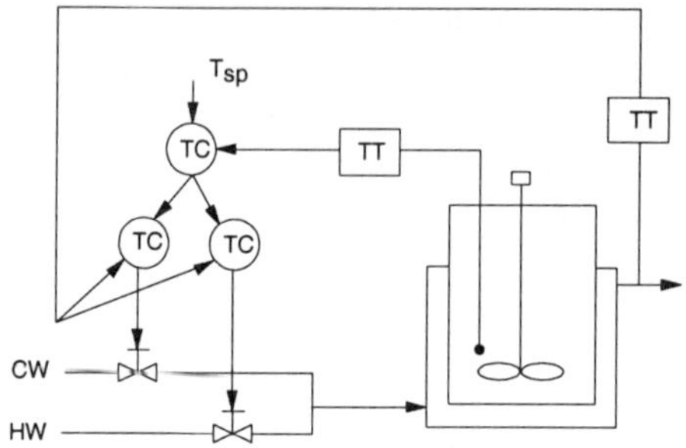

Figure 2. A batch polymerisation reactor

3. A Batch Polymerisation Reactor

The batch polymerisation reactor studied in this paper is a simulation of the pilot scale polymerisation reactor developed in the Department of Chemical Engineering, Aristotle University of Thessaloniki, Greece. The batch polymerisation reactor is shown in Fig. 2. The free-radical solution polymerisation of methyl methacrylate (MMA) is considered in this paper. The solvent used is water and the initiator used is benzoyl peroxide. The jacketed reactor is provided with a stirrer for thorough mixing of the reactants. Heating and cooling of the reaction mixture is achieved by circulating water at appropriate temperature through the reactor jacket. The reactor temperature is controlled by a cascade control system consisting of a primary PID and two secondary PID controllers. The reactor temperature is fed back to the primary controller whose output is taken as the set-point of the two secondary controllers. The manipulated variables for the two secondary controllers are hot and cold water flow rates. The hot and cold water streams are mixed before entering the reactor jacket and provide heating or cooling for the reactor. The jacket outlet temperature is fed back to the two secondary controllers.

A general description of the reactions during the free radical solution polymerisation of MMA initiated by benzoyl peroxide is as follows:

Initiator decomposition

$$I \xrightarrow{k_d} 2R_0$$

Initiation

$$R_0 + M \xrightarrow{k_i} R_1$$

Propagation

$$R_x + M \xrightarrow{k_p} R_{x+1}$$

Transfer to monomer

$$R_x + M \xrightarrow{k_m} P_x + R_1$$

Transfer to solvent

$$R_x + S \xrightarrow{k_s} P_x + R_1$$

Termination by disproportionation

$$R_x + R_y \xrightarrow{k_{td}} P_x + P_y$$

Termination by combination

$$R_x + R_y \xrightarrow{k_{tc}} P_{x+y}$$

In the polymerisation process, initiator I is decomposed into an initiator radical R_0. The initiator radical R_0 reacts with monomer M and a radical R_1 of length 1 is generated. Monomer M is added onto the end of the radical R_x of length x, forming a new radical R_{x+1} of length x+1. The chain of radical R_x is transferred to monomer M and solvent S, forming dead polymers P_x and radicals R_1 of length 1. Termination by disproportionation generates polymers P_x and P_y, while termination by combination generates polymers P_{x+y}.

A detailed mathematical model covering reaction kinetics and heat and mass balances has been developed (Penlidis et al., 1992). Based on this model, a rigorous simulation programme is developed and serves as a test bed for testing different polymerisation control and monitoring techniques before they are implemented on the real reactor.

4. Inferential Estimation Of Polymer Quality

In this reactor, the on-line measured process variables include reactor temperature, jacket inlet temperature, jacket outlet temperature, coolant flow rate, and monomer conversion (*X*) which is measured through a densometer. Polymer quality variables and reactor operation variables include number average molecular weight (*Mn*) and weight average molecular weight (*Mw*). These variables are not measured and are to be estimated from the on-line measurements.

The polymer quality variables during the course of polymerisation is mainly determined by the batch recipe, i.e. the reactor temperature set-point and the initial initiator concentration. Different batch recipes will lead to different polymer growth profiles coupled with different heat generation profiles. Correlation analysis of the reactor operation data indicates that there is a linkage between polymer quality variables and the reactor and jacket temperatures and the coolant flow rate. The reactor temperature setpoint, the initial initiator weight, the jacket inlet and outlet temperatures, the reaction time, and the coolant flow rate through the reactor jacket are used here to estimate the polymer quality variables. The nominal batch time for this reactor is 180 minutes. In this study, data from nine batches were used to develop neural network based inferential estimators. In each of the nine batches, off-line polymer quality "measurements" (simulated) are taken at a 10 minute interval. Thus each batch gives 18 data points. Two additional batches, with different batch recipes from the nine batches, were used as unseen validation data to validate the neural network based inferential estimators. In the two validation batches, polymer quality variables are estimated at every minute and compared with the true values from simulation. Batch recipes for the eleven batches are shown in Table 1. Differences in the recipes in Table 1 reflect different grades of products. Noises with normal distribution are added to the simulated measurements to represent the effects of measurement noise. The means of the noises are zero and the standard deviations of the noises are 10% of those of the corresponding measured variable variations.

Two bootstrap aggregated neural networks were developed to estimate *Mn* and *Mw*. Each of the stacked networks contains *n* neural networks which are in the following form.

$$Mn(t)=f_1(T_{sp},I_0,t,T_i(t),T_o(t),F_c(t),X(t)) \quad (2)$$

$$Mw(t)=f_2(T_{sp},I_0,t,T_i(t),T_o(t),F_c(t),X(t)) \quad (3)$$

where T_{sp} is the reactor temperature set-point, I_0 is the initial initiator weight, T_i is the jacket inlet temperature, T_o is the jacket outlet temperature, F_c is the coolant flow rate, t is the time from the beginning of a batch, and $f_1(.)$ and $f_2(.)$ are nonlinear functions represented by neural networks. In this study, the number of neural networks, n, was selected as 30. Our experience shows that the performance of a stacked network usually settles dwon after stacking about 20 networks. The benefit of selecting a larger n, for eaxample 30, is the improved accuracy in estimating the prediction confidence bonds.

Table 1. Batch recipes

Batch No.	T_{sp} (K)	I_0 (g)
1	343	2.5
2	348	3.0
3	338	2.0
4	343	2.8
5	346	2.0
6	350	1.8
7	332	3.5
8	340	2.6
9	345	2.6
10	342	2.2
11	335	2.4

Data from batches 1, 2, 3, 5, and 7 in Table 1 were used as the training data while data from batches 4, 6, 8, and 9 were used as the testing data. Data from the 10th and 11th batches were used as the unseen validation data. The training data were re-sampled using bootstrap re-sampling with replacement (Efron and Tibshirani, 1993) to form 30 sets of training data. A neural network model was developed for each set of training data. The number of hidden neurons in each individual network was determined by considering a number of neural networks with different numbers of hidden neurons and selecting the one giving the least error on the testing data. Most of the selected networks have around 10 hidden neurons. Each neural network was trained using the Levenberg-Marquardt optimisation algorithm (Marquardt, 1963) together with an "early stopping" mechanism (Sjoberg *et al.*, 1995) to prevent over-fitting. During network training, the training algorithm continuously checks the network error on the testing data. Training is terminated at the point where the network error on the testing data is at its minimum. Network weights were all initialised as random numbers in the range (-0.1, 0.1). The weights for combining individual neural networks were determined through PCR (Zhang *et al.*, 1997).

For the purpose of comparison, single neural network models for estimating *Mn* and *Mw* were also developed. Two single hidden layer feed forward neural networks were used to estimate *Mn* and *Mw*. The number of hidden neurons in each network was determined by studying a number of neural network architectures and selecting the one with the smallest error on the testing data. It was found that the best conventional neural network structures for estimating *Mn* and *Mw* have 16 and 17 hidden neurons respectively. Once again network weights were all initialised as random numbers in the range (-0.1, 0.1) and were trained using the Levenberg-Marquardt optimisation algorithm with "early stopping".

Root mean squared errors (RMSE) from the stacked neural network models and the single neural network models are shown in Table 2. It can be seen that the RMSE from the stacked neural network models are much smaller than the corresponding RMSE from the single neural network models. Note that the seemingly big numbers in the table are due to the fact that *Mn* and *Mw* have large mangitudes, in the order of 10^5 and 10^6 Respectively. Estimations of *Mn* and *Mw* from the single neural network models on the validation data are plotted in Fig. 3 while those from the stacked network models are plotted in Fig. 4. It can be seen that the estimation accuracy has been significantly improved by using bootstrap aggregated neural networks.

Table 2. Estimation errors from different models

Models	RMSE (training & testing)	RMSE (validation)
Single network model *Mn*	4.6330×10^3	7.6613×10^3
Single network model *Mw*	2.1321×10^4	4.1999×10^4
Stacked network model *Mn*	2.4662×10^3	4.0134×10^3
Stacked network model *Mw*	1.1458×10^4	1.7388×10^4

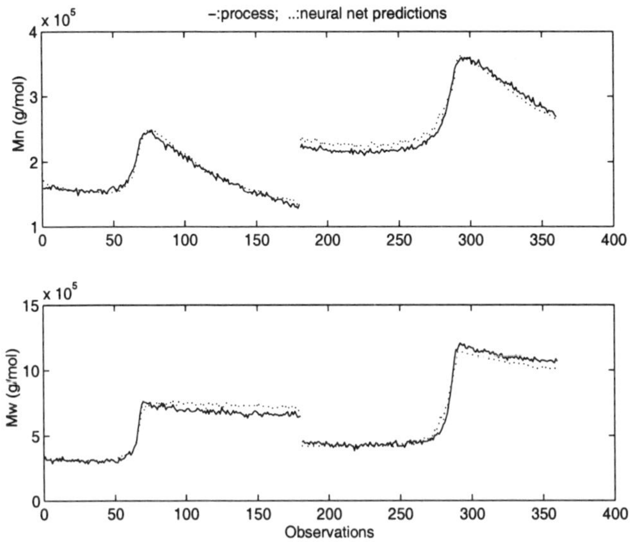

Figure 3. Estimation from the single neural network models (Batch 10: Observations 1 to 180; Batch 11: Observations 181 to 360)

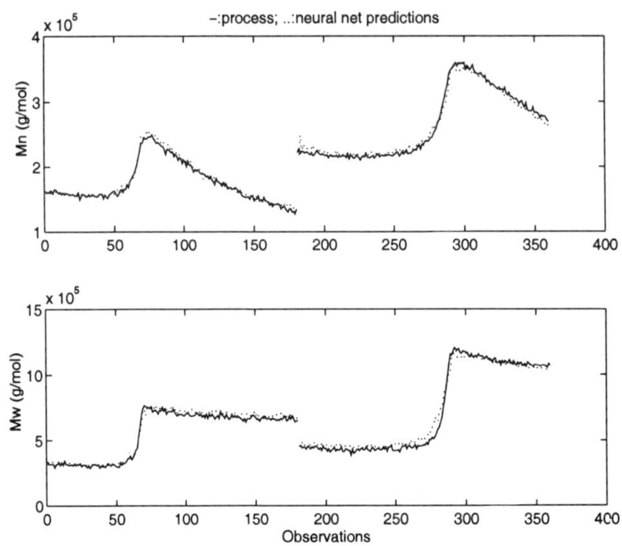

Figure 4. Estimation from the stacked neural network models (Batch 10: Observations 1 to 180; Batch 11: Observations 181 to 360)

5. Estimation of Reactive Impurities and Reactor Fouling

5.1. Neural Network Based Inverse Model

It is also possible to develop a neural network based inverse model which maps the polymerisation trajectories to their corresponding initial conditions. Given a polymerisation trajectory, the neural network model can be used to estimate the effective initial initiator weight and the effective reactor heat transfer coefficient. In this case, the amount of impurities is estimated as the difference between the gross initial initiator weight and the estimated effective initial initiator weight. The amount of reactor fouling is estimated as the difference between the nominal reactor heat transfer coefficient and the estimated effective reactor heat transfer coefficient. The neural network models take the following forms.

$$I_0 = f(T_{sp}, X(t_1), X(t_2), \ldots, X(t_n)) \tag{4}$$

$$U_0 = f(T_{sp}, T_i(t_1), T_i(t_2), \ldots, T_i(t_n), T_o(t_1), T_o(t_2), \ldots, T_o(t_n), F_c(t_1), \ldots, F_c(t_n)) \tag{5}$$

where I_0 and U_0 are the effective initial initiator weight and the effective reactor heat transfer coefficient respectively, T_{sp} is the temperature set-point of the reactor, $X(t_n)$, $T_i(t_n)$, $T_o(t_n)$, and $F_c(t_n)$ are the monomer conversion, the reactor jacket inlet temperature, the reactor jacket outlet temperature, and the coolant flow rate at time t_n respectively. The n points on the polymerisation trajectories and the reactor temperature set-point are used to estimate the effective initial initiator weight and the effective reactor heat transfer coefficient.

To build neural network based inverse models for the batch polymerisation reactor, training data covering various initial conditions should be generated. In this study, 40 different batches of polymerisation are simulated using initial conditions obtained from Monte-Carlo simulation. In this reactor, the nominal values for reactor temperature set-point, initial initiator weight, and reactor wall heat transfer coefficient are 343K, 2.5g, and 0.25B.t.u/m^2-min-K respectively. In the Monte-Carlo simulation, the reactor temperature set points are in the range [323K, 363K]; the initial initiator weights are in the range [0.5g, 2.5g]; and the reactor wall heat transfer coefficients are in the range [0.05, 0.25]B.t.u/m^2-min-K. A further 15 batches were simulated and the resulting data serve as unseen validation data.

The nominal batch time for this reactor is about two to three hours. Since the objective here is to estimate the amount of impurities and fouling at an earlier stage of polymerisation, on-line measurements covering the first 30 minutes of each batch are used. Noises are added to simulated measurements of conversion, temperatures, and coolant flow rate. Noise ranges for temperature, conversion, and coolant flow rate are [-0.5K, 0.5K], [-0.5%, 0.5%], and [-0.1cm^3/min., 0.1cm^3/min.] respectively.

5.2. Impurities Estimation

A stacked neural network model is developed to estimate the effective initiator concentration from the initial monomer conversion trajectory. Discrete monomer conversion measurements during the first 30 minutes of polymerisation were taken. The effect of the number of sampling points on the impurity estimation accuracy has been studied. Table 3 gives the sum of squared errors (SSE) on the 15 unseen validation batches. It can be seen that the estimation accuracy increases with the number of conversion measurements. Monomer conversion can be measured using several different methods such as densometer and gas chromatography. Table 3 indicates that it is a trade off between estimation accuracy and the number of conversion measurements. If conversion measurements are obtained from laboratory analysis, then additional conversion measurements represent additional labour cost. However, the benefit is the improved accuracy in the estimation of impurities which will lead to more appropriate corrective actions. An industrial judgement should be taken here. In this study, conversion measurements at 15, 20, 25, and 30 minutes of each batch are used to estimate reactive impurities. The model for effective initial initiator estimation is of the following form:

$$I_0 = f(T_{sp}, X_{15}, X_{20}, X_{25}, X_{30}) \tag{6}$$

where X_{15} to X_{30} are monomer conversions at times 15 to 30 minutes. Data for building neural network models were re-sampled through bootstrap re-sampling with replacement to form 30 different data sets. For each re-sampled data set, 60% of the data were randomly selected as training data and the remaining severs as testing data. A neural network model is then developed for each re-sampled data set. Each network was trained using the Levenberg-Marquardt optimisation algorithm with "early stopping". Network weights were initialised as random numbers uniformly distributed in the range (-0.1, 0.1). The number of hidden neurons is determined by considering a number of networks with hidden neurons from 5 to 25 and selecting the one giving the least errors on the testing data. The individual networks are then combined together using PCR.

Table 3. Impurity estimation errors with different conversion measurements

No. of conversion measurements	SSE on validation batches
1 (15min.)	0.7553
2 (15, 20min.)	0.3925
3 (15, 20, 25min.)	0.2481
4 (15, 20, 25, 30min.)	0.1147

Figure 5 shows the estimated amount of impurities and the 95% estimation confidence bounds for the 15 unseen validation batches. It can be seen that estimations from the stacked neural network are very accurate. The confidence bounds indicate how confident an estimation is. The narrower the confidence bounds are, the higher the confidence of the estimation is. The neural network prediction confidence bounds are indications of extrapolation.

Figure 6 shows the SSE of the 30 individual neural networks for impurity estimation on training, testing, and validation data. It can be seen that these individual neural networks give various performance. Figure 6 also shows that a single neural network model can give inconsistent performance on training, testing, and validation data. For example, both the 6th and the 24th neural networks have large errors on the training and testing data. However, their performance on the validation data is pretty good. This indicates the non-robustness of single neural network models. Figure 7 shows the SSE of stacked neural networks for impurity estimation on training, testing, and validation data. The x-axis in each plot of Fig. 7 is the number of neural networks in a stacked neural network model. The model errors of stacked neural networks on training, testing, and validation data are very consistent. This is in sharp contrast to the single neural network performance shown in Fig. 6. This clearly demonstrates that stacked neural network models are more robust than single neural network models.

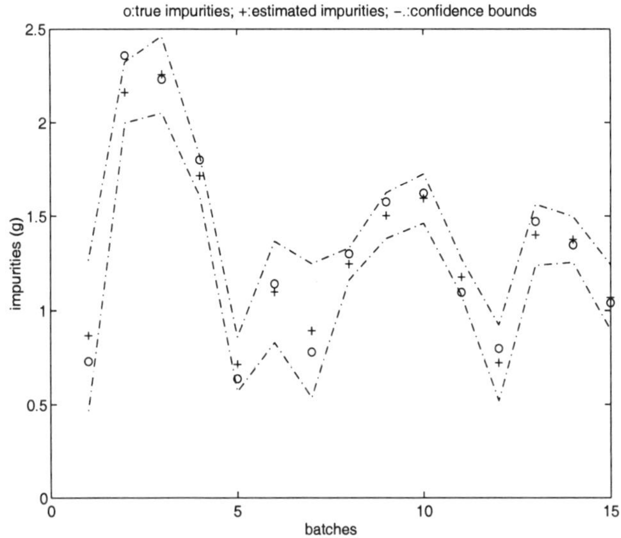

Figure 5. Impurity estimation on validation batches

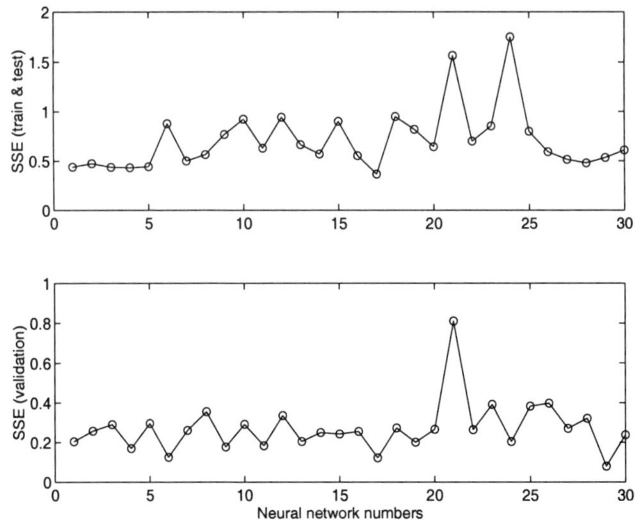

Figure 6. Errors of single neural network models for impurity estimation

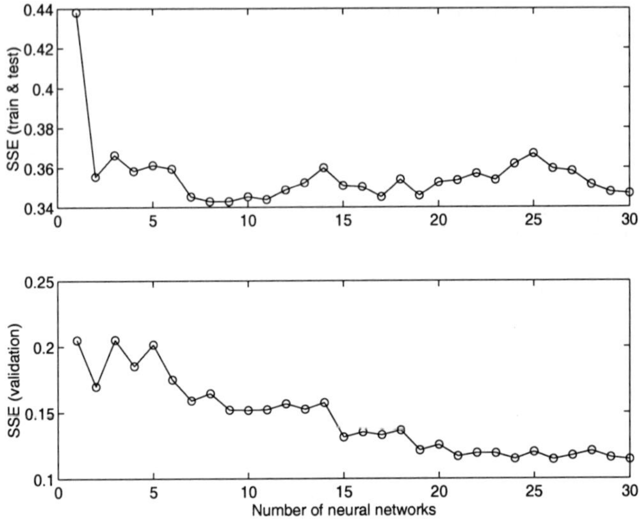

Figure 7. Errors of stacked neural network models for impurity estimation

5.3. Fouling Estimation

A stacked neural network model is developed to estimate the heat transfer coefficients of the reactor wall from temperature and coolant flow measurements. Here, the temperature and flow measurements at 15, 20, 25, and 30 minutes from the start of a batch are used to estimate the effective reactor wall heat transfer coefficient. The amount of reactor fouling is calculated as the difference between the nominal heat transfer coefficient and the estimated heat transfer coefficient. The network model has the following form:

$$U_0 = f(T_{sp}, T_{i15}, \cdots, T_{i30}, T_{o15}, \cdots, T_{o30}, F_{15}, \cdots, F_{30}) \quad (7)$$

Training data are re-sampled through bootstrap re-sampling with replacement to form 30 different training data sets. For each re-sampled training data set, a neural network model is developed. Network training, weight initialisation, and network structure determinations are as outlined before. The individual networks are then combined together using PCR.

Figure 8 shows the estimated amount of fouling and the 95% estimation confidence bounds for the 15 unseen validation batches. It can be seen that the estimations from the stacked neural network are very accurate. The SSE on the validation batches is 0.00067.

6. Robust Neural Network Model Based Optimal Control of The Batch Reactor

In this next study, we consider the following modelling and control scheme. The nominal batch time for this reactor is about 180 minutes. Samples of the monomer conversion and the number average and weight average molecular weights are collected from 60 minutes at a 20 minutes interval. Thus during a batch up to 7 samples of molecular weights are collected. The control variables considered here are the initial reactor temperature set-point and the reactor temperature set-points at 40, 60, 80, 100, 120, 140, and 160 minutes. These reactor temperature set-points provide a control trajectory for the reactor.

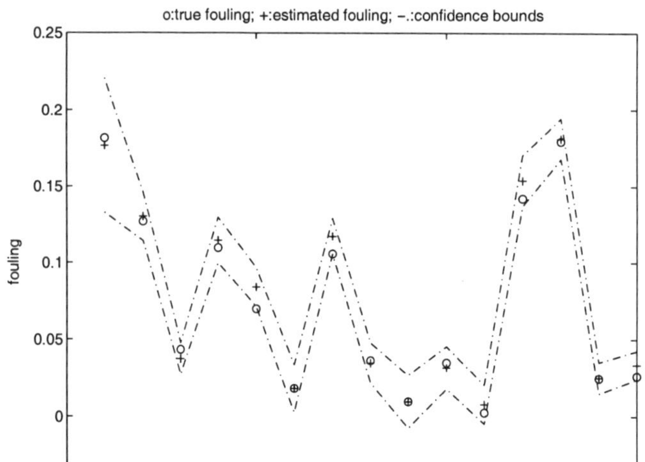

Figure 8. Fouling estimation on validation batches

A neural network model for predicting polymer quality variables at time t_N is then of the following form:

$$Y(t_N) = f(I_0, U(t_N)) \tag{8}$$

where

$$Y(t_N) = [X(t_N)\ Mn(t_N)\ Mw(t_N)]^T$$
$$U(t_N) = [T_{sp0}\ T_{sp1}\ T_{sp2}\ \ldots\ T_{spN}]^T$$

In the above equations, T_{sp0} to T_{spN} are the trajectory of reactor temperature set-points, $X(t_N)$, $Mn(t_N)$, and $Mw(t_N)$ are the monomer conversion, the number average molecular weight, and weight average molecular weight at time t_N respectively.

In order to "simulate" the building of neural network models in an industrial environment, 50 batches were simulated with controls generated from Monte-Carlo simulation. The sampled data were corrupted with typical measurement noises. From the generated data, bootstrap re-sampling with replacement was used to generate 30 sets of replica of the data. For each re-sampled data set, a neural network model is developed. The neural network contains 10 hidden neurons and the network weights were initialised as random numbers in the range (-0.1, 0.1). The networks were trained using the Levenberg-Marquardt optimisation algorithm with regularisation (Zhang and Morris, 1999). The objective to include a regularisation term is to improve the generalisation capability of the networks. The individual networks were then combined together through PCR. A further 20 batches were simulated to generate a set of unseen data to validate the developed neural network models.

Figure 9 shows the scaled SSE of the individual networks on training and validation data sets. It can be observed that the performance of these networks on the training and validation data sets is not consistent. A network having small errors on the training data set may have quite large errors on the validation set. The minimum SSEs of individual networks on the training and validation data sets are about 18 and 19 respectively. The SSEs from the stacked network on the training and validation data sets are 9.8 and 13.8 respectively. Thus the model accuracy is significantly improved by combining multiple non-perfect models.

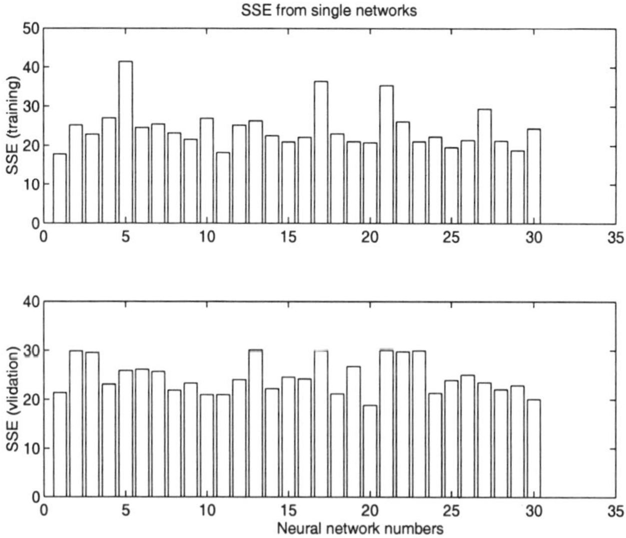

Figure 9. Model errors of individual networks

The objective in optimum batch polymerisation operation is to produce polymers with desired quality and efficiency within a short time. This is achieved by solving the following optimisation problem:

$$\min_{U,\, t_f} J = (1-X)^2 + wt_f$$
$$\text{s.t. } 0.85 \le Mn/Mnd \le 1.15$$
$$2 \le Pd \le 3$$

where U is a vector of control actions (i.e. reactor temperature setpoints), X is the monomer conversion predicted from the stacked neural network model, t_f is the batch ending time, w is a weighting factor for batch time, Mn is the number average molecular weight predicted from the stacked neural network model, Mnd is the desired value of Mn, and Pd is the polydispersity, which is defined as Mw/Mn. In this study, w is selected as 0.001 hour^{-1}. This objective function represent a trade-off between maximising monomer conversion and minimising batch time subject to the desired molecular weight distribution constraints. Since the neural network models only predict polymer qualities at a 20 minute interval starting from the 60th minutes into reaction (since the reaction is not likely to finish within 60 minutes), the batch ending time can only take one of the following values: 60, 80, ..., and 180 minutes. The optimisation problem is solved by considering each of the possible batch ending time and selecting the one resulting in the smallest objective function value. By such means, the above free-terminal-time optimisation problem is converted into several fixed-terminal-time optimisation problems.

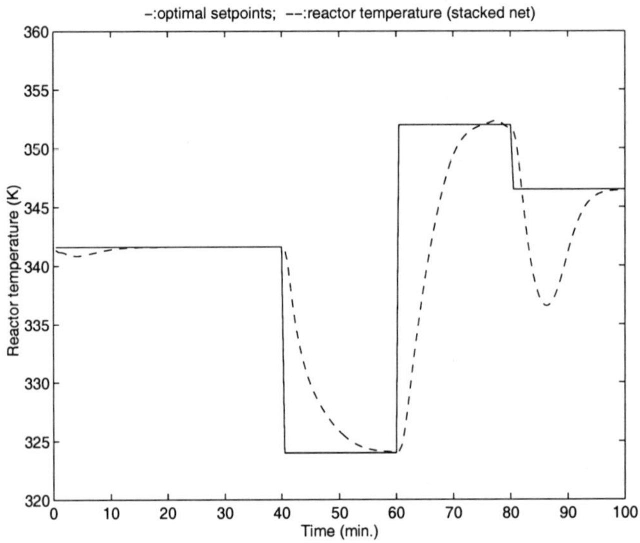

Figure 10. Optimal reactor temperature profile calculated from a stacked network

The following example demonstrates that optimum reactor temperature control strategies can be calculated from empirical models developed from minimal plant data and that the optimal trajectories calculated improve product quality and production efficiency. In this example, *Mnd* is taken as 2×10^5 g/mole corresponding to a specific grade of product. Optimum batch recipe and reactor temperature control profile can be obtained by solving the above optimum control problem. The optimum batch time is found to be 100 minutes, the optimum reactor temperatures set-points for the time intervals 0-40, 40-60, 60-80, and 80-100 minutes are found to be 341.6K, 324K, 352K, and 346.5K respectively. The optimal reactor temperature set-points and the reactor temperature are shown in Fig. 10. Under this optimum control strategy, the following final product quality variables were obtained from simulation: Mn=1.79×10^5 g/mole, Pd=2.7 and X=84%. These quality variables are within their constraints indicating that the product quality is satisfactory. The monomer conversion is also quite high under this control strategy.

For the purpose of comparison, a single neural network model is also used to calculate the optimal control actions. The optimum batch ending time is also found to be 100 minutes and the optimum reactor temperature set-points for the time intervals 0-40, 40-60, 60-80, and 80-100 minutes are 337.7K, 249.3K, 325.2K, and 352K respectively. The optimal reactor set-points and the reactor temperature are shown in Fig. 11. Under this optimum control strategy, the following final product quality variables were obtained from simulation: Mn=1.96×10^5 g/mole, Pd=8.17 and X=85.7%. However, the polydispersity is seen to be well above its upper constraint 3.0. This indicates that model-plant mis-matches can have a significant impact on the calculated optimal control strategies. The "optimum" control actions calculated from an inaccurate single neural network model can turn out to be significantly "non-optimal".

Although not a direct comparison, it is interesting to observe that the optimal control strategy obtained form a stacked neural network is qualitatively similar to that obtained from a mechanistic model. Thomas and Kiparissides (1984) calculated near-optimal temperature policies for a batch MMA polymerisation process using a mechanistic model. The results shown in Fig. 10 is qualitatively similar to those presented in Thomas and Kiparissides (1984). The control strategy obtained from a single neural network shown in Fig. 11, however, is very different from those obtained from a mechanistic model. This observation is very encouraging and indicates that it may be possible to build robust neural network representations and make use of a stacked neural network based optimal control strategy for real process applications.

Figure 12 shows the polydispersity under the two optimum control strategies. Under the optimum control strategy calculated from a stacked neural network model, Pd is always within its constraints. However, under the optimum control strategy calculated from a single neural network model, Pd significantly overshoot its upper constraint after 70 minutes. This is mainly due to the poor generalisation capability of the single neural network model. When calculating the "optimal" control actions based on this model, the model predicted polydispersity is within the constraints. However, when the calculated "optimal" control actions are applied to the reactor, the actual polydispersity moves outside its constraints.

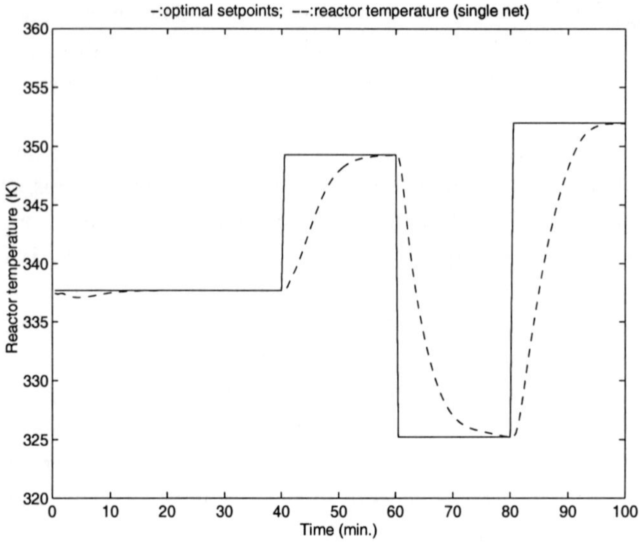

Figure 11. Optimal reactor temperature profile calculated from a single network

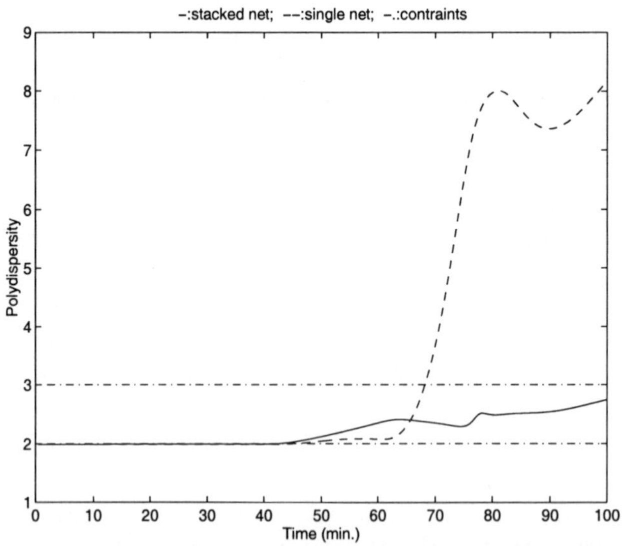

Figure 12. Polydispersities under two optimal control strategies

7. Conclusions

Studies in this paper have demonstrated that combining multiple neural networks can improve model generalisation capability and provides an attractive approach to developing robust empirical models from a limited amount of process operational data. Robust neural network based techniques for inferential polymer quality estimation, estimation of reactive impurities and reactor fouling during the early stage of a batch, and optimal control of batch polymerisation processes are developed and successfully demonstrated in simulation studies. These techniques have significantly potentials in agile batch manufacturing where modelling and control based on detailed mechanistic model is usually not feasible due to the frequent change in product designs and process operations.

References

Breiman, L., *Technical Report No. 367*, (Department of statistics, University of California at Berkeley, USA, 1992).

Breiman, L., *Technical Report No. 421*, (Department of statistics, University of California at Berkeley, USA, 1994).

Bulsari, A. B., (Ed), in *Computer-Aided Chemical Engineering, Volume 6: Neural Networks for Chemical Engineers* (Amsterdam, Elsevier, 1995).

Cybenko, G., *Math. Cont. Signal Sys.* **2** (1989), 303-314.

Dimitrators, J., Georgakis, C., El-Aasser, M. S., and Klein, A., *Comput. Chem. Engng.* **13** (1989), 21-33.

Efron, B., and Tibshirani, R., *An Introduction to Bootstrap* (Chapman and Hall, London, 1993).

Ellis, M. F., Taylor, T. W., Gonzalez, V., and Jensen, K. F., *AIChE Journal.* **34** (1998), 1341-1353.

Girosi, F., and Poggio, T., *Biological Cybernetics.* **63** (1990), 169-179.

Hashem, S., *Neural Networks.* **10** (1997), 599-614.

Jordan, M. I., and Jacobs, R. A., *Neural Computation.* **6** (1994), 181-214.

Kiparissides, C. (1996). *Chem. Eng. Sci.* **51** (1996), 1637-1659.

Kozub, D. J., and MacGregor, J. F., *Chem. Eng. Sci.* **47** (1992), 1047-1062.

Marquardt, D., *SIAM J. Appl. Math.* **11** (1963), 431-441.

Morris, A. J., Montague, G. A., and Willis, M. J., *Trans. IChemE*, Part A. **72** (1994), 3-19.

Park, J., and Sandberg, I. W., *Neural Computation.* **3** (1991), 246-257.

Penlidis, A., Ponnuswamy, S. R., Kiparissides, C., and O'Driscoll, K. F., *Chem. Eng. Journal.* **50** (1992), 95-107.
Raviv, Y., and Intrator, N., *Connection Science.* **8** (1996), 355-372.
Schuler, H., and Zhang, S., *Chem. Eng. Sci.* **40** (1985), 1891-1904.
Sjoberg, J., Zhang, Q., Ljung, L., Benveniste, A., Delyon, B., Glorennec, P. Y., Hjalmarsson, H., and Judisky, A., *Automatica.* **31** (1995), 1691-1724.
Sridhar, D. V., Seagrave, R. C., and Bartlett, E. B., *AIChE J.* 42 (1996), 2529-2539.
Taniguchi, M., and Tresp, V., *Neural Computation.* **9** (1997), 1163-1178.
Thomas, I. M., and Kiparissides C., *Canad. J. Chem. Eng.* **62** (1984), 284-291.
Tibshirani, R., *Neural Computation.* **8** (1996), 152-163.
Wolpert, D. H., (1992)., *Neural Networks.* **5** (1992), 241-259.
Zhang, J., Martin, E. B., Morris, A. J., and Kiparissides, C., *Comput. Chem. Engng.* **21** (1997), s1025-s1030.
Zhang, J., Morris, A. J., and Martin, E. B., *Comput. Chem. Engng.* **22** (1998), 1051-1063.
Zhang, J., Morris, A. J., Martin, E. B., and Kiparissides, C. (1999). *Comput. Chem. Engng.* **23** (1999), 301-314.
Zhang, J., and Morris, A.J., *IEEE Trans. on Neural Networks.* **10** (1999), 313-326.

Acknowledgements

The work is supported by the European Community under the grant BRITE EURAM Project No. 7009. The authors thank Prof. C. Kiparissides for providing the polymerisation reactor simulation programme.

Part IV New Learning Technologies

12. REINFORCEMENT LEARNING IN BATCH PROCESSES

J. A. WILSON
School of Chemical, Environmental and Mining Engineering
University of Nottingham, University Park, Nottingham, NG7 2RD, UK

E. C. MARTINEZ
INGAR-CONICET, Avellaneda
3000 Santa Fe, Argentina

Conventional methods for batch chemical process optimisation and control depend on having both perfect process models and measurements available. Here, to avoid this, we apply a novel methodology centred on *reinforcement learning* (RL) whereby, unlike most forms of machine learning, an autonomous agent is not instructed on how to act by example but instead learns directly by trying control actions and seeking for those giving maximum reward. A central notion is the *performance* or *value function* that, in a given current state, signifies the contribution a specific action will make towards maximising the final performance or reward over an entire batch. For batch-to-batch, incremental learning and control, the initially unknown value function is here represented using *wire fitting* and a neural network. This is a simple yet powerful means of simultaneously learning and fitting the value function. The performance achieved in each completed batch can be propagated from the end point back through the intermediate states. With echoes of dynamic programming, this allows calculation of *Bellman=s errors* which can be minimised in neural network fitting. The higher level optimisation and control problem in batch processing thus fits neatly into this framework and some results of a case study illustrate the potential of the approach.

1. Introduction

A recent shift in the attention of the chemical industry has been towards fine and speciality chemicals and bioprocess products, which are normally produced batchwise. For many batch processes, continuous human intervention is still the key to success in achieving products of high and reproducible quality. In the current economic climate, where global markets impose intense competition, a shorter product life cycle and an ever-increasing number of products, such a dependency is unsatisfactory.

In the typical industrial batch process environment, where control action based on observation of progress can be taken at discrete intervals during the course of a batch (i.e. intra-batch actions), optimising the batch operation represents a challenging decision problem. Firstly, information on end-product quality and process performance is often delayed until after a batch is completed. Secondly, key measurements during the course of a batch are often scarce and also delayed. Moreover, even in the rare cases where a first-principles model is available, ever-present process uncertainties make the final outcome of a batch run difficult to predict accurately using a model alone (Terwiesh, Agarwal and Rippin, 1994). For all of these reasons, conventional optimal control methods are rarely part of everyday practice in industrial batch processing. However, many batch processes are still operated on a day-to-day basis with acceptable levels of performance, thanks to the availability of that scarce resource - experienced human operators. The success achieved can be attributed to the ability of an operator to learn incrementally from experience, batch-to-batch. After completing each batch the benefit of hindsight allows an operator to update his strategy for the next batch to come. Figure 1 shows this schematically. The work presented here is part of a research project aimed at developing performance and quality control methodologies that implement this type of learning in a computer. Artificial neural networks (ANNs) lie at the heart of the learning approach.

2. Incremental Learning Control

The basic problem of learning a control strategy from examples has been defined as >learning what to do= (Sutton and Barto, 1997), i.e. how to map sensed situations or process states into control actions, so as to maximise some externally provided (often delayed) scalar reward signal. According to this definition, the learner is not instructed to act under the tutelage of an exemplar teacher, as in most forms of machine learning, but instead must try control actions whilst always seeking for those that provide the maximum reward. This is broadly termed *Reinforcement Learning*, where in psychology *to reinforce* is to >reward an action or response so that it becomes more likely to occur again=. The learning process, as shown in Fig. 1, emphasises the *interaction* between an active decision-making agent, or controller, and its target system (Sutton and Barto, 1997). A final state or *goal* is sought for the system, despite imperfect knowledge of its behaviour and the influence of external disturbances.

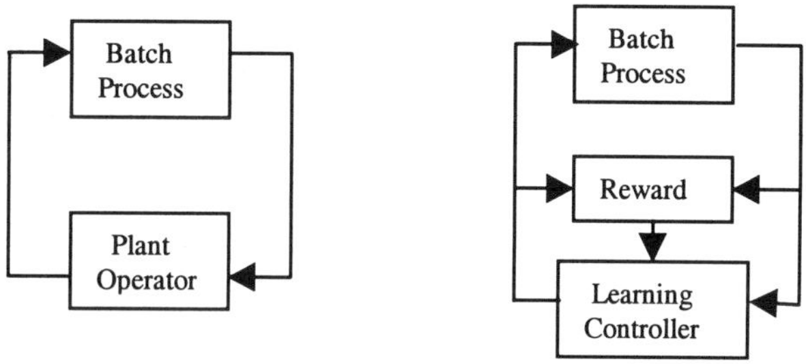

Figure 1. Reinforcement learning paradigm showing interaction between the human or computer controller and the plant.

For batch process optimisation the final batch condition is often the goal for control and each time the controller chooses a given action during the course of a batch all ensuing states will be affected, thereby constraining the degrees of freedom available at later times in the decision sequence. Thus, the long-term influence of every chosen action during a batch is of outstanding importance. This is shown schematically in Fig.2 where, in addition to the goal, the reward signal also incorporates information on one or more *preference* indices associated with each run outcome (Wilson and Martinez, 1997). Normally, the goal specification expresses hard constraints on end-product quality, whereas preferences are used for softer operational objectives like reducing end-time and energy consumption, or increasing reactant conversion.

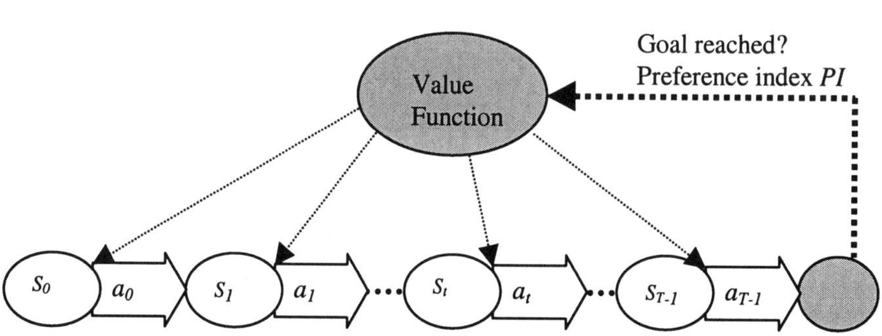

Figure 2. Multi-stage decision making with delayed rewards.

Goal achievement and preference optimisation both demand foresight to account for the indirect, delayed consequences of each individual control action. This is particularly critical for most batch processes. To reflect the long-term impact of control actions, and hence to give guidance in selecting good actions, a mathematical device is needed to assign them rewards or penalties as appropriate. Here, for this purpose, a value function is proposed which can be incrementally learned on-line.

3. The Value Function

The objective of learning a value function is to establish an explicit strategy for the selection of intra-batch control actions that, if applied, lead to achieving the process goal and maximising the value of *PI* a scalar preference index. *PI* embodies the resulting values of the preferences associated with the outcome of each batch run.

The process goal is defined to be a subset of end states that meet the necessary constraints on product quality, safety and operational performance. Thus the goal embodies conditions that must be met, otherwise the batch counts as >bad= and potentially must be rejected or reprocessed. Preferences and the preference index *PI*, on the other hand, are used to express the relative desirability associated with different paths towards the process goal. Thus they register a degree of success which if not maximised represents a marginal economic penalty rather than a catastrophic loss.

At any instant during the progress of a batch, the value function, to be denoted here by •, maps the current measured state *s0S* and an action *a0•* to a real number representing the goodness or badness of the action from the point of view of achieving the goal and maximising *PI*. Thus, when picking action *a* given the process state *s*, the larger and the more positive the corresponding value of •, the better. The importance of the value function • is that it contains, in an implicit form, the knowledge of a good control policy. That is, *Q* is a good policy, at best the optimal policy, if actions are selected for each state according to

$$\text{Policy } Q: \quad a_t{}^* = \arg\max_{a\in\Omega} \pi(s_t, a_t) \tag{1}$$

where • represents the set of feasible control actions and $a_t{}^*(s_t)$ is the optimum action in state s_t. However, at the outset the value function itself is not explicitly known. In order to construct an approximation to it inductively, examples of the form $\{(s_t, a_t), \bullet\}$ need to be generated by practical experience during batch production by exercising decision making in different process states so as to enable a distinction to be drawn between Agood≅ actions and Abad≅ actions. By considering a given number of batch runs, and the intra-batch actions taken, sampled values for the value function are calculated using the following relationship:

$$\pi(s,a) = \begin{cases} \leftarrow PI \text{ if } a_t \text{ is a final action and the goal has been met} \\ \leftarrow -1 \text{ if } a_t \text{ is a final action and the goal has not been met} \\ \leftarrow \max_{a_{t+1}\in\Omega} \pi(s_{t+1},a) \text{ otherwise} \end{cases} \tag{2}$$

Here s_{t+1} is the state reached at the next decision stage as a result of taking action a_t from state s_t.

Once each batch run has been completed and the outcome is known, the benefit of hindsight provides room to understand the goodness of each control action that was taken. To allow this, the value function is defined in Eq. 2 to approximate the maximum final reward (or penalty) the controller is expected to receive on completing the batch by executing action a_t when the process state s_t is observed, and then *acting optimally* for the remainder of the batch. Hence, Eq. 2 requires a backward recursive calculation along the sequence of decision stages during the batch. The reader can easily recognise the underlying Dynamic Programming (DP) style.

Note that Eqs. 1 and 2 are strongly linked, which initially impedes making a good approximation to • when there are only a few batch runs to learn from. But, as enough batch-to-batch data accumulate, and/or are artificially augmented (with the aid of model prediction, as explained later), the approximation to the value function, and along with it the optimum control policy of Eq. 1, can be sensibly improved.

4. The Value Function and Optimum Operation

A nice mathematical property of the value function as defined in Eq. 2 is that it can be recursively expressed as

$$\pi(s_t, a_t) = E\left\{\max_{b\in\Omega} \pi(s_{t+1}, b)\right\} \tag{3}$$

where *E{.}* is the expected value operator over all sources of randomness. Again, this implies that the value of taking action a_t in state s_t, which will carry the batch to state s_{t+1} as a result, is the value of subsequently taking optimal actions at all remaining stages through to completing the batch.

Equation 3 is the well-known Bellman=s criterion for DP (Bertsekas, 1995), written over the continuum of states and feasible control actions. The solution to this infinite dimensional set of equations is the value function but an exact solution as demanded by conventional DP is almost impossible to find. Classically, DP consists of forward sweeping through the entire state-action space and backing up each state-action pair once per sweep. However, in many problems of batch process optimization the vast majority of the state space is irrelevant because either there are regions of states that are never visited or they can be >visited= only under very poor control policies. So, the curse of dimensionality can be eased by focusing backing up only where it is needed. This can be done by combining (forward) state sweeping, which is made using a process model, with selective backups that update the current approximation to the value function (Bertsekas and Tsitsiklis, 1996). In the following sections we will look both at building a suitable approximation to the value function and at using it, in conjunction with predictive models learned on-line, to control operation of future batches.

5. Learning an Approximation to the Value Function

Equation 3 represents the system of so called *Bellman Optimality Equations* (Bertsekas, 1995), one for each possible state-action pair, the solution to which is the value function •. But remember, • is unknown at the outset and an approximation to it must be learned progressively, batch-to-batch, by interaction with the plant. An approximation scheme which facilitates the learning process is therefore needed.

5.1. Approximation Using a Neural Network

As a basis for approximating the value function • let us first consider a neural network scheme with states and actions as inputs and the value function as scalar output. For a given set of weights w in the ANN approximation $•(s_t, a_t | w)$, and a given state-action pair (s_t, a_t), the Bellman residual is defined to be the difference between the two sides of the Bellman Equation (Eq. 3). Accordingly, for a batch process involving a sequence of n decision stages, the *mean squared Bellman error,* for all the data accumulated, is defined to be:

$$E_B = \frac{1}{n}\sum_{s,a}\frac{1}{2}\left[E\left\{\max_{b\in\Omega}\pi\left(s_{t+1},b\mid w\right)\right\}-\pi\left(s_t,a_t\mid w\right)\right]^2 \tag{4}$$

If the Bellman Error E_B is non-zero, then the fitted ANN approximation to the value function will provide a sub-optimal control policy. This suggests it might be reasonable to change the weights w in the ANN approximation, e.g. by performing backpropagation and gradient descent on E_B. Accordingly, a specific weight update rule is

$$\Delta w = -\frac{\alpha}{n}\sum_{s,a}\left[E\left\{\max_{b\in\Omega}\pi\left(s_{t+1},b\middle|w\right)\right\}-\pi\left(s_t,a_t\middle|w\right)\right] \times\sum_{s,a}\frac{\partial}{\partial w}\left[E\left\{\max_{b\in\Omega}\pi\left(s_{t+1},b\middle|w\right)\right\}-\frac{\partial}{\partial w}\pi\left(s_t,a_t\middle|w\right)\right] \tag{5}$$

where w is the vector of neural network weights and • is the learning rate. If, for the sequence of decisions in a batch run, E_B is zero, then the value function is locally optimal for the sampled data, as will also be the control policy Q derived from it through Eq. 1. Therefore, performing gradient descent on the Bellman Error E_B guarantees that Q will eventually converge, at least locally, to an optimal control policy.

The speed of ANN training using Eq. 4 depends heavily on the presentation sequence of examples followed during training. As expected, the fastest training speed is obtained when state-action pairs are stratified in accord with the batch decision sequence $s_0, \ldots s_T$, and stage-wise backward training is used. Assuming that the neural network has enough hidden neurons, the following stage-wise procedure provides good results. First consider only state-action pairs associated with the last decision stage, that is pairs where states are indicated by s_{T-1}. According to Eq. 2, for these pairs the value function can be directly calculated from the corresponding final outcome of the batch. Once training is achieved for this subset of state-action pairs, add to the training set those pairs involving s_{T-2} and repeat. Continue the backward inclusion of training pairs and re-training until the training set includes the pairs associated with initial state s_0 (i.e. the training set includes all data accumulated to date). Figure 3 illustrates this stage-wise backward training scheme. Note that in Eq. 4 the optimum trajectory onwards from state s_{t+1} to the end point is always already known when training w to minimise E_B from states s_t. Each time a new experimental batch is completed and the data from it become available, this whole training procedure is repeated.

```
2=T
repeat
    2=2-1
    add experimental data pairs (s2,a2) to training set
    repeat
        repeat for every pair (s_t,a_t) in training set
            ANN for a_i^+ and forward sweep for B_i
            B(s_t,a_t|w) from Equ. 7
            max_bOΣ B(s_t+1,b|w)
            square Belman error contribution
        until training set exhausted
        Belman error E_B by summation across n batches
        backpropagate E_B
        update ANN weights w
    until E_B minimised
until 2=0
```

Figure 3. Stagewise backward training strategy for the wire fitting/neural network based approximation to the value function.

Within this scheme, solving the optimisations embedded in Eq. 4 under a pure neural network approximation is computationally inefficient in involving searches for the optimum action across large parts of the state-action space. For this reason a modified approximation to the value function is attractive.

5.2. Approximation Using Wire Fitting and a Neural Network

Wire fitting (Baird and Klopf, 1993) is a function approximation method specifically designed for self-learning control problems where, as here, a given function needs to be *simultaneously* learned and fitted. Significantly, it also allows the maximum of the function to be found very quickly. First consider a new approximation to the value function $\bullet(s_t,a)$ for a given state s_t which uses a number, m, of so called >support points= $(a_i^+,\bullet_i)$. Here the value $\bullet_i$ corresponds to action a_i^+

and the actions $a_1^+, \ldots a_m^+$ are free parameters that can be adjusted as long as every $a_i^+ 0 \bullet$. The function approximation is given as

$$\pi(s_t, a_t) = \frac{\sum_{i=1}^{m} \pi_i \left[\left| a_t - a_i^+ \right| + \pi_{\max} - \pi_i \right]^{-1}}{\sum_{i=1}^{m} \left[\left| a_t - a_i^+ \right| + \pi_{\max} - \pi_i \right]^{-1}} \tag{6}$$

where $\bullet(s_t, a_t)$ for a given control action a_t is defined as a weighted average of the m values of $\bullet_i$, weighted by the distance between a_t and a_i^+, and also by the distance between $\bullet_i$ and $\bullet_{max}$ (=max $\bullet_i$). This approximation $\bullet(s_t, a_t)$ may not go through every support point, but, most importantly, it is guaranteed to pass through the one that provides the maximum value $\bullet_{max}$. Thus, for optimisation purposes, the action that maximises $\bullet(s_t, a_t)$ is simply that action a_i^+ amongst the support points which has subscript corresponding to the maximum value $\bullet_{max}$. Thus, optimisation reduces to choosing the optimum action from the set of m possible support points.

Now consider the problem of learning this approximation to the value function $\bullet(s_t, a_t)$ for a given state s_t. It must be learned from the accumulated batch data according to the Bellman Error criterion, as already described, but this time by adjustment of the parameters a_i^+. As training samples are observed, the parameters a_i^+ and $\bullet_i$ must be adjusted so that $\bullet(s_t, a_t)$ becomes a good fit to the training data.

These ideas on actions at a single state can be extended to the general rule for action at all states by replacing the parameters a_i^+ and $\bullet_i$ with state-dependent functions $a_i^+(s)$ and $\bullet_i(s)$. With this change, the support points ($a_i^+, \bullet_i$) become *support wires* $(a_i^+(s), \bullet_i(s))$ in a higher-dimensional (state-action-value) space where the value function is a surface >supported= by those wires. This is illustrated in Fig. 4 where m=3 and thus three support wires, which in the case shown are straight i.e. state-independent, shape a notional value function surface. The maximum • at a given state s always lies on one of the three support wires. On that basis Eq. 6 can be generalised into Eq. 7, where the additional constants $c_i > 0$ can be used to fix the smoothness of the approximation. When all $c_i = 0$, then the approximation is forced to pass through all the wires, potentially giving rise to abrupt changes in the value function. Otherwise, the interpolation is smoother, but may not go exactly through all the wires.

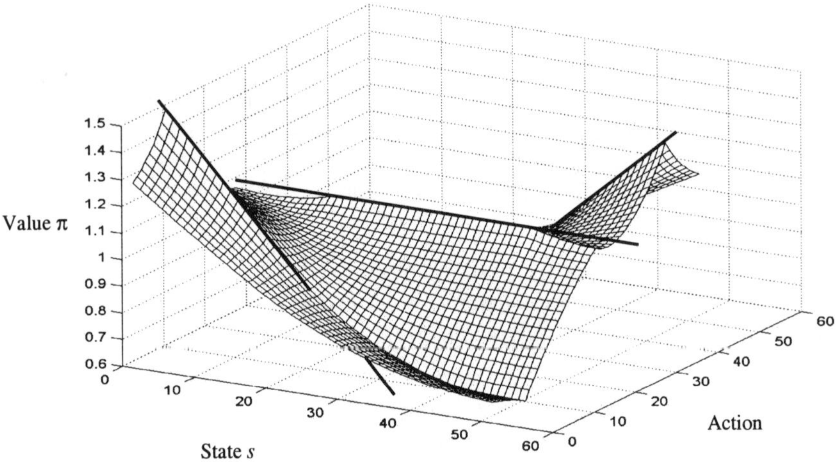

Figure 4. Notional wire fitted approximation to the value function having three straight support wires. Notice that the maximum value at any state lies on one of the wires (e.g. the horizontal wire for states between 13 and 41).

$$\pi(s_t, a_t) = \frac{\sum_{i=1}^{m} \pi_i(s)\left[\left|a_t - a_i^+(s)\right| + c_i\left(\pi_{\max}(s) - \pi_i(s)\right)\right]^{-1}}{\sum_{i=1}^{m}\left[\left|a_t - a_i^+(s)\right| + \pi_{\max}(s) - \pi_i(s)\right]^{-1}} \tag{7}$$

In either case, the most attractive property of the approximation given by Eq. 7 is that, no matter what values the vectors associated with states take, it is guaranteed that:

$$\max_a \pi(s, a) = \max_i \pi_i(s) \equiv \pi_{\max}(s) \tag{8}$$

The general approach proposed here for learning this approximation to the value function is to use a neural network to learn positioning for the support wires $(a_i^+(s), \bullet_i(s))$ in order to minimise the Bellman Error criterion in Eq. 4. The neural network has s_t as input and $a_1^+, \ldots a_m^+$ as outputs. Thus for a given w, the wires are wholly defined and an approximation to the value function is then obtained through Eq. 7. AWire fitting≅ is accomplished through adjusting the vector of neural network weights w.

Thus, using this wire fitting/neural network approximation, the Bellman Error in Eq. 4 can be calculated for any state s_t and its successor s_{t+1} along the state sequence in a batch run. To introduce changes in the neural network weight components w, the error found is backpropagated as before through Eq. 5 but this time with partial derivatives evaluated from Eq. 7.

6. Model-based Learning and Optimisation

As explained in Sec. 4, the exact solution to the infinite dimensional equation set in Eq. 3 demands forward sweeping through the entire state-action space and backing up each state-action pair once per sweep. To reduce the dimensionality of this problem we search for an approximation to the value function that focuses backing up only where it is needed. This can be done by combining (forward) state sweeping, which is made using predictive models, with selective backups according to the Bellman Error criterion Eq. 4. Figure 5 indicates how the local predictive models M_i and the neural network/wire fitting approximation to the value function are combined in making a forward sweep from an experimental state s_t. A more detailed discussion is given elsewhere (Martinez and Wilson, 1998). Where n samples and control actions are taken during a batch run, n local predictive models will be required. Each predictive model represents the state transition from one sample period to the next, i.e. it predicts (or simulates) the next most immediate measured state s_{t+1} to be expected on executing a given control action a_t at the state s_t according to

$$s_{t+1} = M_t\,(s_t, a_t) \qquad (9)$$

The only exception is the model for the final transition in a batch, i.e. from state s_{T-1} which as output yields the terminal value function as defined in Eq. 2. We here of course assume that these predictive models are unknown at the outset and must therefore be fitted on-line using the data observed from batch-to-batch. Any inductive (black box) approximation technique (e.g. neural networks, local weighted regression) could be used for this purpose. As the batch-to-batch data accumulate the quality of these models will, like the approximation to the value function itself, improve incrementally.

The forward sweep illustrated in Fig. 5 is also used to implement the control strategy, learned (and embedded in •) from the experience accumulated during the previous batches, in making a new batch of product. The most recent approximations to the value function and predictive models are both employed in identifying the optimum state-action trajectory from each successive on-line measurement of batch state s_t as it is reached.

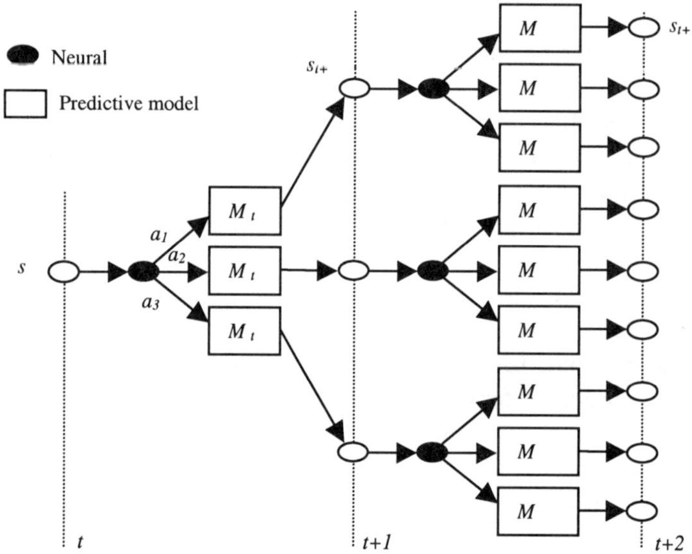

Figure 5. A forward sweep from the measured state at time t using the neural network and predictive models.

Because the optimum value function is assured to arise only from the neural network generated actions (i.e. the support wires) it is a trivial task to work back from the best predicted terminal outcome to fix the best action a_t^*. This is illustrated in Fig. 6 where the best action a_{T-3}^* follows from the maximum *PI*. Having taken this first step along the optimal trajectory we then await arrival of the next plant measurement of resulting state s_{t+1} before repeating the cycle. In this mode the proposed strategy echoes the model predictive control approach which has proved so successful in continuous process control applications. Once having completed the new production batch the data collected are added to the accumulated data set as a basis for updating the predictive models and retraining the value function, as previously described in Sec. 5.

During value function learning, the predictive models are instrumental in providing a base for artificially augmenting the amount of batch-to-batch data by means of forward simulations. Backpropagation of the corresponding Bellman Errors provides corrections to the fitting weights w. Using wire fitting, the best control action is found from the m support wires in constant time after only a few evaluations of the value function (e.g. for the case in Fig. 6 the optimum is one amongst only 27 outcomes). Moreover, wire fitting of the value function provides an optimisation framework that can respond quickly to process changes and unmeasured disturbances.

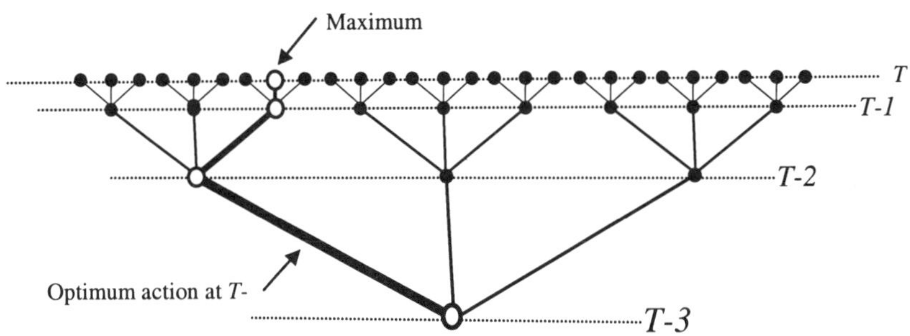

Figure 6. A forward sweep across the last three decision steps to batch completion (each node contains the neural network and predictive model as shown in Fig. 5 and the optimum action and value follow by backing up from the maximum *PI* amongst the 27 final values reached when using three support wires).

7. An Implementation Example

As an example of how the proposed approach can be applied, consider the case of a semi-batch reactor where the main product *B* is formed according to an autocatalytic reaction scheme, experiencing a slower irreversible decay. The exact kinetic mechanism is assumed unknown but for simulation purposes use is made of the following scheme.

$$A + 2\,B 6 3\,B, \qquad r_1 = k_1\,C_A\,(C_B)^2$$

$$B\ 6\ impurities, \qquad r_2 = k_2\,C_B \qquad (10)$$

For the purpose of control during a batch, only the concentration of *B* can be measured fast enough to be useful. The analysis for the accumulated concentration of impurities is both costly and time-consuming, so this will only be analysed for in the final product. Final product is either "on-spec" if less than 2% of *B* is lost to impurities (the process *goal*) or "off-spec" otherwise. A minimum conversion of 90% of the reactant *A* fed is expected within a 5 hour time scale. Thus, the *preference* is to achieve maximum possible conversion with a lower reaction time. To control the final level of impurities, both reactor temperature and feed flow rate can be altered or profiled during the batch. During each production batch three samples are taken to measure the concentration of *B* at intervals corresponding to $V=0.2V_f$, $V=0.4V_f$ and $V=0.6V_f$ (i.e. $n=3$). The analysis result from each sample is available after a delay of 30 minutes. Other relevant data for the example are given in Table 1.

Table 1. Data used in the semi-batch reactor case study.

Initial reactor charge:	Vol V=0.5m^3; C_A=1.92 kmol m^{-3}; C_B=0.55 kmol m^{-3};
Reactor feed:	C_A=1.42 kmol m^{-3}; C_B=0.75 kmol m^{-3};
Kinetic parameters:	k_1=10.5 exp(-985/(• +273) m^6 h^{-1} kmol^{-3}; k_2=2.1e15 exp(-13600/(• +273)) h^{-1}
Operating constraints:	Feed rate *F*# 1.5 m^3 h^{-1}; Temp • # 80°C; Volume V_f# 5 m^3

Thus the objective here can be stated as >to produce a product within specification preferably in less than 5 hours and with a conversion above 90% of all reactant fed=. If the goal is achieved, the preference index *PI* is defined to have 3 units for each additional percent conversion obtained over 90% plus 1 unit for each hour reduction within the maximum reaction time. For example, if an on-spec product is obtained in 3.3 hours with 91.5% conversion then *PI*=6.2.

The predictive, local model at each of the three sample periods was taken as linear. A preliminary set of 6 batches was run to provide data for first setting up the predictive models and then training the value function approximation. The reinforcement learning strategy already described was then applied to a sequence of simulated batches. The performance of the learned optimisation strategy as it evolved can be compared with that obtained independently using a Aperfect≡ model with known kinetic parameters. The results obtained are summarised in Figs. 7 to 9. Figure 7 shows the time profile of process state under the optimum control policy learned for F and •, which is itself presented in Fig. 8. Figure 9 shows the incremental performance shift as batch-to-batch data accumulate and quality of the local predictive models and the value function approximation improve. Initially the speed of improvement is slow, but as soon as a reasonable approximation to the predictive models is obtained the improvement rate increases dramatically.

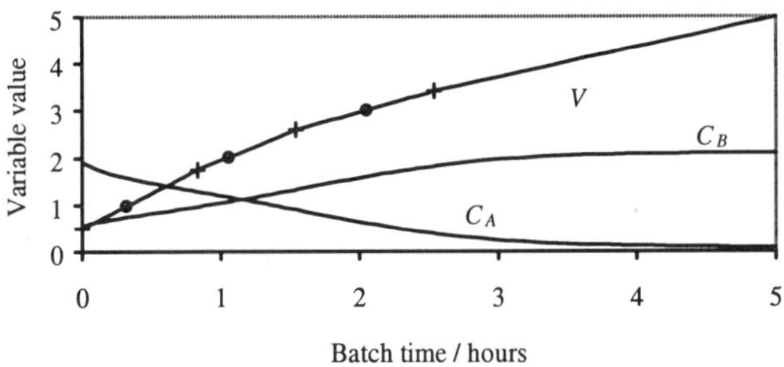

Figure 7. Batch reactor case study: Variable profiles under the optimum control policy learned (o=sample taken, +=control action taken based on analysis result)

8. Closing Remarks

In the batch processing context we address here, there is a strong pressure to work with scarce plant data (i.e. to learn quickly from very few production batches). Under the strategy we have presented our experience is that the value function can be learned quickly, provided good predictive models are available. The speed of convergence is heavily linked to the model fidelity. When working from very sparse experimental data there is a strong incentive to improve the predictive model quality by introducing enhancements based on any information available about the process behaviour. Rigorous first principles models are rarely available but there is nearly always some knowledge, perhaps from process research or development, which could be of use. How to efficiently encapsulate available process knowledge, both qualitative and quantitative, into suitable predictive model forms is a central topic of on-going research.

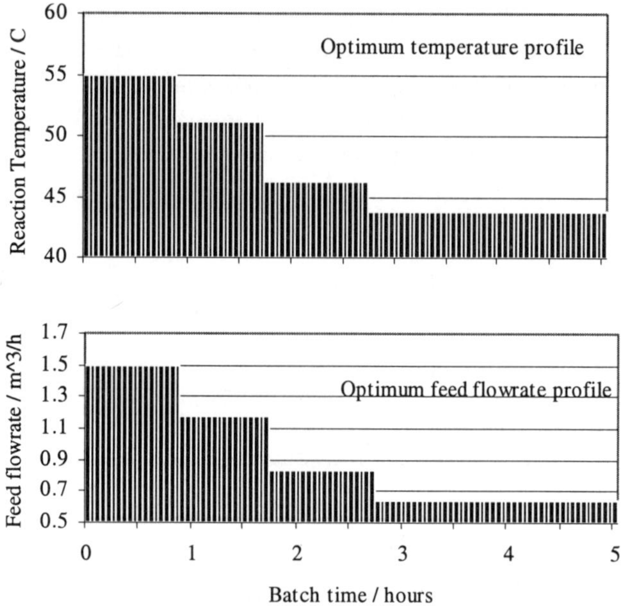

Figure 8. Batch reactor case study: Optimal profiling policy learned for temperature and feed flowrate.

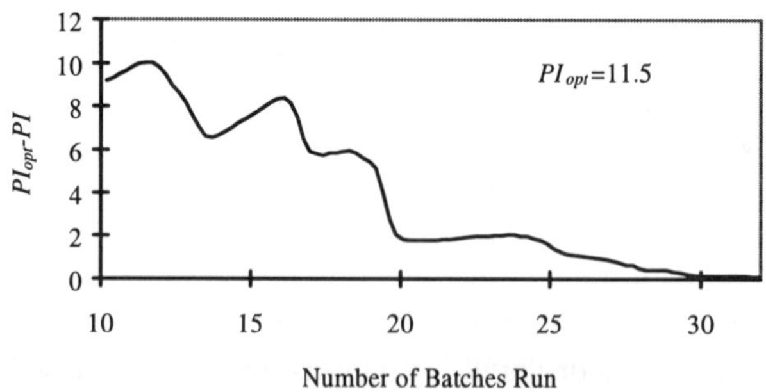

Figure 9. Batch reactor case study: Convergence towards the optimum performance index PI_{opt} for the strategy learned

9. Conclusion

An incremental learning approach, based on reinforcement learning, has been presented as a novel methodology capable of automatic optimisation of a batch process in the face of information uncertainty and modelling imperfections. Improved operation is achieved through a value function that is incrementally learned using wire fitting, with an embedded neural network, and Bellman=s Error backpropagation. Location of optimum control actions is greatly facilitated as an important by-product of the wire fitting technique and the use of on-line fitted predictive models is shown to be a promising way to build upon observed batch-to-batch data to enhance the ability to learn from scarce information. Implementing the strategy in a production environment involves use of the value function/predictive model combination in a scheme which echoes the successful Model Predictive Control strategy for continuous processes. However, convergence towards the optimum batch operating policy is linked closely to the fidelity and >quality of fit= of the predictive models in use and this is a key area in further development of the approach.

Nomenclature

a_t	control action taken at time t during the batch cycle
a_t^*	optimum control action taken at time t during the batch cycle
a_i^+	action parameter in wire fitted function approximation
ANN	artificial neural network
C	component concentration (kmol m^{-3})
E_B	error between observed and fitted performance (the Bellman error)
F	flowrate of reactant into the semi-batch reactor (m^3 h^{-1})
k	reaction rate constant
m	number of support wires used in •($s,a\|w$) the wire-fitted approximation to the value function
M_t	predictive model for state transition s_t to s_{t+1}
n	number of decision stages (samples) during a batch cycle
PI	preference index for a complete batch
Q	control policy
r	specific reaction rate (kmol m^{-3} h^{-1})
s_t	state of the process at time t during a batch cycle

t	time during the batch cycle when the state of the process is measured
T	terminal time for a batch
V_f	maximum volume of liquid in the batch reactor (m^3)
V	volume of liquid (reaction mixture) in the batch reactor (m^3)
w	weights in the neural network representation of •(s,a), the value function
•	neural network learning rate
•(s_t,a_t)	the value function (value of control action a_t in state s_t during the batch cycle)
•($s_t,a\|w$)	a neural network based approximation to the value function
•	set of feasible control actions
•	temperature of reactor contents (C)

References

Baird, L. C. and Klopf, A. H., *Technical Report WL-TR-93-1147* (Wright Laboratory, Wright-Patterson Air Force Base, OH 45433-7301, 1993).

Bertsekas, D., *Dynamic programming and optimal control.* **I and II** (Athena-Scientific, Belmon, MA, 1995).

Bertsekas, D. And Tsitsiklis, J., *Neuro-dynamic programming* (Athena-Scientific, Belmon, MA, 1996).

Martinez, E. C. And Wilson, J. A., *Comput. Chem. Engng.*, **22** (1998), S893-S896.

Sutton, R. S. and Barto, A. G., *An Introduction to reinforcement learning* (MIT Press, Boston, MA, 1997).

Terwiesch, P. , Agarwal, M. and Rippin, D. W., *J. Proc. Cont.* **4** (1994), 238-259.

Wilson, J. A. and Martinez, E. C., *Comput. Chem. Engng.* **21** (1997), S1233-S1238.

Acknowledgement

The authors gratefully acknowledge the support of EPSRC in conducting the work reported here under Visiting Fellowship Research Grant No. GR/K88132.

13. KNOWLEDGE DISCOVERY THROUGH MINING PROCESS OPERATIONAL DATA

X.Z. WANG
Department of Chemical Engineering, The University of Leeds, Leeds LS2 9JT, UK

In process plant operation and control, modern computer control and automatic data logging systems create large volumes of data, which contain valuable information about normal and abnormal operations, significant disturbances and changes in operational and control strategies. The data unquestionably provide a useful source of information for supervisors and engineers to monitor the performance of the plant and identify opportunities for improvement and causes of poor performance. This contribution describes the use of data mining and knowledge discovery techniques for automatic analysis and interpretation of process operational data both in real time and over the operating history. Techniques studied include data pre-processing using wavelets and principal component analysis, multivariate statistical analysis, and unsupervised machine learning approaches as well as inductive learning for conceptual clustering. Examples and industrial case studies are used to illustrate these methods.

1. Introduction

Modern computer-based control systems are often designed with automatic data logging systems. Being able to collect and display to operators a large amount of information is regarded as one of the most important advances provided in distributed control (DCS) over earlier analogue and direct digital control systems. The data are used by plant operators and supervisors to develop an understanding of plant operations through interpretation and analysis. It is this understanding which can then be used to identify problems in current operations and find better operational regions which result in improved products or in operating efficiency.

It has long been recognised that the information collected by DCS systems tends to overwhelm operators and so makes it difficult to take quick and correct decisions, especially on critical occasions. For example, olefin plants typically have more than 5000 measurements to be monitored, with up to 600 trend diagrams[23]. Clearly there is a need to develop methodologies and tools to automate data interpretation and analysis, and not simply rely on providing the operators with large volumes of multivariate data. The role of the acquisition system should be to provide the operators with information, knowledge, assessment of states of the plant and guidance in how to make adjustments. Operators are more concerned with the current status of the process and possible future behaviour rather than the current values of individual variables.

Process monitoring tends to be conducted at two levels. Apart from immediate safe operation of the plant, there is also the need to deal with the long term performance which has been the responsibility of supervisors and engineers. The databases created by automatic data logging provide potentially useful sources of insight for engineers and supervisors to identify causes of poor performance and opportunities for improvement. Despite a number of recent efforts to develop computer-aided technologies for analysing the operational data, including multivariate statistical analysis and inductive and analogical machine learning, such data sources have not been adequately exploited.

This contribution introduces development in automatic analysis and interpretation of process operational data both in real-time and over the operational history, and describes new concepts and methodologies for developing intelligent, state-space-based systems for process monitoring, control and diagnosis. It is now possible to exploit data mining and knowledge discovery technologies to the analysis, representation, and feature extraction of real-time and historical operational data to give deeper insight into systems' behaviour. The emphasis is on addressing challenges facing interpretation of process plant operational data, including the multivariate dependencies which determine process dynamics, noise and uncertainty, diversity of data types, changing conditions, unknown but feasible conditions, undetected sensor failures and uncalibrated and misplaced sensors, without being overwhelmed by the volume of data.

2. Data Mining and Knowledge Discovery in Databases

The emerging of data mining (DM) and knowledge discovery in databases (KDD) as a new technology is due to the fast development and wide application of information and database technologies. With the increasing use of databases the need to be able to digest large volumes of data being generated is now critical. It is accepted that database technology has been successful in recording and managing data but failed in the sense of moving from data processing to making it a key strategic weapon for enhancing business competition. The large volume and high dimensionality of databases leads to the breakdown of traditional human analysis. DM and KDD are aimed at developing methodologies and tools to automate the data analysis process and create useful information and knowledge from data to help in decision-making. It is a non-trivial process of identifying valid, novel, potentially useful, and ultimately understandable patterns in data[8]. It draws upon methods, algorithms and technologies from these diverse fields, and the unifying goal is extracting knowledge from data. DM and KDD methods and tools can be categorised in different ways. According to application purposes, they can be divided into pattern discovery and cluster analysis, regression, dependency modelling, sequence analysis, link analysis and trend prediction. DM and KDD are complex procedures involving a number of steps as shown in Fig. 1.

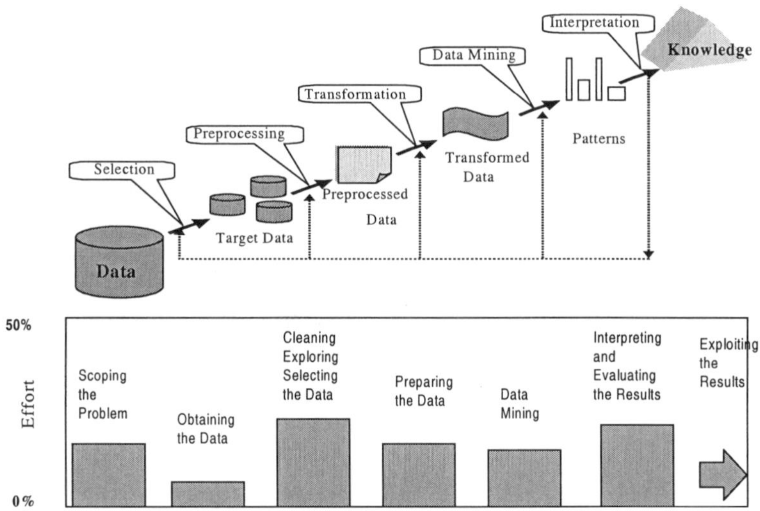

Figure 1. An overview of the steps comprising the DM and KDD processes

2.1. Characteristics of Process Operational Data

The major challenge in applying DM and KD techniques to process operational data analysis arises from the characteristics of the data, which are summarised as follows:[22]

- Large volume. A DCS automatic data logging system continuously stores data. The large volume makes manual probing almost impossible. Large volumes of data also demand large computer memory and high speed.

- High dimensionality. The behaviour of a process is usually defined by a large number of correlated variables. As a result it is difficult to visualise the behaviour without dimension reduction.
- Process uncertainty and noise. Uncertainty and noise emphases the need for good data pre-processing techniques.
- Dynamics. In operational status identification, it is very important to take account of the dynamic trends. In other words, the values of variables are dynamic trends. Many data mining and knowledge discovery tools, such as the well-known inductive machine learning system C5.0[17,18,19], are mainly designed to handle categorical values such as a colour being red or green. They are not effective in dealing with continuous-valued variables. These tools are not able to handle variables that take values as dynamic trends.
- Difference in the sampling time of variables. On-line measurements and laboratory analyses have variable sampling periods.
- Incomplete data. Some important data may not be recorded.
- Small and stale data. Sometimes, data analysis is used to identify abnormal operations. The data corresponding to abnormal operations might be buried in a huge database. Some tools are not effective in identifying small patterns in a large database.
- Complex interactions between process variables. Many techniques require that attributes be independent. However, many process variables are interrelated.
- Redundant measurements. Sometimes several sensors are used to measure the same variable, which gives rise to redundant measurements.

Current methods only address some of these issues, certainly not all, and the following observations can be made:

(1). Data pre-processing is critical for various reasons including noise removal, data reconciliation, dimension reduction and concept formation.

(2). Effective integration of the tools is needed. It means combining various tools for data preparation for other tools or for validation.

(3). Validation of discoveries from the data and presentation of the result is essential. Many times, because of lack of knowledge about the data, interpretation becomes a major issue.

(4). Windowing and sampling from a large database for analysis. This is necessary particularly for analysis of historical operational data.

3. Integrated Data Mining System

Sometimes, it is clear what we would like to discover from the data, but some other times we are not sure what we want to find out though we might expect useful information in the data. The integrated data mining prototype is designed to be able to provide some basic functions and is flexible enough to be tailored for other special subjective mining purposes. The basic functions include:

- Pattern discovery. Grouping data records into clusters and then analysing the similarities of data within a cluster and dissimilarities between clusters is a useful way starting the analysis. The most obvious application is abnormal operation identification as well as new operational states identification.
- Trend and deviation analysis. There have been various technologies for trend and deviation analysis including statistics, calculation of mean and standard deviation, as well as drawing.
- Link and dependency analysis. The linkage and dependency between variables, variables and performance metrics are important for understanding the process behaviour and improving performance. Some existing tools such as C5.0 as well as many graphical tools can not be directly used due to the real-valued dynamic trends as well as interactions between variables.
- Summarising. Summarising provides a compact description of a subset of data, for example, the mean and standard deviation of all fields. More sophisticated tools involve summary rules, multivariate visualisation techniques, and functional relationships between variables.
- Sequence analysis. Sequence analysis models sequential patterns (e.g., in data with time dependence, such as time series analysis). The goal is to model the states of the process, generating the sequence or to extract report deviations and trends over time. A typical application area is in batch process operations.
- Regression for predictive model development.

It is important to notice that one of the main features of DM and KDD is that they promise to discover novel and previous unknown knowledge in data. It is important to develop the system with great flexibility so that it can be tailored to any specific purpose-oriented systems. Fig. 2 illustrates the components involved in the integrated prototype system.

3.1. Data Pre-processing

Process data often contains noise and erroneous components and has missing values. There is also the possibility that redundant or irrelevant variables are recorded, while important features are missing. Data pre-processing includes provision for correcting inaccuracies, removing anomalies and eliminating duplicate records, and filling holes in the data and checking entries for consistency. It also requires making the necessary transformation of the original to put it in the format suitable for data mining tools.

The other important requirement with KDD process is feature selection. KDD is a complicated task and often depends on the proper selection of features. Feature selection is the process of choosing features which are necessary and sufficient to represent the data. There are several issues influencing feature selection, such as, masking variables, the number of variables employed in the analysis and relevancy of the variables[9].

Masking variables hide or disguise patterns in data. Numerous studies have shown that inclusion of irrelevant variables can hide real clustering of the data so only those variables which help discriminate the clustering should be included in the analysis[9].

The number of variables used in data mining is also an important consideration. There is generally a tendency to use many variables. However, increased dimensionality has an adverse effect because, for a fixed number of data patterns, increased dimensionality makes the multidimensional data space sparse. However, failing to include relevant variables also causes failure in identifying the clusters. A practical difficulty in mining some industrial data is to know if all important variables have been included in the data records.

Prior knowledge should be used if it is available. Otherwise, mathematical approaches need to be employed. Feature extraction shares many approaches with data mining. For example, principal component analysis (PCA), which is a useful tool in data mining, is also very useful for reducing dimensions. However, PCA is only suitable for dealing with real-valued attributes. Mining of association rules is also an effective approach in identifying the links between variables which take only categorical values. Sensitivity studies using feedforward neural networks (FFNNs) are also an effective way of identifying important and less important variables.

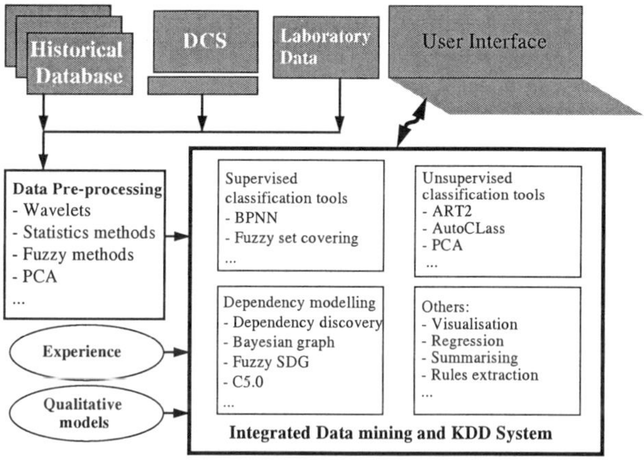

Figure 2. The integrated data mining system

3.2. DM and KDD Tools

Figure 2 shows the tools included in the prototype system. The various tools are loosely integrated in a way that they can be used independently or co-operatively. There is an unified interface for managing the data. Though the efficiency is lower compared with a fully integrated system, it provides very high flexibility for users and for future development.

The four parts in the block "Integrated Data Mining System" in Fig. 2 just mean functional grouping. In practice, individual tools are independent and loosely integrated together. The tools are categorised as follows.

- Supervised classification refers to tools that can learn from data cases with known classification to predict the assignments of new data cases. It is therefore a kind of technology that learns from the known to predict the unknown.
- Unsupervised classification is a technology that can automatically or semi-automatically group a set of unclassified data cases into clusters in a way that cases within a cluster are similar according to certain measures, and are unlike those in a different cluster. So unsupervised tools can learn from the unknown. Normally supervised classification gives more accurate predictions. An obvious advantage of tools integration is that unsupervised tools can be used to first classify the data prior supervised tools to using. Another advantage comes with a property of clustering approaches. Clustering approaches may give different

classification schemes if they start from different initial states. Different classification tools can provide cross-validation of discoveries.

- Clustering can also be divided into similarity- (or distance) based and conceptual clustering tools. The majority of methods studied for process operational state identification belong to the former. Similarity-based approaches though give predictions of states, they do not provide causal and qualitative explanations. Conceptual clustering, on the other hand, is able to give both predictions and a language describing the causal knowledge for the predictions.
- Graphical models can transform a complex problem into an easy understandable form so can be used for representation of discovered knowledge. Dependency discovery or link analysis tools are used to identify variables responsible for observed operational states, as well as links between variables.
- Other tools such as automatic extraction of knowledge in the form of rules.

4. Signal Pre-processing for Feature Extraction, Dimension Reduction and Concept Extraction

Data pre-processing is used to

(1) Filter out the noise components otherwise this may result in wrong conclusions being reached from the data.

(2) Extract features, reduce the dimensionality of the original signal and retain as much relevant information as possible. The main reasons for feature extraction are, first of all to minimise the dependencies between attributes, and secondly to reduce dimensionality.

(3) Deal with the problem of variable sampling periods for data, such as on-line real-time signals and laboratory analytical data.

(4) Develop concept formation because some data mining and KDD tools have been developed only for dealing with discrete-valued attributes and are not effective in dealing with continuous-valued variables. It is not possible to use variables represented by a trend without pre-processing the data.

It is worth noting that data pre-processing has many features in common with data mining, such as principal component analysis, and supervised and unsupervised classification using statistical and neural network algorithms. The following discussion focuses on pre-processing of dynamic trend signals.

4.1. Use of Principal Component Analysis

The method of principal component analysis (PCA) was originally developed in the 1900's[10,16], and has now re-emerged as an important technique in data analysis[12]. The central idea is to reduce the dimensionality of a data set consisting of a large number of interrelated variables, while retaining as much as possible of the variation present in the data set. Multiple regression and discrimination analysis use variable selection procedures to reduce the dimension but result in the loss of one or more important dimensions. The PCA approach uses all of the original variables to obtain a smaller set of new variables (principal components - PCs) that they can be used to approximate the original variables. The greater the degree of correlation between the original variables, the fewer the number of new variables required. PCs are uncorrelated and are ordered so that the first few retain most of the variation present in the original set. PCA has mainly been used as a clustering tool to identify deviation of process operation from normal state and in developing multivariate monitoring systems. In this section, PCA is used to extract features from dynamic trends.

In computer control systems such as DCS, nearly all important process variables are recorded as dynamic trends. Dynamic trends can be more important than the actual real time values in evaluating the current operational status of the process and in anticipating possible future developments. Figure 3 shows the trends of a variable under different operating conditions. The eigenvalues of the first 20 principal components are summarised in Fig. 4. It is apparent that the eigenvalues of the first few principal components can be used as a concise representation of the original dynamic trend.

Since the first two principal components can capture the main feature of a dynamic trend, this can be displayed graphically by plotting the eigenvalues on a two-dimensional plane. Figure 5 shows such a plot of the eigenvalues of the first two principal components of a variable.

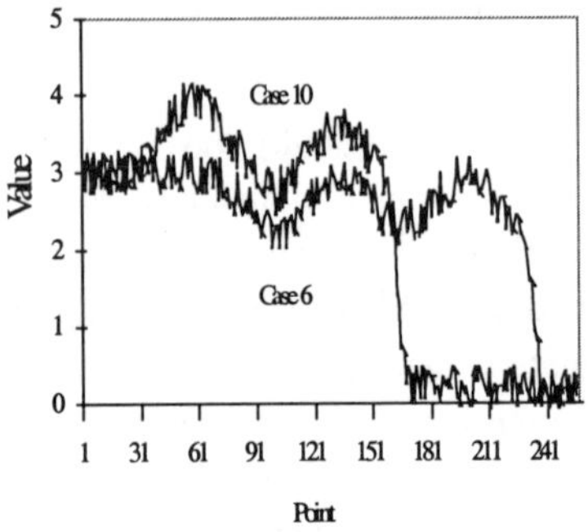

Figure 3. The dynamic trends of a variable

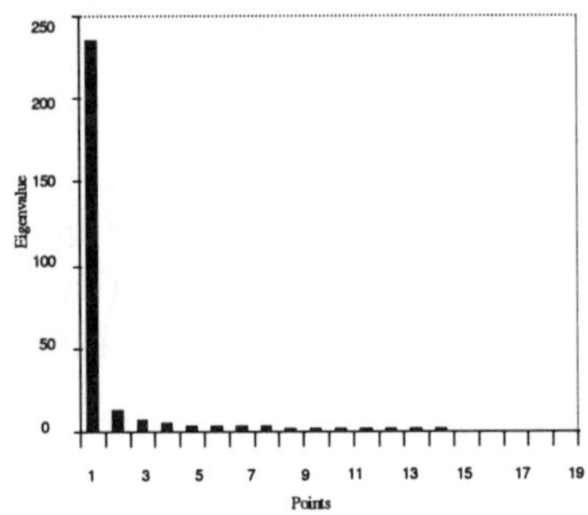

Figure 4. The first 20 eigenvalues

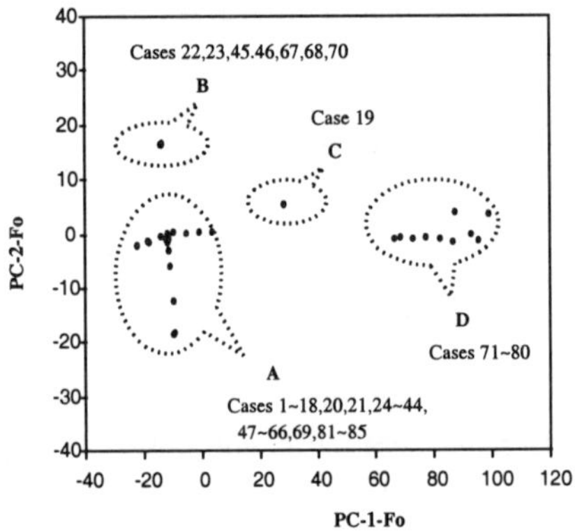

Figure 5 The PCA two dimensional plane of a variable Fo

The fact that a two-dimensional plot is able to capture the features can be seen from Figs. 6 and 7. Figure 6 shows the dynamic responses of the variable T_MTBE for seven data cases. After being processed using PCA (actually the seven data cases are processed using PCA together with another 93 data cases, but here only the seven are shown for illustrative purpose), the results are shown on the two-dimensional PCA plane in Fig. 7. It is clear that the dynamic trends of data cases 1 and 2 are more alike than with the others in Fig. 6 and they are grouped closer in Fig. 7. Similar observations can be made for data cases 40 and 80, as well as 14 and 15.

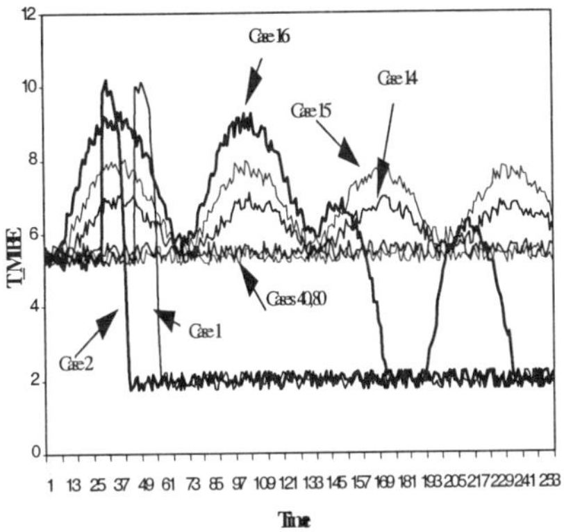

Figure 6. The dynamic trends of T_MTBE

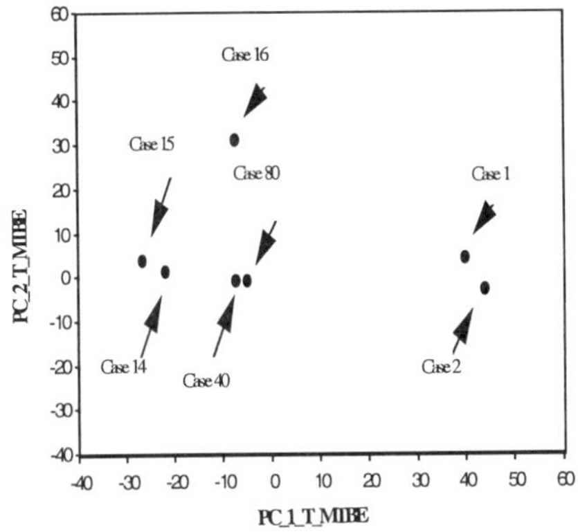

Figure 7. Projection of Fig. 6 on the two-dimensional PCA plane.

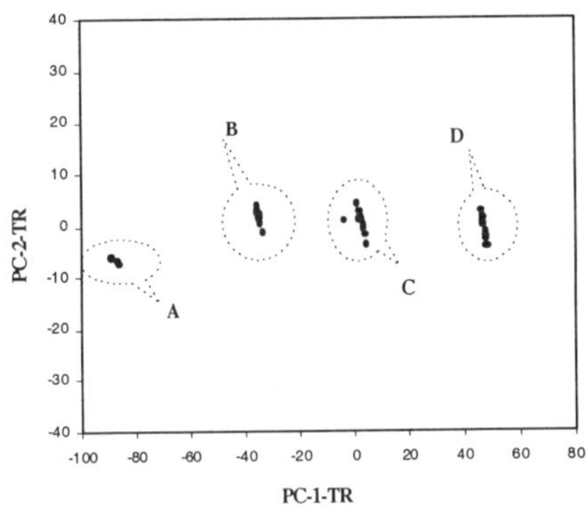

Figure 8. PCA plane of the variable TR

The system is able to give a conceptual clustering language as production rules,

IF	TR is in region C of Fig. 8
AND	Fo is in region D of Fig. 5
THEN	The operation will be in region ABN-1 of Fig. 9.

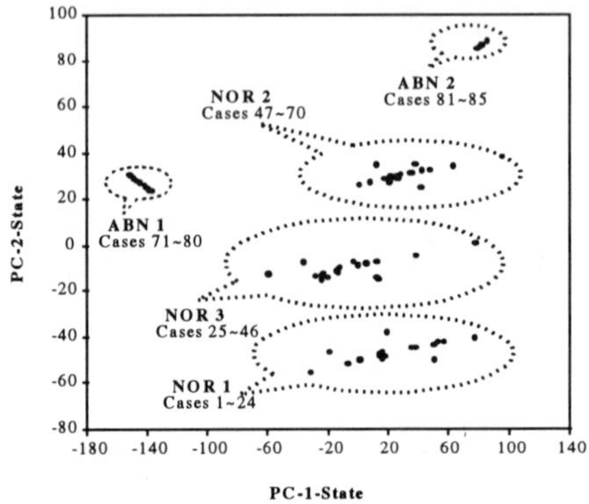

Figure 9. PCA plane of operational states

4.2. Signal Feature Extraction Using Wavelets

Signal feature extraction using wavelets is based on the fact that irregularities and singularities contain the most important information of trend signals. Since the extrema of wavelet transform of signals are able to capture all the irregularities and singularities of a signal when a filter bank and wavelet function are selected properly, they are regarded as the features of the trend. Mathematically, the local singularity of a function is measured by Lipschitz exponents[14]. Mallat and Hwang[14] proved that the local maxima of the wavelet transform modulus detect the locations of irregular structures and provided numerical procedures for computing the Lipschitz exponents. Within the framework of scale-space filtering, inflexion points of f(t) appear as extrema for $\partial f(t) / \partial t$ and zero crossing for $\partial^2 f(t) / \partial t^2$, so Mallat and Zhong[15] suggested using a wavelet which is the first derivative of a scaling function $\Phi(t)$,

$$\psi(t) = \frac{d\phi(t)}{dt}$$

with a cubic spline being used for the scaling function.

The wavelet modulus maxima and zero-crossing representations were developed from underlying continuous-time theory. For computer implementation, this has to be cast in the discrete - time domain. Berman and Baras[2] proved that wavelet transform extrema / zero-crossing provide stable representations of finite length discrete-time signals. Cvetkovic and Vetterli[7] have developed a more complete discrete-time framework for the representation of the wavelet transform. They designed a non-subsampled multi-resolution analysis filter bank to implement the wavelet transform for the representation. Using this filter bank, the wavelet function can be selected from a wider range than the B-spline in Mallat's method.

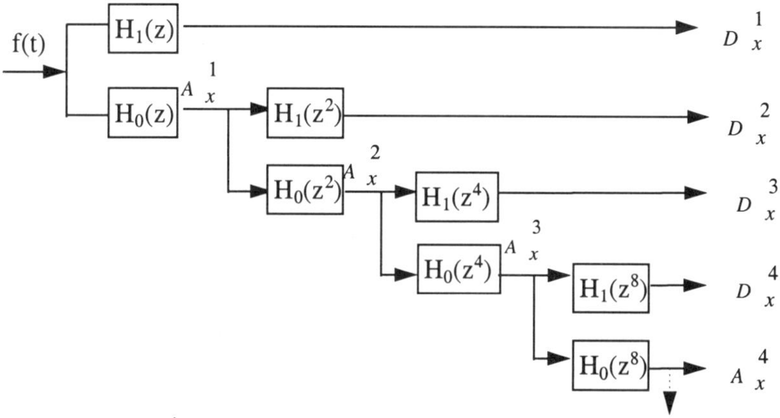

D_x^i -- detail on the ith decomposition

A_x^i -- approximation on ith decomposition

H_0, H_1 -- low-pass and high-pass filters

Figure 10. An octave band non-subsampled filter bank.

Non-subsampled multi-resolution analysis can then be used to detect singularities of a signal. An octave band non-subsampled filter bank with analysis filters $H_0(z)$ and $H_1(z)$ is shown in Fig. 10. In this method, a wavelet transform is defined in terms of the bounded linear operators $W_j: l^2(Z) \rightarrow l^2(Z)$, j = 1, 2, ..J+1. The operators W_j are the convolution operators with the impulse responses of the filters:

$$V_1(z) = H_1(z)$$
$$V_2(z) = H_0(z)H_1(z^2)$$
$$\bullet\bullet\bullet\bullet\bullet\bullet\bullet\bullet\bullet$$
$$V_J(z) = H_0(z)\cdots H_0(z^{2^{J-2}})H_1(z^{2^{J-1}})$$
$$V_{J+1}(z) = H_0(z)\cdots H_0(z^{2^{J-2}})H_0(z^{2^{J-1}})$$

The multi-resolution procedure depicted in Fig. 10 can be described less rigorously. Figure 10 shows four steps, or four scales analysis. In the first step, the original signal is split into *approximation* A_x^1 and *detail* D_x^1. The *detail* D_x^1 is assumed to be mainly the noise components of the original signal and the *approximation* A_x^1 represents mainly the trend of the original signal. A_x^1 is further decomposed into *approximation* A_x^2 and *detail* D_x^2, A_x^2 to A_x^3 and D_x^3, and A_x^3 to A_x^4 and D_x^4. In each step we find the extrema of the *detail*. In the first few steps, the extrema are due to both the noise and the trend of the noise-free signal. As the scale increases, the noise extrema are gradually removed while the extrema of the noise-free signal remain. In this way, using multi-scale analysis and extrema determination, the extrema of the noise-free signal can be found, which represent the features of the signal. Multi-resolution analysis of an example of a signal with noise components is shown in Fig. 11.

The wavelet approach for signal feature extraction has a number of advantages. Firstly, the extrema of wavelet multi-scale analysis can completely capture the distinguished points of a trend signal, because the original signals can be reconstructed. Secondly, the method is robust in the sense that the features captured do not change with the change of scales of analysis. Thirdly, the episode representation of a trend is primitive, there are no *a priori* measurements of compactness for the representation of extrema of wavelet multi-scale decomposition. In addition, a wavelet-based noise component removal procedure has been included so that noise effects can be filtered out.

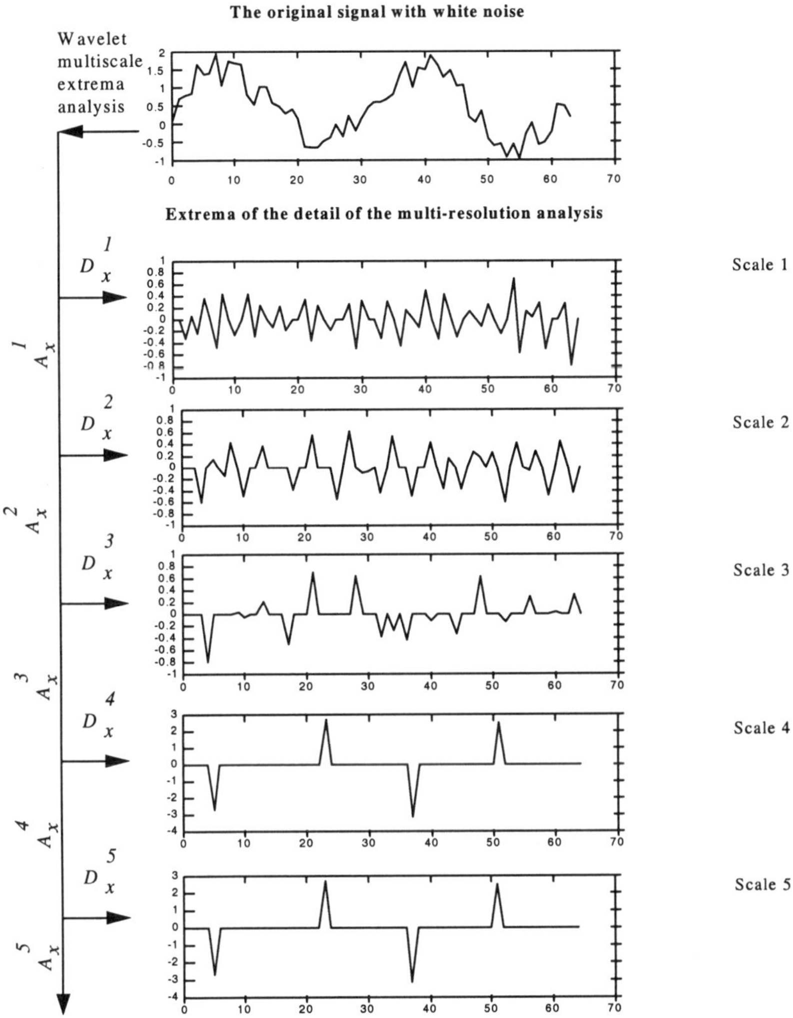

Figure 11. Noise signal and its multi-resolution analysis.
(A_x^i - approximation of multi-resolution analysis, D_x^i - detail.)

5. Multivariate Statistical Analysis for Operational Data Analysis

Multivariate statistics have recently been widely studied for designing multivariate statistical monitoring and control systems[12,11]. In this section we give an industrial example to demonstrate how multivariate data analysis can be used to get insight into past operational records.

5.1. The FCC Main Fractionator and Product Quality

The fluid catalytic cracking process (FCC) of a refinery converts a mixture of heavy oils into more valuable products. The relevant section of the process is shown in Fig. 12, where the oil gas mixture leaving the reactor goes into the main fractionator to be separated into various products. The individual side draw products are further processed by down-stream units before being sent to blending units.

One of the products is light diesel whose quality is typically characterised by the temperature of condensation. Traditionally the temperature of condensation has been monitored by off-line laboratory analysis, which caused time delays because the interval between two samples is between four to six hours. As a result a software sensor has been developed using 303 data patterns spanning over nearly a year for predicting the condensation point using fourteen process variables which are measured on-line. The fourteen variables are listed in Table 1.

An interesting problem with the process is that it is required to produce three product grades according to seasons and market demand, namely -10#, 0# and 5# defined by the ranges of condensation temperature. Because there are more than one process variable the operators use their experience through trial-and-error to adjust process variables to move the operation from producing one product grade to another. There is a clear need to minimise the time of change over because off-specification product may be produced during transition.

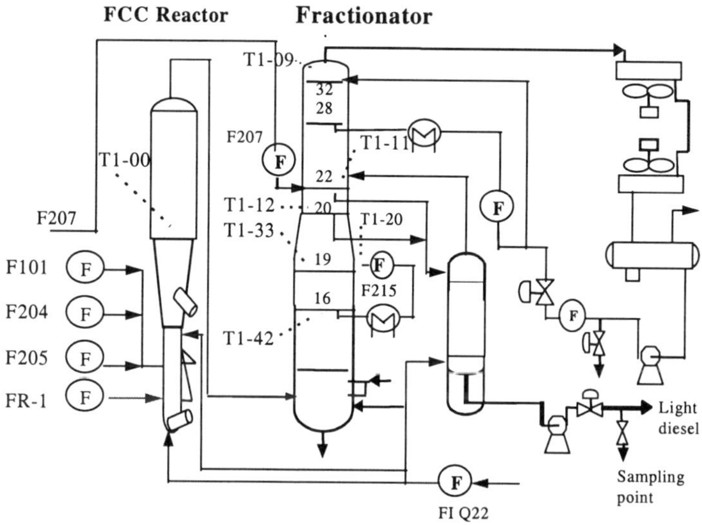

Figure 12. The main fractionator of the FCC process.

Table 1. The fourteen variables used as input to the FFNN model.

T1-11	- the temperature on tray 22 where the light diesel is withdrawn
T1-12	- the temperature on tray 20 where the light diesel is withdrawn
T1-33	- the temperature on tray 19
T1-42	- the temperature on tray 16
T1-20	- the return temperature of the pumparound
F215	- the flowrate of the pumparound
T1-09	- column top temperature
T1-00	- reaction temperature
F205	- fresh feed flowrate to the reactor
F204	- flowrate of the recycle oil
F101	- steam flowrate
FR-1	- steam flowrate
FIQ22	- flowrate of the over-heated steam
F207	- flowrate of the rich-absorbent oil

5.2. Knowledge Discovery Using PCA

The difficulty of the problem comes from the fact that there are fourteen process variables to consider. Application of PCA to the database of the size 303×14 (number of data patterns×number of process variables) found that the first seven variables account for about 93% of the variance. The PC1 and PC2 two-dimensional plot is shown in Fig. 13. It was found that the 303 data patterns are grouped into four clusters. Three clusters correspond to three products -10#, 5# and 0# and the cluster at the bottom-right corner is found to be a cluster that has a high probability of product off-specification.

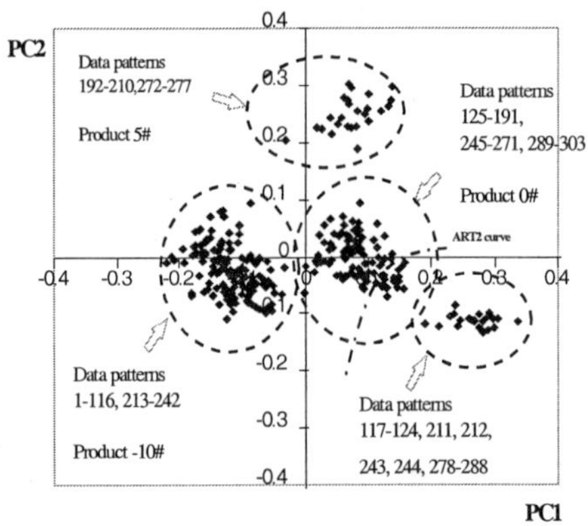

Figure 13. PC1 and PC2 plot

Therefore the strategy for operation and product design should be to operate the process in the region of the bottom-left if the desired product is -10#, or the region at the top if the desire product is 5#, or the region at the middle if the desired product is 0#, and try to avoid the region at the bottom-right corner. Another point is that to move from producing -10# to 0#, adjusting PC1 is more important than changing PC2. To switch from producing 0# to 5#, PC2 is more important than PC1. Both PC1 and PC2 are important in avoiding the region at the bottom-right corner which produces off-specification product.

However, PC1 and PC2 are latent variables. To link PC1 and PC2 to the original variables, contribution plots are used. The contribution plot of PC1 is shown in Fig. 14, from which it is found that the most important variables are TI12 (the temperature on tray 20 where the product is withdrawn) and TI42 (the temperature on tray 16 close to the flashing zone). Some other variables are not important such as FR-1. The above discovery is confirmed by looking at the change of TI-12 over the 303 data patterns (Fig. 15). It clearly shows that TI-12 can distinguish product -10# from 0# and 5#, but can not distinguish 0# and 5#.

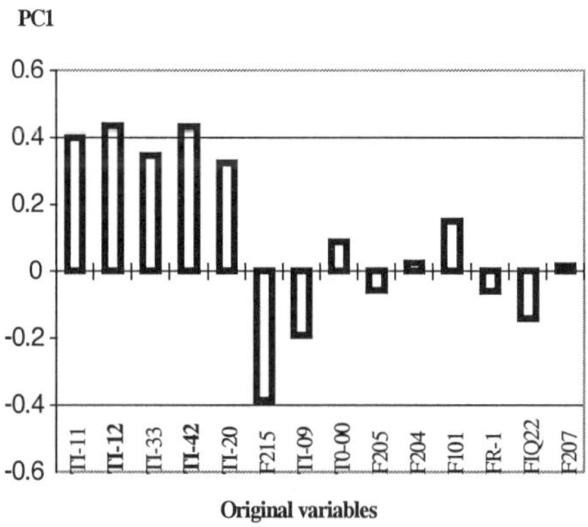

Figure 14. The contributing plot of PC1.

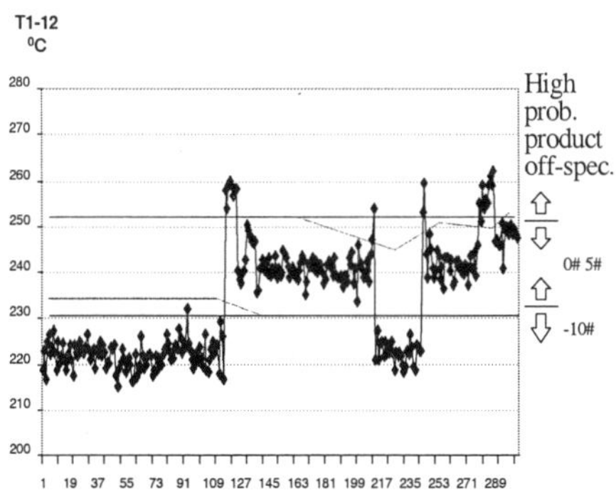

Figure 15. TI-12

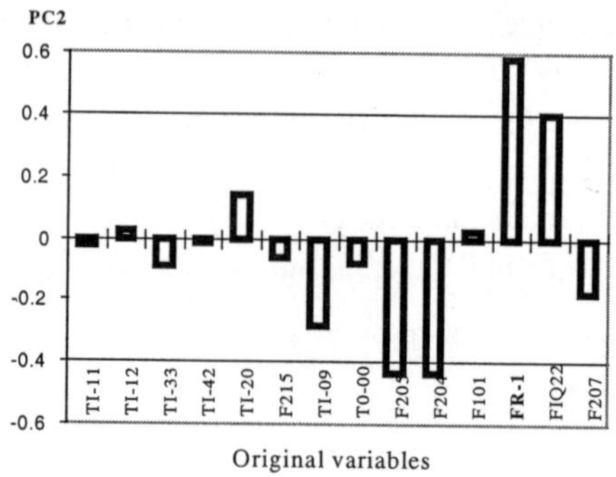

Figure 16. Contribution plot for PC2

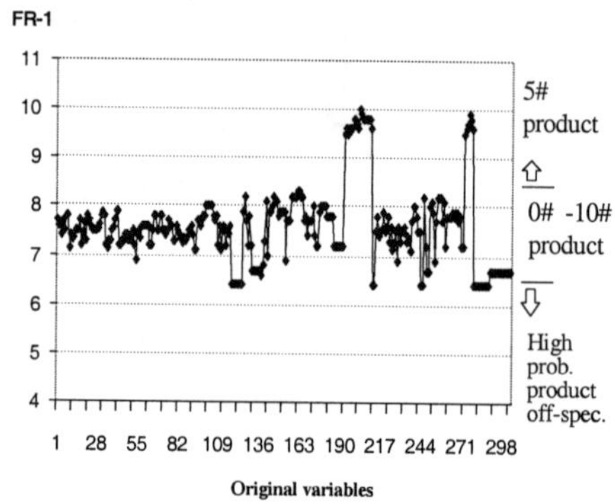

Figure 17. The changing profile of FR-1

The contribution plot of PC2 is shown in Fig. 16 which indicates that FR-1 is the most important variable. The changing profile of FR-1 for the 303 data patterns are shown in Fig. 17. It clearly shows that FR-1 can distinguish product 5# from 0# and -10#, but not 0# from -10#. The figure also confirms that FR-1 is not important to PC1.

Therefore the operational strategy for product design should be that if we want to change from producing -10# to 5#, we should increase TI-12 and TR-42 and then increase FR-1. In order to avoid off-specification product we should carefully monitor TI-12, TR-42 and FR-1 to avoid the region at the bottom-right corner. Of course it is important to be aware that fine-tuning of all the variables is necessary but this guidance can help operators to move the process from producing one product quickly to another product.

Close examination gives a more interesting discovery, that is, the region at the bottom-right corner of Fig. 13 - the region of high possibility of product off-specification is very likely caused during product change-over. For example, 117-124 at the bottom-right corner were due to transition from the region of -10# (1-166) to the region of product 0# (125-191). Other data cases in the bottom right corner can be explained similarly. 211-212 were due to transition from 5# region (192-210) to -10# region (213-242); 243-244 due to transition from -10# region (213-242) to 0# region (245-271); 278-288 due to transition from 5# (272-277) to 0# (289-303). This shows that some transitions took a long time. If the knowledge discovered had been known, together with an on-line sensor, the transition time could have been reduced.

5.3. General Observations

PCA and PLS have proved to be powerful tools for operational data analysis and statistical process control. However they still have limitations. PCA and PLS-based data analysis for statistical process control has the assumption that the first few PCs can capture most of the variations in a multivariate database. This assumption may be violated in some cases, e.g., when the dimensions of the original variables are very large. Multiblock PCA and PLS can tackle this problem for some applications, however, dividing variables into blocks may not always be possible. In such cases alternative approaches may have to be used such as the unsupervised machine learning approaches, including neural network and Bayesian automatic classification methods. However, PCA and PLS may still be a useful approach for pre-processing the data to eliminate the linear dependencies in the data.

The variable contributing plots may not be applicable in cases where the contributions of the original variables to the PCs are not equally distributed. Use of other approaches to compensate this limit of PCA can be a good alternative. For example, neural network models can be developed and used as sensitivity study tools to identify the contributions of variables.

In the above applications, PCA and PLS are used mainly for statistical process control for long term performance monitoring and the data dealt with are averaged over hours or days. PCA and PLS are also potentially useful for on-line real-time data analysis. As already discussed in section four, PCA is also useful for feature extraction and concept formation from dynamic trend signals. Bakshi[1] combined wavelet multiscale analysis with PCA for developing on-line monitoring systems.

PCA can also be categorised as an unsupervised learning approach. However its learning is not recursive or incremental. For on-line real time use, it is useful for PCA to be able to learn incrementally, i.e., learn from a single example when it is presented. There has been a report on such on-line learning for principle component analysis[3].

6. Operational State Identification using Unsupervised Methods

Data encountered can be broadly divided into the following four categories:
(1) Part of the database is known, i.e., the number and descriptions of classes as well as the assignments of individual data patterns are known. The task is to assign unknown data patterns to the established classes.
(2) Both the number and descriptions of classes are known, but the assignment of individual data patterns is not known. The task is then to assign all data patterns to the known classes.
(3) The number of classes are known but the descriptions and the assignment of individual data patterns are not known. The problem is to develop a description for each class and assign all data patterns to them.
(4) Both the number and descriptions of classes are not known and it is necessary to determine the number and descriptions of classes as well as the assignments of the data patterns.

For the first type of data where the objective is to assign new data patterns to previously established classes, supervised methods, such as feedforward neural networks can be used. Clearly supervised methods are not appropriate for the last three types of data, since training data are not available. In these cases unsupervised learning approaches are needed and the goal is to group data into clusters in a way such that intraclass similarity is high and interclass similarity is low. In other words, supervised approaches can learn from known to predict unknown while unsupervised approaches learn from unknown in order to predict unknown. Supervised learning can generally give more accurate predictions, but can not be extrapolated: when new data are not in the range of training data, predictions will not generally be reliable. For process operational state identification and diagnosis, supervised learning needs both symptoms and faults. Therefore the routine data collected by computer control systems can not be used directly for training. Faults are unlikely to be deliberately introduced to an industrial process in order to generate training data.

Grouping of data patterns using unsupervised learning is often based on a similarity or distance measure, which is then compared with a threshold value. The degree of autonomy will depend on whether the threshold value is given by the users or determined automatically by the system. In this section three representative approaches are studied, the adaptive resonance theory (ART2), a modified version of it named ARTnet, and Bayesian automatic classification (AutoClass). ART2 and ARTnet, though requiring a pre-defined threshold value, are able to deal with both the third and fourth types of data. AutoClass is a completely automatic clustering approach without the need to pre define a threshold value and the number and descriptions of classes, so is able to deal with the fourth type of data.

6.1. An Integrated Framework ARTnet and its Application

We have developed an integrated framework named ARTnet (Fig. 18) which combines wavelet for feature extraction from dynamic transient signals and adaptive resonance theory[4]. In ARTnet the data pre-processing part uses wavelets for pre-processing the data for feature extraction[6,21,22]. In order to introduce ARTnet it is helpful to first examine the mechanism of ART2 for noise removal. ART2 has a data pre-processing unit which is very complicated but the mechanism for removing noise uses a simple activation function A(x),

$$A(x) = \begin{cases} x & x > \theta \\ 0 & x < \theta \end{cases} \tag{1}$$

where θ is a threshold value. If an input signal is less than θ, it is considered to be a noise component and set to zero. This has proved to be inappropriate for removing noise components contained in process dynamic transient signals which are often of high frequencies and of certain magnitude.

In the ARTnet architecture, wavelets are used to pre-process the dynamic trend signals. The extracted features are used as inputs to the kernel of ARTnet for clustering. A pattern feature vector $(x_1, x_2, \cdots, x_N)$ is fed to the input layer of the ARTnet kernel and weighted by b_{ij}, bottom-up weights. The extrema of wavelet multiscale analysis should be regarded as the features of dynamic transient signals.

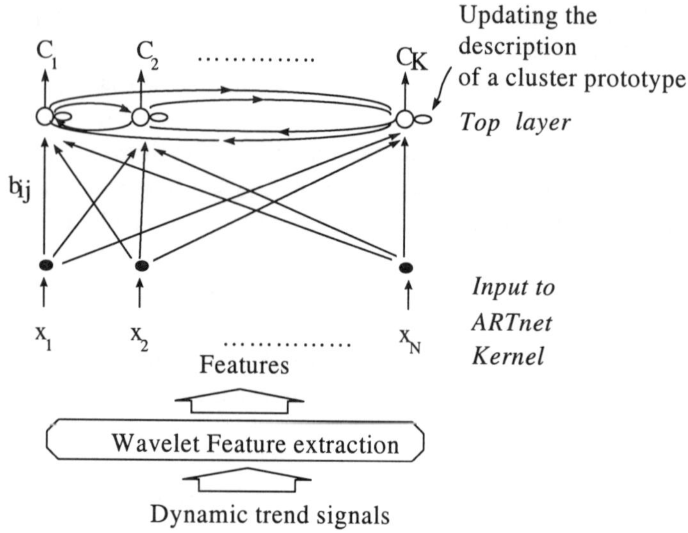

Figure 18. The conceptual architecture of ARTnet.

The weighted input vector is then compared with existing clusters in the top layer by calculating the distance between the input and existing clusters. The existing cluster prototype, which has a smaller distance than the input is called the winner. By considering this input the description or knowledge of the wining cluster is updated. Whether or not a winning cluster prototype is allowed to learn from an input data pattern depends on how similar the input is to the cluster. If the similarity measure exceeds a predetermined value, called the vigilance parameter, learning is enabled. If the similarity measure is less than the required vigilance parameter, a new cluster unit is then created which reflects the input. Clearly this is unsupervised and a recursive learning process.

It is apparent that the learning process is concerned with the extent to which how similar two vectors are. The Euclidean distance between two vectors **x** and **y** is defined as the root sum-squared error,

$$\|\boldsymbol{x}-\boldsymbol{y}\|_2 = \left\{\sum_{n=1}^{N}(x_n - y_n)^2\right\}^{1/2} \tag{2}$$

Suppose there are K existing cluster prototypes. The kth cluster prototype consists of a number of data patterns and is also described by a vector, denoted as $z^{(k)}$, which has considered all data patterns belonging to it. Clearly, if there is only one data pattern in the cluster, $z^{(k)}$, it is equal to that data pattern. When a new input data pattern **x** is received, a distance between **x** and $z^{(k)}$ is calculated according to the expression,

$$\sigma^2(k) = \sum_{(x \in Cluster\ k)} \left(\left\| x - z^{(k)} \right\|_2 \right)^2 \tag{3}$$

Since the distance between **x** and all existing cluster prototypes is calculated, the cluster prototype with the smallest distance is the winner. If the distance measure for the winner is smaller than a pre-set distance threshold, ρ, then the input **x** is assigned to the winning cluster and the description of the cluster is then updated,

$$z_i^{(k)} = z_i^{(k)} + \frac{1}{NF} x_i b_{ij} \qquad i = 1...N,\ j = 1...K \tag{4}$$

where $z_i^{(k)}$ refers to the ith attribute of the vector **z** for the cluster k. b_{ij} is the weight between the ith attribute of the input and the jth existing cluster prototype. NF is the number of features.

6.2. Application of ARTnet to the FCC Process

The FCC process shown in Fig. 19 has been described in detail by Wang[22] and Wang et al.[21]. To demonstrate the procedure, 64 data patterns are used which include the following faults or disturbances:

- fresh flow rate is increased or decreased
- preheat temperature for the mixed feed increases or decreases
- recycle slurry flow rate increases or decreases
- opening of the hand valve V20 increases or decreases
- air flow rate increases or decreases
- the opening of the fully open valve 401-ST decreases
- cooling water pump fails
- compressor fails
- double faults occur

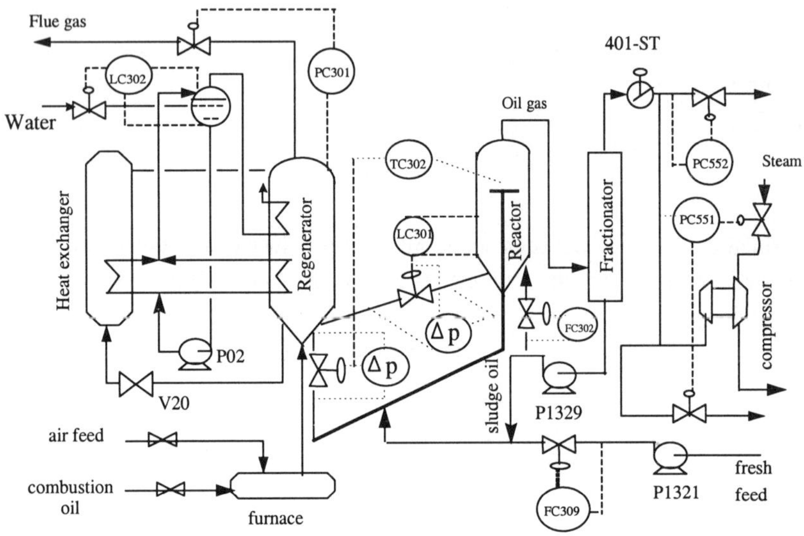

Figure 19. The simplified flowsheet of the R-FCC process.

The sixty four data patterns were obtained from a customised dynamic training simulator, to which random noise was added using a zero-mean noise generator (MATLAB®). In the following discussions, the term "data patterns" refers to these sixty four data patterns and "identified patterns" to the patterns estimated by ARTnet.

As stated previously, the extrema that are mostly influenced by noise fluctuations are those (1) where the amplitude decreases on average as the decomposition scale increases and (2) do not propagate to large scales. Using these criteria, noise extrema are removed.

It is important that a suitable threshold for pattern recognition is used when applying ARTnet. For a threshold $\rho = 0.8$, all 64 data patterns are identified as individual patterns. A more suitable threshold is obtained by analysing clustering results for increased threshold values.

Table 2. ATRnet identified clusters (when the distance threshold is 4.5) and the corresponding data patterns[a].

Identified clusters	Corresponding data patterns	Identified clusters	Corresponding data patterns	Identified clusters	Corresponding data patterns
1	1	**19**	32	**37**	51
2	2	**20**	33	**38**	52
3	[3 4 5 6 7 8 9]	**21**	34	**39**	53
4	10	**22**	[35 36]	**40**	54
5	11	**23**	37	**41**	55
6	12	**24**	38	**42**	[56 57]
7	13	**25**	39	**43**	58
8	14	**26**	40	**44**	59
9	15	**27**	41	**45**	60
10	16	**28**	42	**46**	61
11	17	**29**	43	**47**	62
12	18	**30**	44	**48**	63
13	[19 20 21 22 23 24]	**31**	45	**49**	64
14	[25 26]	**32**	46		
15	[27 28]	**33**	47		
16	29	**34**	48		
17	30	**35**	49		
18	31	**36**	50		

a - [3 4 5 6 7 8 9] means data patterns 3 to 9 are identified in the same cluster

When the threshold value is 4.5, the groupings are [3,4,5,6,7,8,9], [19 20 21 22 23 24], [25 26], [27, 28], [35,36] and [56 57]. The pairing of identified patterns and original data patterns are shown in Table 2. The clustering is justified by inspecting the results in detail.

However, any further increase in threshold is not useful because some data patterns that are significantly different are grouped in the same cluster. For instance, when the threshold value is 5, data pattern 29 (opening ratio of the hand-valve V20 increasing by 5%) is merged with the clusters representing increase and decrease in the preheat temperature of the mixed feed. Therefore, the threshold $\rho = 4.5$ is considered as the most appropriate value for this case.

6.3. Comparison Between ARTnet and ART2

It is apparent that the data pre-processing part of ARTnet is able to effectively reduce the dimension of the dynamic trend signals using wavelet feature extraction and piece-wise processing. ARTnet has also shown other advantages over ART2 in operational data analysis. These include the determination of threshold values, the ability to deal with noise and computational speed. In the comparison following only the first fifty-seven data patterns were used.

6.3.1. Threshold Determination

In this case, only 57 data patterns are used to compare the distance threshold for using ARTnet and the vigilance value in ART2 using noise-free data. For noise-free data, ARTnet and ART2 give the same results if the ARTnet distance threshold and the ART2 vigilance are appropriately adjusted, as shown in Table 3. From Table 3, for the same groupings, the ARTnet distance threshold changes from 0.8 to 4.5 while the vigilance of ART2 varies from 0.9998 down to 0.9985. So the distance threshold for ARTnet is less sensitive than the vigilance of ART2. The ART2 clustering is too sensitive to the vigilance value, making it difficult to set a value.

6.3.2. Robustness with Respect to Noise

The following demonstrates that ARTnet gives a consistent clustering result regardless of the magnitude of noise-to-signal ratio, providing it is in a reasonable range. ART2 gives fewer clusters at a low noise-to-signal ratio and more clusters at a larger ratio. 57 data patterns are considered with white noise added. A constant C_{noise} is introduced to control the magnitude of noise defined by

$$\text{The magnitude of noise} = \frac{\begin{array}{c}\text{The magnitude of noise from}\\ \text{the noise generator}\end{array}}{C_{noise}} \tag{5}$$

Table 3. Comparison of the value ranges of the distance threshold of ARTnet and the vigilance value of ART2, for the same grouping schemes[a,b,c].

ARTnet distance threshold	ART2 vigilance value	Grouping of data samples
0.8	0.9998	
1.0	0.9996	[56 57]
2.0	0.9992	[5 7] [25 26] [27 28] [56 57]
3.0	0.9990	[5 7] [19 20 23 24] [25 26] [27 28] [56 57]
4.0	0.9987	[5 6 7 8] [19 20 21 23 24] [25 26] [27 28] [56 57]
4.5	0.9985	[3 4 5 6 7 8 9] [19 20 21 22 23 24] [25 26] [27 28] [35 36] [56 57]

[a] [56 57] means that data patterns 56 and 57 are grouped in the same cluster, [b] Only the first 57 data patterns are considered and the data are noise-free, [c]The ARTnet distance threshold changes in a wider range while ART2 vigilance is too sensitive making it difficult to set a value.

In Eq. 5, C_{noise} changes ranging from 0.001 to 100 are examined: in what follows the smaller the C_{noise}, the larger the noise-to-signal ratio.

The best clustering results are obtained when the distance threshold of ARTnet is 4.5. This result is not affected by changing C_{noise} from 0.001 to 100, as can be seen in Table 4. For ART2, the best value of the vigilance is 0.9985 and C_{noise} = 100, and is the same result as ARTnet (Table 4). However, as C_{noise} decreases to 10, i.e., larger noise-to-signal ratio, ART2 splits the cluster [3 4 5 6 7 8 9] into two [3 4 5 6 7] and [8 9]. As C_{noise} decreases to 0.001, i.e., a much larger noise-to-signal ratio, there are further new groupings, [20 42] and [29 51]. The new groups can not be satisfactorily explained. Although the inappropriate groupings [20 42] and [29 51] can be avoided by changing the vigilance value, other unreasonable groupings are generated.

Table 4. Clusters predicted by ARTnet when the distance threshold is 4.5 and C_{noise} varies over a wide range, from 0.001 to 100[a].

Identified patterns	Corresponding data patterns	Identified patterns	Corresponding data patterns	Identified patterns	Corresponding data patterns
1	1	**15**	[27, 28]	**29**	43
2	2	**16**	29	**30**	44
3	[3 4 5 6 7 8 9]	**17**	30	**31**	45
4	10	**18**	31	**32**	46
5	11	**19**	32	**33**	47
6	12	**20**	33	**34**	48
7	13	**21**	34	**35**	49
8	14	**22**	[35 36]	**36**	50
9	15	**23**	37	**37**	51
10	16	**24**	38	**38**	52
11	17	**25**	39	**39**	53
12	18	**26**	40	**40**	54
13	[19 20 21 22 23 24]	**27**	41	**41**	55
14	[25 26]	**28**	42	**42**	[56 57]

[a][3 4 5 6 7 8 9] means that data patterns 3 to 9 are grouped in the same cluster.

6.3.3. Computational Speed

It is found that ARTnet is faster than ART2. After optimum values of the distance threshold of ARTnet and the vigilance of ART2 are found, for the same data, ARTnet is typically two times faster than ART2.

6.4. Bayesian Automatic Classification

Both ARTnet and ART2 require the user to give a threshold value (though ARTnet is much superior over ART2 in this aspect). A Bayesian method termed AutoClass which was developed by NASA[5] is described in this section. For a given number of data patterns (some times called cases, observations, samples, instances, objects or individuals), each of which is described by a set of attributes, AutoClass can devise an automatic procedure for grouping the data patterns into a number of classes such that instances within a class are similar, in some respect, but distinct from those in other classes. The approach has several advantages over other clustering methods.

- The number of classes is determined automatically. Deciding when to stop forming classes is a fundamental problem in classification. More classes can often explain the data better, so it is necessary to limit the number of classes. Many systems rely on an *ad hoc* convergence criterion. For example, ART2 (Adaptive Resonance Theory) is strongly influenced by a vigilance or threshold value which is set by users based on trial and error. The Kohonen network requires the number of classes to be determined beforehand. The Bayesian solution to the problem is based on the use of prior knowledge. It assumes that simpler class hypotheses (e.g., those with fewer classes) are more likely than complex ones, in advance of acquiring any data, and the *prior probability* of the hypothesis reflects this preference. The prior probability term prefers fewer classes, while the likelihood of the data prefers more, so both effects balance at the most probable number of classes. Because of this, AutoClass finds only one class in random data.
- Objects are not assigned to a class absolutely. AutoClass calculates the probability of membership of an object in each class, providing a more intuitive classification than absolute partitioning techniques. An object described equally well by two class descriptions should not be assigned to either class with certainty, because the evidence cannot support such an assertion.
- All attributes are potentially significant. Classification can be based on any or all attributes simultaneously, not just the most important one. This represents an advantage of the Bayesian method over human classification. In many applications, classes are distinguished not by one or even by several attributes, but by many small differences. Humans often have difficulty in taking more than a few attributes into account. The Bayesian approach utilises all attributes simultaneously, permitting uniform consideration of all the data. At the end of learning, AutoClass gives the contributing factors to class formation.
- Data can be real or discrete. Many methods have difficulty in analysing mixed data. Some methods insist on real valued data, while others accept only discrete data. The Bayesian approach can utilise the data exactly as they are given.
- It allows missing attribute values.

AutoClass has been studied for clustering process operational data produced by operating a refinery fluid catalytic cracking process[22,24]. It was found that that it is able to automatically convert data into clusters that represent significantly different operational modes. Most of the classified results are what would have been expected, some are certainly not. It is only after detailed thought and inspection that they can be seen to be valid classes.

6.5. General Comments

The above discussion has introduced unsupervised machine learning as a powerful method for process operational state identification. The data pre-processing methods described in section IV. have been used to reduce the dimensionality of the data and remove noise before analysis of data using unsupervised machine learning. There are several issues that are important but have not been fully addressed. First, for on-line process monitoring, it is important for the approach to be recursive. ART2 is an recursive method, but AutoClass is not. Second, although unsupervised procedures do not need training data, they are usually not as accurate as supervised methods, therefore interpretation and validation of results becomes an important issue. Furthermore when adapted to on-line monitoring the speed is obviously critical as is the selection of variables used for classification. Process variables tend to be interrelated so it is necessary to remove redundant variables without losing the important ones.

7. Conceptual Clustering for Process Monitoring

Multivariate statistics, supervised and unsupervised machine learning approaches all depend on calculating a similarity or distance measure for identifying clusters in data. Apart from giving predictions, however they are not able to provide causal explanations for why a specific set of data is assigned to a particular cluster. Conceptual clustering is distinguished from similarity or distance based clustering and able to generate conceptual knowledge about the major variables which are responsible for clustering, as well as predicting operational states. The resulting knowledge is expressed in the form of production rules or decision trees.

Inductive learning attempts to acquire a conceptual language for describing an object by drawing inductive inference from observations. The focus is on deriving rules or decision trees from unordered sets of examples, especially attributes-based induction, a formalism where examples are described in terms of a fixed collection of attributes. It is relatively easy for human experts to document cases rather than for them to articulate the expertise explicitly and clearly. The conceptual clustering approach used in C5.0 was developed by Quinlan[17,18,19]. A database of objects (or in other words data sets) is described in terms of a collection of attributes, which measure some important feature of an object. Each object belongs to one of a set of mutually exclusive classes; the task is to develop a classification rule that can determine the class of any object from its values of the attributes. The decision tree generated can be used for conceptual clustering. The procedure is iterative and can be summarised as follows[17,18]:

(1) Select a random subset of the given training examples (called the window)
(2) *Repeat* (a) to (c)
  (a) Develop a decision tree which correctly classifies all objects in the window
  (b) Find exceptions to this decision tree in the remaining examples
  (c) Form a new window by adding incorrectly classified objects to the window
Until there are no exceptions to the decision tree.

The crux of the problem is how to develop a decision tree for an arbitrary collection of objects in the window. To form a decision tree requires selecting the root attribute. To do this, assume that there are only two classes representing all the data, P and N (extension to any number of classes is not difficult). The method of finding the root attribute is adopted from an information based method that depends on two assumptions. Suppose the window C contains p objects of class P and n objects of class N. The assumptions are:

(1) Any correct decision tree for the window C will classify objects in the same proportion as their representation in C. An arbitrary object will be determined as belonging to class P with probability p/(p+n) and to class N with probability n/(p+n).

(2) When a decision tree is used to classify an object, it returns a class. A decision tree can then be regarded as the source of a message 'P' or 'N'. with the expected information needed to generate this message given by

$$I(p, n) = -\frac{p}{p+n}\log_2\frac{p}{p+n} - \frac{n}{p+n}\log_2\frac{n}{p+n} \quad (6)$$

If attribute, A, having values $\{A_1, A_2, \ldots A_v\}$, is used for the root of the decision tree, it will partition the window C into $\{C_1, C_2, \ldots C_v\}$ where C_i contains those objects in C that have values A_i of A. Suppose C_i contains p_i objects of class P and n_i of class N. The expected information required for the subtree for C_i is $I(p_i, n_i)$ and for the tree with A as root is then obtained as the weighted average given by

$$E(A) = \sum_{i=1}^{v} \frac{p_i + n_i}{p + n} I(p_i, n_i) \tag{7}$$

where the weight for the ith branch is the proportion of the objects in C that belong to C_i. The information gained by branching on A is therefore

$$\text{gain}(A) = I(p, n) - E(A) \tag{8}$$

The approach calculates the gain for all attributes and chooses the attribute having the biggest gain as the root node. The root node will have as many branches as its values. The branches divide the database into a number of subsets. For each subset, the root node is obtained following the same procedure.

The approach has been used in the commercial software C5.0[17], which has evolved from the earlier versions C4.5 and ID3[19]. A major limitation of ID3 was that it assumed that the values of all attributes are discrete, for instance a colour being red or green. C4.5 claimed to be able to deal with continuous-valued attributes, but is still weak considering the way it deals with discrete-valued attributes, as noted by Quinlan[18]. Though Quinlan[18] made a further effort to improve the method so that it could deal with continuous-valued attributes, the outcome is still not very satisfactory. Nevertheless, C5.0 has become one of the most well-known tools for use in data mining and knowledge discovery, especially in domains involving only discrete values.

Like most of the available inductive learning methods, C5.0 was developed for problem domains where attributes only take discrete values. The methods have proved to perform remarkably well with discrete-valued attributes. However, when the problem domains contain real numbers, the performance usually decreases in terms of accuracy. Using inductive learning based on continuous-valued attributes requires discretisation of the values into a number of intervals. To deal with this, a number of approaches have been proposed.

In process monitoring and control, dynamic trends of variables might be more important than the instantaneous values[13,22]. The differences for the seven dynamic trends for a variable shown in Fig. 6 can have important implications. The issue of dealing with this kind of problem has not previously been considered. An approach using principal component analysis to extract qualitative concepts from dynamic trend signals was given in Section IV, so will not be repeated here. In this section it will be shown how the concept formation can be used in inductive learning to develop conceptual clustering systems for process monitoring and diagnosis.

Saraiva[20] presented examples to extract decision trees from process operational data which has been averaged.

7.1. Inductive Learning for Conceptual Clustering and Real-time Monitoring

In this section, we present our work on application of inductive learning to the analysis of data collected on-line by computer-based control systems. A conceptual clustering approach is thus developed for designing state-space-based on-line process monitoring systems. The approach is illustrated by reference to a simple case study based on a CSTR reactor. It is concerned with analysis of an operational database consisting of eighty five data patterns obtained in operating the CSTR. For each data pattern seven variables are recorded including, reaction temperature TR, reaction mixture flow out of the reactor F_o, cooling water flowrate F_w, feed flowrate F_i, feed inlet temperature T_i, feed concentration C_i, and cooling water temperature T_w. Each variable is recorded as a dynamic trend comprising 150 sample points. The goal is to identify operational states using a conceptual clustering approach.

The approach basically comprises the following procedures: (1) concept extraction from dynamic trend signals using PCA. (2) identification of operational states using an unsupervised machine learning approach, and (3) application of the inductive machine learning system to develop decision trees and rules for process monitoring.

7.1.1. Concept Extraction from Dynamic Trend Signals

This has been discussed in detail in section IV, so only a brief review is presented here. For a specific set of data, the value of a variable represents a dynamic trend, consisting of tens to hundreds of sampled points. In inductive learning, it is the shape of the trend that matters so for a specific variable, when the trends of all the data sets are considered and processed using PCA, the first two principal components (PCs) can be plotted in a two dimensional plane. Figure 8 showing PC-1-TR and PC-2-TR corresponds to the first two PCs of the reaction temperature TR. The data sets are grouped into clusters in this two dimensional plane. This permits a dynamic trend to be abstracted as a concept as typically by PC-1-TR in region A. The following sections will show how this process can be used for conceptual clustering using inductive learning.

7.1.2. Identification of Operational States

The next step is identification of operational states. In this case, this can be done using PCA because there are only eight variables. For more complex processes, more sophisticated approaches need to be used, which will be described later in the case study of MTBE. The PC1-PC2 two-dimensional plots for TR and Fo are shown in Figs. 8 and 5. The PC1-PC2 plots for Fw, Fi, Ti, Ci, Twi and L are given in Fig. 20 and 21(a)-(e). The first two PCs of the eight variables (TR, Fo, Fw, Fi, Ti, Ci, Twi, L) are plotted in Fig. 9. The five groups which are identified represent the 85 data cases as five clusters corresponding to five distinct operational modes. Detailed examination of the clusters shows that these groups are reasonable.

7.1.3. Conceptual Clustering

Having characterised the dynamic trend signals and identified the operational states, it is necessary to find out how to generate knowledge which correlates the variables and operational states. To do this requires generating a file as shown in Table 5. In fact, each data set in Table 5 can be interpreted as a production rule. Thus, the first case is equivalent to the following rule,

IF PC-L = C AND PC-TR = D AND PC-Fo = A
AND PC-Fw = D AND PC-Twi = B AND PC-Ci = A
AND PC-Ti = D AND PC-Fi = B
THEN States = NOR1

Obviously this is simply an explanation of the database and the decision tree developed will be very complex. C5.0 makes it possible to develop a simpler tree. A simple tree is preferable because it can usually perform better than a complex tree for data cases outside the training data set.

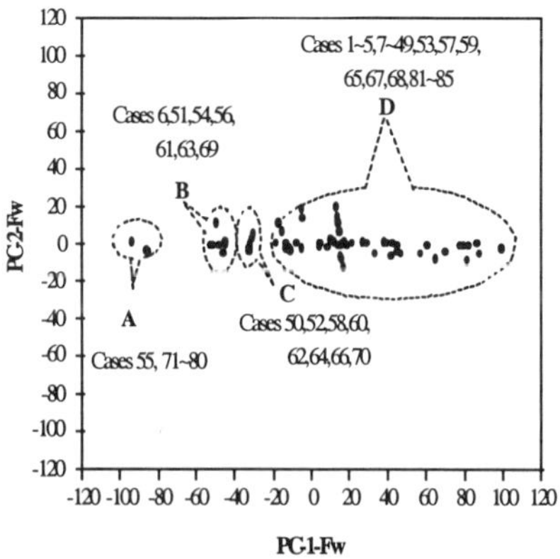

Figure 20. PC1-PC2 plot for F_w.

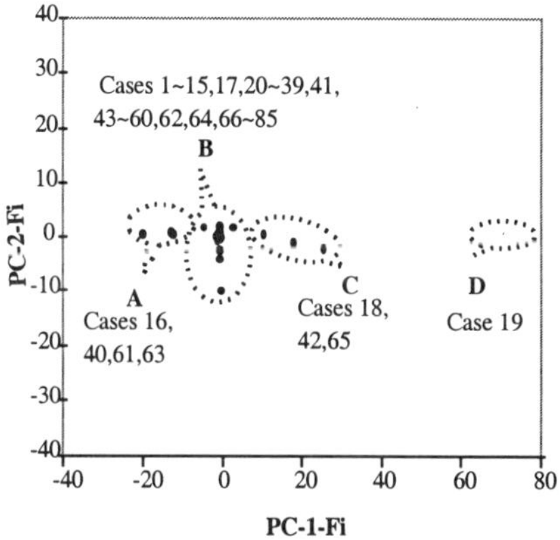

Figure 21(a). PC1-PC2 plot for F_i.

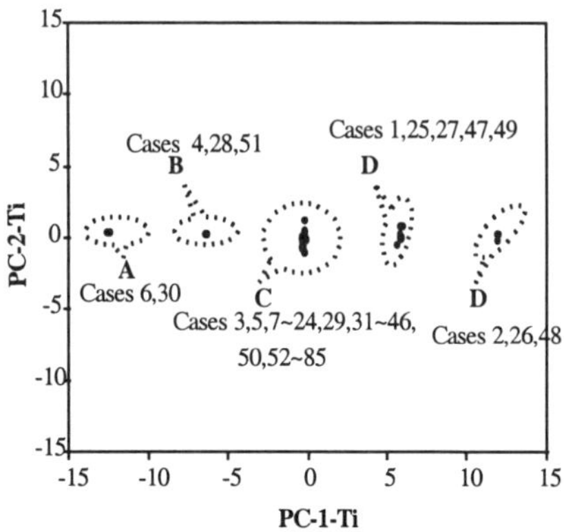

Figure 21(b). PC1-PC2 plot for T_i.

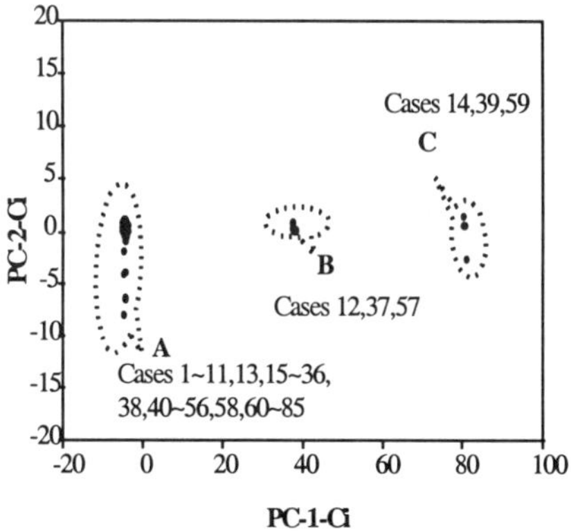

Figure 21(c). PC1-PC2 plot for C_i.

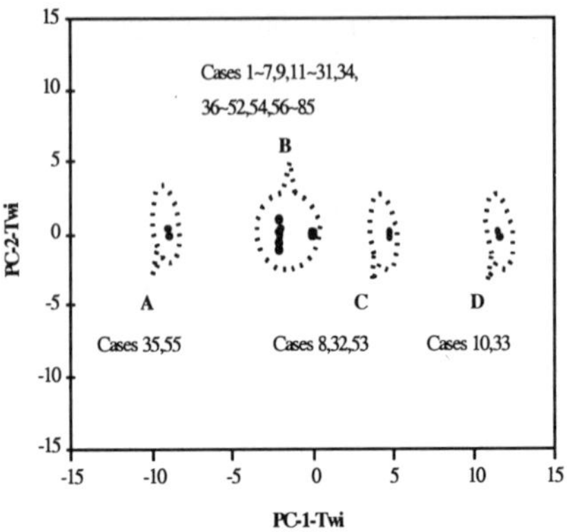

Figure 21(d). PC1-PC2 plot for T_{wi}.

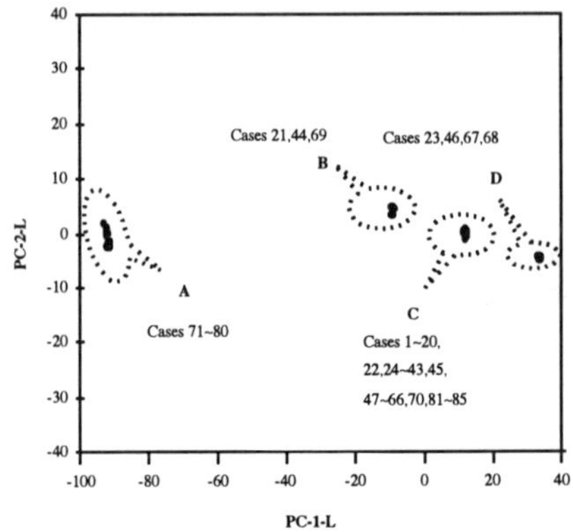

Figure 21(e). PC1-PC2 plot for L.

Table 5. The data structure used by C5.0 for conceptual clustering

PC_L	PC_TR	PC_Fo	PC_Fw	PC_Twi	PC_Ci	PC_Ti	PC_Fi	States
C	D	A	D	B	A	D	B	NOR1
C	D	A	D	B	A	E	B	NOR1
...	...	...	...	...	...	...	...	...
A	C	D	A	B	A	C	B	ABN1
A	C	D	A	B	A	C	B	ABN1
...	...	...	...	...	...	...	...	...

The decision tree developed for the CSTR case study is shown in Fig. 22 and can be converted to production rules, as shown in Table 6. C5.0 identifies the reactor temperature as the root node and states that if TR is in the region of A, B or D of Fig. 8, then the operation will be in regions ABN2 (abnormal mode 2), NOR2 (Normal operation mode 2), or NOR1 (Normal operation mode 1) of Fig. 9. If TR is in the region C in Fig. 8, then there are three possible situations depending on Fo. If Fo is in the region D of Fig. 5, then the operation will correspond to ABN1 (Abnormal operation 1): if Fo is in A or B of Fig. 5, then the operation will be NOR3 (Normal operation 3). The result effectively states that it is possible to focus on monitoring TR in Fig. 8. If TR is in the region C, then Fo in Fig. 5 should be examined. It also shows the variables responsible for the location placing the operation in a specific region of Fig. 9.

The decision tree shown in Fig. 22 and the rules in Table 6 provide guidance for operation clustering which is transparent. The approach has also been applied to a more complicated case study, the production of methyl tertiary butyl ether (MTBE)[22,13].

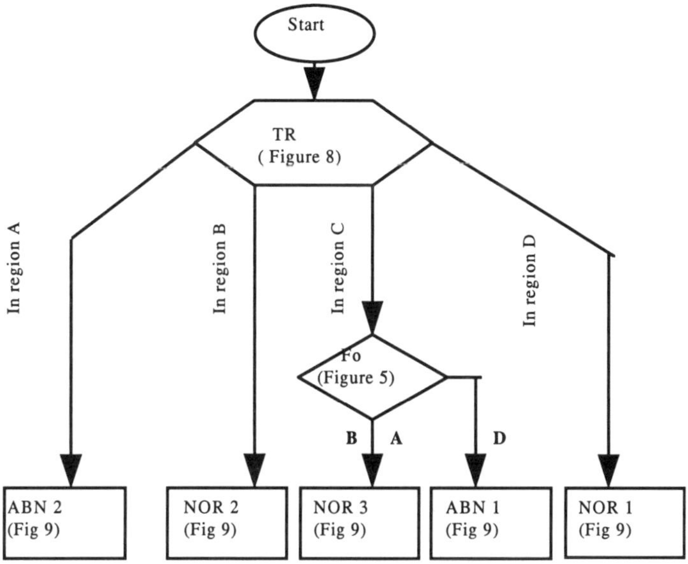

Figure 22. The decision tree developed for the CSTR.

Table 6. The production rules converted from the decision tree in Figure 22.

Rule 1:	IF TR = A in Fig. 8 THEN		Operational state = ABN 2 in Fig. 9
Rule 2:	IF TR = B in Fig. 8 THEN		Operational state = NOR 2 in Fig. 9
Rule 3:	IF TR = C in Fig. 8	AND	Fo = A or B in Fig. 5
		THEN	Operational state = NOR 3 in Fig. 9
Rule 4:	IF TR = C in Fig. 8	AND	Fo = D in Fig. 5
		THEN	Operational state = ABN 1 in Fig. 9
Rule 5:	IF TR = D in Fig. 8	THEN	Operational state = NOR 1 in Fig. 9

7.2. General Review

Inductive learning has been introduced as a method for analysis of data records averaged over days or weeks and a conceptual clustering tool for developing on-line operational monitoring systems. It can learn from a large number of examples to develop explicit and transparent knowledge in the form of decision trees and production rules. It is also able to identify the most important variables that contribute to clustering, which is clearly valuable for analysing process operational data and process monitoring. There are several issues that need to be addressed. Most inductive learning systems are not recursive. In addition though PCA has proved to be an effective way for concept extraction from dynamic trend signals, it is expected that the combination of PCA and Wavelet will deliver more effective pre-processing methods. Compared with similarity or distance-based methods which have been widely studied, conceptual clustering clearly needs more research attention.

8. Final Remarks

This contribution has examined the use of data mining technology in process operational data analysis and knowledge discovery. A critical issue is the pre-processing of on-line signals of measurements which are interrelated, contain noise components and change with time. Methods have been developed based on principal component analysis and wavelet multiscale analysis for dimension reduction, removal of noise components, feature extraction and concept extraction from dynamic trends. Multiscale wavelet analysis was used to replace the data pre-processing part of the adaptive resonance theory and an integrated framework ARTnet was thus developed which demonstrates much improved performance in dealing with noise. Multivariate statistical analysis based on principal component analysis was also used in discovering knowledge from averaged operational data and consequently developing operational strategies.

Multivariate statistics and unsupervised machine learning often depend on calculating a similarity or distance measure to group data sets into clusters. Apart from giving predictions, they are not able to give causal explanations on why a specific set of data is assigned to a particular cluster. A conceptual clustering approach has been developed which is able to generate conceptual knowledge on the major variables which are responsible for clustering, as well as projecting the operation to a specific operational state. A critical issue in this approach is how to conceptually represent dynamic trend signals. For this purpose, principal component analysis is used for concept extraction from real-time dynamic trend signals.

References

1. Bakshi, BR., *AIChE J.* **44** (1998), 1596-1610.
2. Berman, Z, Baras J.S., *IEEE Trans. Signal Processing.* **41** (1993), 3216-3231.
3. Biehl, M. and Schlosser, E., *J. Phys. A: Math. Gen.* **31** (1998), L97-L103.
4. Carpenter, G.A. and Grossberg, S., *Appl Opt.* **26** (1978b), 4919-4930.
5. Cheeseman, P, Stutz, J., in *Advances in Knowledge Discovery and Data Mining.* eds. Fayyad, U.M., Piatetsky-Shapiro, G., Smyth, P., Uthurusamy, R. (AAAI Press/MIT Press, 1996), 153-180.
6. Chen, B.H., Wang, X.Z., Yang, S.H. and McGreavy, C. *Comput. Chem. Engng.* **23** (1999), 899-906.
7. Cvetkovic, Z. and Vetterli, M., *IEEE Trans. Signal Processing.* **43** (1995), 681-693.
8. Fayyad, U.M., Piatetsky-Shapiro, G., Smyth, P., in *Advances in Knowledge Discovery and Data Mining.* eds. Fayyad, U.M., Piatetsky-Shapiro, G., Smyth, P., Uthurusamy, R. (AAAI Press/MIT Press, 1996), 1-36.
9. Graco, W. and Cooksey, R.W. in *Proc. PADD98 - The second int. conf. on the practical application of knowl. discovery and data mining*, London, March 1998, 111-130.
10. Hotelling, H., *J. Educ. Psychol.* **24** (1933) 417-441, 498-520.
11. Zhang, J., Martin, E.B. and Morris, A.J., *Trans. IChemE.* **74A** (1996), 89-96.
12. Kourti, T. and MacGregor, J.F. *Chemometrics and Intell. Lab. Systems.* **28** (1995), 3-21.
13. Wang, X.Z. and Li, R.F. *Ind. Eng. Chem. Res.* **38** (1999), 4345-4358.
14. Mallat, S. and Hwang, W.L. *IEEE Trans. on Inf. Theory.* **38** (1992), 617-643.
15. Mallat, S. and Zhong, S. *IEEE Trans. Pattern Analysis and Machine Intelligence.* **14** (1992), 710-732.
16. Pearson, K. *Phil Mag.* **2** (1901), 559-572.
17. Quinlan, J.R. *C4.5: programs for machine learning* (Morgan Kauffman, 1993).
18. Quinlan, J.R. *J. Artif. Intell. Res.* **4** (1996), 77-90.
19. Quinlan, J.R., *Machine learning.* **1** (1986), 81-106.

20. Saraiva, P.M., in *Intelligent Systems in Process Engineering: Paradigms from Design and Operations.* eds. Stephanopoulos, G. and Han, C. (Academic Press, Inc., San Diego, California, 1996), 377-435.
21. Wang, X.Z., Chen, B.H., Yang, S.H. and McGreavy, C., *Comput. Chem. Engng.* **23** (1999), 945-954.
22. Wang, X.Z., *Data mining and knowledge discovery for process monitoring and control* (Springer, London, 1999).
23. Yamanaka, F. and Nishiya, T., *Comput Chem Engng.* **21** (1997), s625-s630.
24. Wang, X.Z. and McGreavy, C., *Ind. Eng. Chem. Res.* **37** (1998), 2215-2222.

PART V EXPERIMENTAL AND INDUSTRIAL APPLICATIONS

14. USE OF NEURAL NETWORKS FOR PROCESS CONTROL. EXPERIMENTAL APPLICATIONS

M. CABASSUD, M.V. LE LANN
Laboratoire de Génie Chimique - UMR CNRS 5503
Ecole Nationale Supérieure d'Ingénieurs de Génie Chimique - INPT
18, chemin de la Loge - 31078 Toulouse-Cedex 4 - France

In this paper the problem of design and elaboration of artificial neural networks as direct process controllers is developed. The neural controller is a feedforward multi-layer network, and the controller design methodology is based on the modelling of the process inverse dynamics. The advantage of this method is that it is not necessary to perform initial closed-loop experiments with a classical controller to generate the learning data base. By this way, multivariable controllers can be easily developed, taking into account the dynamics and the interactions of the different control loops. The efficiency of such a control methodology is exemplified through its application to different chemical processes :

- a semi-batch pilot plant chemical reactor
- a liquid-liquid extraction column
- a low pressure chemical vapour deposition reactor

1. Introduction

In the last few years, a new approach for process control has been proposed in the literature based on the use of artificial neural networks. ANN are computing tools made up of many highly interconnected processing elements. They are able to model a wide range of complex and non-linear problems. Their design is based on an auto-organisation of their parameters during a learning phase. These parameters are optimised in order to model the functionalities between input and output vectors ; these two vectors form the learning data base. The principal fields of application within chemical engineering are: modelling, prediction, fault detection and diagnosis, and process control [Bulsari, 1995 ; Morris et al., 1994]

Different methods can be considered for the design of a controller based on artificial neural networks. In a very simple way the neural controller can be obtained from a learning data base provided by an other "controller" or by control values delivered by an human operator [Dirion, 1993]. In both cases, it has been shown that as far as the evolution of the process output is included in the learning data set, the neural controller gives good control performances[Dirion et al., 1996]. Moreover, the neural controller is able to generalise to new situations. Nevertheless, in this case, a reference control system has been used for creating the learning data base and the neural network models the functioning of this reference control system.

In an other approach, the neural controller is still a classical feedforward multi-layer network, but the controller design methodology is based on the modelling of the process inverse dynamics. The advantage of this method is that it is not necessary to perform initial closed-loop experiments with a classical controller implemented on the process in order to generate the learning data base.

This chapter is devoted to the application of this methodology to the design and the implementation of direct neural controllers for the experimental control of different chemical processes :

- a semi-batch pilot plant chemical reactor [Dirion, 1993]
- a liquid-liquid extraction column [Chouai, 1999]
- a low pressure chemical vapour deposition (LPCVD) reactor [Fakhr-Eddine, 1998]

2. Design of Neural Networks for Process Control

2.1. Introduction

Because of their intrinsic nonlinearity, neural networks appear as useful tools for process control [Thibault et al., 1991].

Due to their capability to realise dynamic modelling of complex processes, they can be used as models within a model based control strategy (internal model control, predictive control, reference model control, ...) [Psichogios et al., 1991 ; Nahas et al., 1992 ; Grondin-Perez et al., 1996].

The approach adopted in this work is rather different and consists in designing an autonomous neural controller. In such a case, a supervised learning is not easy to carry out because the solutions (optimal command variables) are a priori unknown: the set-point is fixed by the user, but the control variable is not known. However, several strategies to solve this problem have been proposed. The most popular one is the inverse modelling for which the neural network is trained to represent the inverse process dynamics which is then considered as the control law [Psichogios et al., 1991]. An other possibility is to explicitly define a control law [Zaldivar et al., 1992].

2.2. Neural Network

Artificial neural networks consist of a large number of computational units connected in a massively parallel structure. The processing units from each layer are linked to the processing units in successive layers by weighted connections. Collectively these connections, as well as the transfer functions of the processing units, can form distributed representations of the relationships between input and output data to some degree of accuracy and even when the information is noisy and imprecise. Neural networks are trained by a self-organisation of their parameters during a learning phase. The parameters are optimised in order to model the relationships between input and output vectors as closely as possible.

In process engineering, artificial neural networks have so far mainly been used in process modelling [Hamachi et al., 1999; Delgrange et al., 1998], process control, fault diagnosis, error detection, data reconciliation and process analysis.

An important aspect of a neural network is the learning process, based on a set of measured numerical values (the learning data base). Representative examples are presented to the network so that it can integrate this knowledge within its structure.

The protocol used to obtain a neural model is relatively simple. The input and output data vectors used to teach the network are scaled into the range 0.9 to 0.1 (preferably to 0 to 1), and the sigmoidal function may be used as an activation function. The first layer of neurones, the input layer, is strictly a pre-processing layer that simply distributes the inputs to the next layer. It does not perform, as the other layers, a non-linear transformation of its input data. An offset, also called bias or reference, is added at each layer except at the output layer.

The data from the input neurones is propagated through the network via the interconnections. Every neurone in a layer is connected to every neurone in adjacent layers. A scalar weight is associated with each interconnection.

Neurones in the hidden layers receive weighted inputs from each of the neurones in the previous layer and perform two tasks: they sum the weighted inputs to the neurone and then pass the resulting summation through a non-linear activation function.

The weighted sum to the kth neurone in the jth layer ($j >= 2$) is given by:

$$S_{j,k} = \sum_{i=1}^{N_{j-1}} \left(w_{j-1,i,k} \, I_{j-1,i} \right) + w_{j-1,N_{j-1}+1,k} \, b_{j,k} \qquad (1)$$

$I_{j-1,i}$ is the information from the ith neurone in the $j\text{-}1^{th}$ layer, $b_{j,k}$ is the bias term and N_{j-1} is the number of neurones in the previous layer (j-1).

The output of the kth neurone in the jth layer (j >= 2) is then :

$$O_{j,k} = \frac{1}{\left(1 + \exp\left(-S_{j,k}\right)\right)} \tag{2}$$

The learning process consists of identifying the weights $w_{j,i,k}$ that produce the best fit of the predicted outputs over the entire training data set. The weights are first set to random values. During the training process, the weights of the network are adjusted continuously based on the error signal generated by the discrepancy between the output of the network (O) and the actual output of the training examples (target vector T). This is accomplished by means of learning algorithms designed to minimise the least square total output error (F).

The errors between networks outputs and targets are summed over the entire data set and updating of the weights is performed after every presentation of the complete data set.

$$F = \frac{1}{2} \sum_{l=1}^{Nd} \sum_{k=1}^{N3} \left(T_k(l) - O_{3,k}(l)\right)^2 \tag{3}$$

Nd is the number of examples in the data set and N3 corresponds to the number of outputs of the neural network.

The topology of the neural network determines the accuracy and the degree of representation of the model. A number of papers have shown that a feedforward network has the potential to approximate any non-linear function. In this paper, only one hidden layer has been considered. The number of neurones in this hidden layer has been chosen by trial and error tests.

Many different network architectures are used. The most popular architecture is the backpropagation multilayer network with sigmoidal activation functions, often called 'the backpropagation network'. However, this procedure converges slowly, which is not surprising since the backpropagation algorithm is essentially a steepest descent method. Consequently, it is restricted to feed-forward layered networks only. Theoretical and numerical results proved that Quasi-Newton algorithms are superior to steepest descent algorithms [Dennis et al., 1983]. Watrous [1987] compared Davidon-Fletcher-Powel (DFP) and Broyden-Fletcher-Goldfarb-Shanno (BFGS) methods with the backpropagation algorithm and showed that (DFP) and

(BFGS) algorithms need less iteration. For this reason, a Quasi-Newton learning algorithm has been used in this work to train the neural nets.

Two data sets are considered for the learning phase. The first one, called the learning data set, is used to calculate F and to update the weights. The second one, called the test data base, is used to determine the optimal weights, which give the minimum error on this test base. By this way, the problem of overlearning which is a main drawback to neural networks, is avoided.

2.3. Neural Network Architecture Design

To build the neural network, several design problems still exist:

- The nature and the list of their inputs and the ouputs. This has been done according to a physical analysis of the process behaviour. Neural networks realise their mappings by auto-organisation of their weights. Therefore, the user must choose cautiously the most relevant information so that the neural network performs its task well. A good knowledge of the covered operating domain is essential.

- The relevance of the examples in the learning set. Once the above problem is resolved, one must ensure that the data in the learning set span the domain of expected operation.

- The choice of the network architecture. How many hidden layers and neurones should be used ? If too few hidden neurones are used, the weights of the neural network will not converge during the learning phase. If there are too many, "over-fitting" will occur, i.e. the network will model the learning examples well, but interpolation and extrapolation will fail. Up to now, there does not exist a clear procedure to determine a priori this number of neurones. Therefore we used a trial and error procedure. A first learning is carried out with a given architecture. Then, the number of neurones is increased and learning is carried out again. If the second neural network leads to a best fitting, the procedure is repeated until there is no more improvements in results. The classical approach adopted here is to determine a minimum and sufficient number of neurones for the task at hand. In general, the number of learning examples must be many times larger than the number of parameters [Baum, 1989].

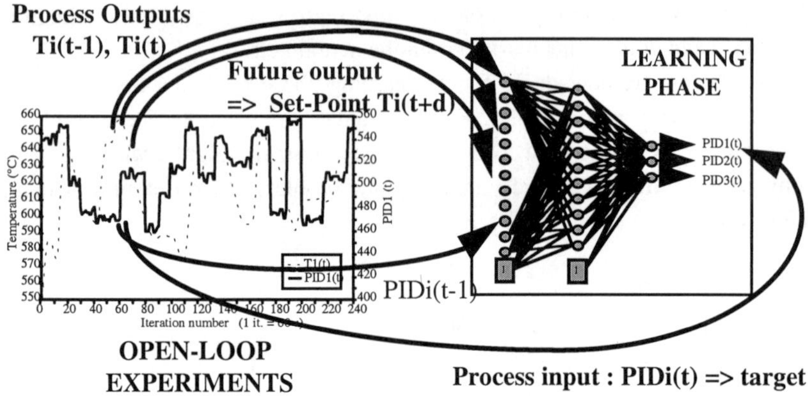

Figure 1. Principle of the inverse modelling methodology

2.4. Design Of The Neural Controller

The objective of the neural network controller is to directly compute the values of the control variables in order to make the plant outputs follow the desired set-points. Given this goal, the learning objective consists in modelling the functionalities between the inputs and the outputs of the plant (see Fig. 1). With the inverse dynamics modelling methodology [Thibault *et al.*, 1991], the learning data base is obtained in applying input values to the plant in an open-loop structure. These inputs can be randomly generated, but they must preferably cover the entire input domain and must contain frequencies to fit with the dynamics of the pilot-plant. The applied inputs and resulting process outputs are recorded during experiments. At the end of the learning phase, the network must be able to model off-line the inverse dynamics of the plant, i.e. to compute the inputs which have been applied to the process. The set of the network inputs is composed of present and future values of the process states over a sliding horizon. The network outputs are the process inputs, which have been applied and which led to these process output values.

Figure 1 gives a schematic global overview of the learning process for inverse dynamics modelling.

After a successful learning, the neural network is integrated in a feedback control loop. At this step, the neural controller must be able to compute the inputs (the manipulative variables) to apply to the process, from the knowledge of its current state and the desired future state. The input units, which coded the future state of the plant during the learning phase, are then replaced by the future desired set-points.

2.5. Conclusion

The section above has presented the general framework for design and elaboration of neural networks for process control applications. In the following, the efficiency of such a control strategy will be exemplified through its application to different complex chemical processes.

3. Application To Batch Reactors

3.1. Introduction

Control of batch or semi-batch reactors remains an open and challenging problem. Such operations are widely encountered in fine chemicals or pharmaceuticals production. The characteristics of this production (flexibility and multipurpose character) necessitate the operation over a wide range of conditions and dynamics. Start-ups and shutdowns are frequently encountered operations. Moreover, the reactor never reaches a steady state but remains in transient state. Often these processes exhibit strong non-linear and time-varying dynamic behaviour. Thus, due to the complex nature of these processes, conventional process control strategies have usually only limited performances. Therefore, in industry, a precise reactor temperature control is essential to ensure a tight control of the kinetics, and then to favour the reaction yield.

In our laboratory, a lot of studies have dealt with the temperature control of semi-batch reactors using advanced control strategies [Le Lann et al., 1995]. In this paper, the design and implementation of a neural network controller are presented. This approach is illustrated by a real-time application of this neural controller for the temperature control of an experimental pilot-plant reactor equipped with monofluid heating-cooling system [Dirion, 1993].

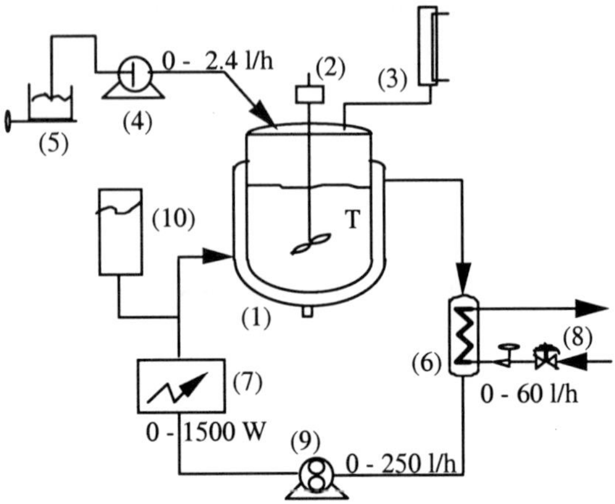

Figure 2. The experimental batch pilot-plant reactor.

3.2. Experimental Configuration

The experimental apparatus is depicted in Fig. 2. It consists of a jacketed glass reactor of 1 litre (1). The reactor is fitted out with a monofluid heating-cooling system. The internal and the external diameters of the reactor are 100 mm and 140 mm respectively. A Rushton turbine (2) is fitted through the central socket at the top of the reactor and its speed is fixed at approximately 300 tr.min^{-1} to ensure a good mixing. A condenser (3) is used to condense possible vapour, which may be produced during a chemical reaction.

For semi-batch operations, liquid reactants can be fed into the reactor by means of a piston pump (4). The inlet reactants mass flow rate is obtained by means of a balance (5), which measures the mass time-evolution of reactants introduced into the reactor. The flow rate of the reactants can be automatically controlled during the feeding operation.

The heating-cooling system consists of a plate heat-exchanger (6) and an electric resistance (7) in order to modulate the inlet jacket temperature of the thermal fluid. The plate exchanger uses cold water (water temperature between 20 and 25°C). In order to cool the monofluid, the flow rate of the cooling water is manipulated by an air-to-open valve (8). Alternatively, heating of the thermofluid is ensured by acting on the electric tension applied to the extremities of the electric resistance. The power produced by the resistance is proportional to the electric tension applied. A constant flow rate of the thermal fluid (250 $l.h^{-1}$) is ensured by a gearing pump (9). An expansion vase (10) is installed to avoid a possible pressure rise in the thermal loop. Moreover, all the pipes of the external monofluid loop are insulated to minimise heat losses to the environment.

Several PT-100 temperature sensors allow the measurement of the temperature inside the reactor, the inlet and outlet jacket temperatures and the inlet and outlet temperatures of the cooling loop. The constant flow rates of the monofluid circulating in the jacket and the cooling flow rate are measured too.

A computer (PC 486) equipped with A/N and N/A converters provides a real-time data acquisition and control.

3.3. Description Of The Control Strategy

In this work, our main goal is to control the temperature of the jacketed semi-batch reactor by directly acting on the different thermal elements (i.e. the plate exchanger and the electric power). The control system computes the inputs from the following information: the measured reactor temperature and the desired time-varying set-point. This single control loop necessitates the manipulation of two different elements in the same time: valve opening and electric tension. The sign of the control variable allows determining the required thermal element. Thus, positive values imply that heating is needed and negative values imply cooling using the plate exchanger. The control variable is bounded between -1 and +1. The electric power and the cooling valve opening are proportional to the control signal in this range.

The objectives of the control system are on the one hand, to ensure temperature set-point tracking and, on the other hand, to ensure satisfactorily rejection of internal (e.g. heat generated by exothermic reaction) and external disturbances (e.g. thermal losses, cooling water temperature fluctuations).

3.4. Experimental Results

The first step consists in generating a learning data base, by carrying out experiments where the inputs are randomly computed and applied to the reactor in an open loop structure. To fit with the dynamics of the pilot plant reactor, a "smooth" pseudo-random input signal with frequency within the appropriate range has been used. Figuress 3 and 4 present the learning data base so generated: the reactor temperature evolves between 22 and 57°C which correspond to the appropriate temperature range according to the capacities of the thermal elements.

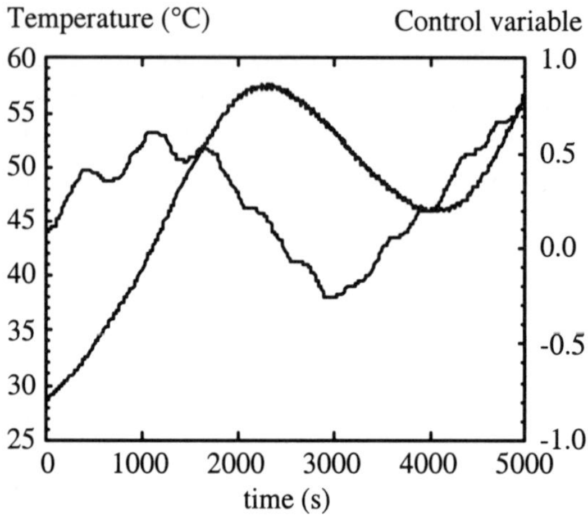

Figure 3. Open-loop experiment performed on the reactor pilot-plant
(▬: Temperature (y) ; —: Control variable (u))

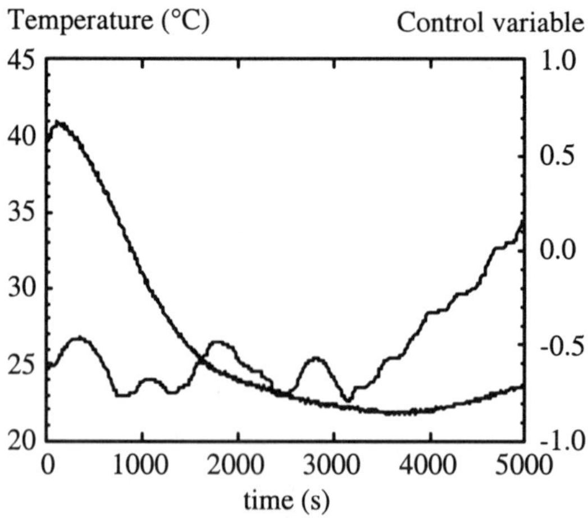

Figure 4. Open-loop experiment performed on the reactor pilot-plant
(▬: Temperature (y) ; —: Control variable (u))

As shown in previous papers [Dirion et al., 1995], perfect knowledge of the process time-delay significantly improves the performance of the controller. The sampling period has been chosen equal to 10 seconds for this reactor. A simple step response experiment has allowed us to approximate the time-delay to 3 sampling periods.

Different architectures have been studied with this value of time-delay introduced in the neural network. In the case of a simple architecture (NN[3, 4, 1]), where the inputs correspond to u s_{k-1} (process input: control variable applied the last sampling period) ; y_k (process output: measured temperature) and s_{k+3} (set-point for the next delayed sampling time, the neural controller gives good results [Dirion et al., 1995]. Tracking is well performed during the heating phase, but oscillations appear during the constant temperature-level step. To reduce these oscillations, a prediction horizon has been considered by adding supplementary inputs coding future set-point values (s_{k+R+1}, s_{k+R+2}, ...) . This prediction horizon is used to make the controller react by anticipation to set-point slope changes. The choice of the number of supplementary set-point values has been made by making a compromise: a small learning error with the minimum number of inputs. The chosen architecture is NN[7, 4, 1] and all the input units are the following ones:

$$N_1 = \{u_{k-1}, y_k, s_{k+3}, s_{k+4}, s_{k+5}, s_{k+6}, s_{k+7}\} \quad (4)$$

For the experiment presented Fig. 5, a good tracking is observed for the different phases. Performances of the neural controller are equivalent for different set-point profiles. The property of disturbance rejection of the controller has been studied: cool water has been poured into the reactor between 1500 and 2000 seconds. It can be observed in Fig. 6 that the controller increases rapidly the control value in order to compensate the cooling of the reactor contents.

To clearly demonstrate the importance of time-delay mismatch, the same architecture has been used, replacing the time-delay in the network by 0 (no time-delay considered). In this case (Fig. 7), an oscillatory behaviour of the controller is observed. This confirms the necessity of a process dynamics behaviour analysis before designing the artificial neural network.

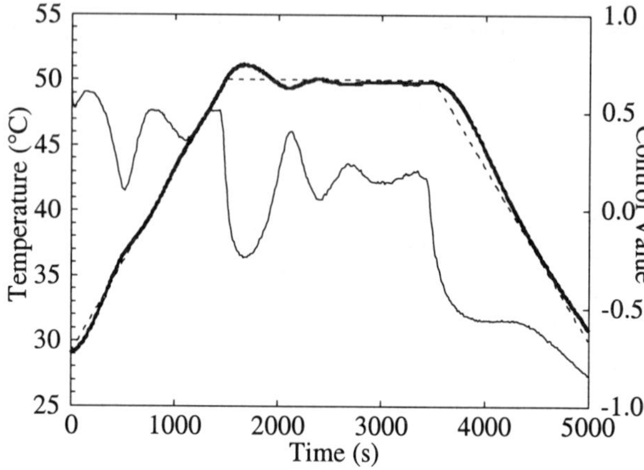

Figure 5. Control by the neural network with prediction horizon.

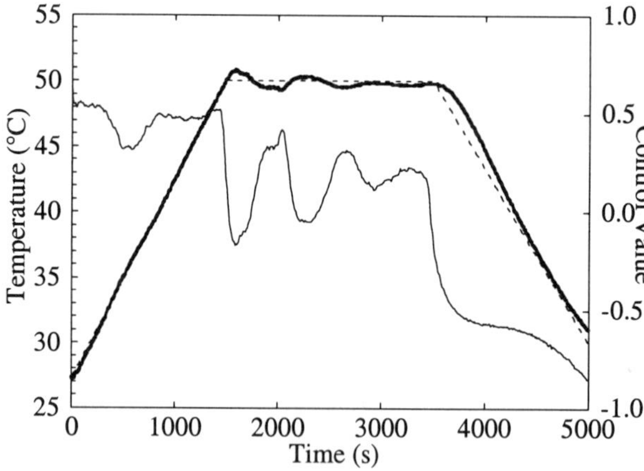

Figure 6. Control by the neural network being faced with thermal disturbance (introduction of cold water into the reactor).

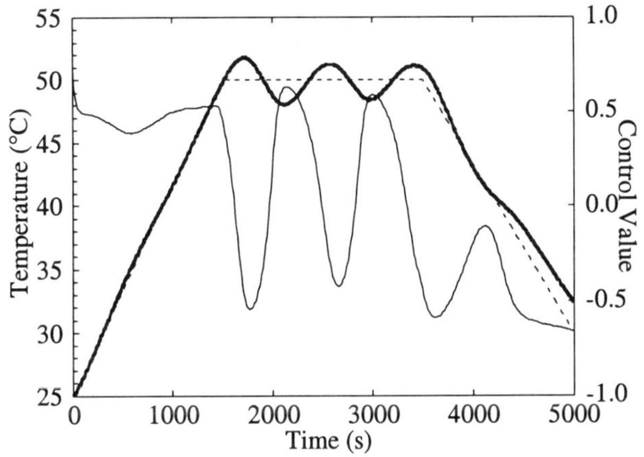

Figure 7. Neural controller without time-delay

3.5. Conclusion

This first application of neural networks as direct controller concerns a single input – single output process even if the output has to be monitored in order to choose the right element of the thermal loop on which the control variable has to be applied. This example shows the importance of the choice of the inputs and particularly the necessity to perform a suitable process behaviour analysis before designing the neural network architecture.

4. Multivariable Control of A Pulsed Liquid-Liquid Extraction Column

4.1. Introduction

Solvent extraction in continuous columns is one of the most important separation processes in chemical engineering. This separation process presents a strongly non-linear behavior and time-varying dynamics. The control of this process can often be problematic, partially due to the difficulty of on-line measurement of output variables, and partially due to the complex behavior of the two-phase flows.

In this chapter, we are interested by an application concerning industrial wastewater treatment. More precisely, the ozonation of poplar sawdust in order to study its enzymatic digestibility produces soluble substances in water especially oxalic acid [Faizal et al., 1991] . This work deals with the separation of this carboxylic acid from aqueous solutions by a liquid-liquid extraction with a mixture of tributylphosphate (60 vol.%) + dodecane (40 vol.%) as selective solvent. The experiments to recover the oxalic acid from wastewater were carried out in a continuous agitated countercurrent discs and doughnuts column. Previous studies have dealt with mass-transfer transients while assuming hydrodynamic steady-state. More realistic models relaying on drop population [Casamatta, 1981 ; Casamatta et al., 1985], describing the hold-up profiles along the column, the drop breakage and coalescence have been developed. Therefore, these approaches need complex mathematical formulations and have not yet been developed for both on-line hydrodynamics and mass transfer control. For this purpose, a new approach of extraction column dynamics multivariable control, relying on neural networks, has been introduced [Chouai, 1999]. This section will present the development and the application of a multivariable controller based on neural networks. The pilot plant to be controlled is a pulsed liquid-liquid extraction column. Previous works have shown that the column could be maintained in its optimal behaviour by means of the control of conductivity by action on the pulse frequency. In the same time, a given product specification can be obtained by the control of the product concentration in the outlet stream by acting on the solvent feed flow rate.

4.2. Experimental Pilot Plant

The extraction pilot plant is represented in Fig. 8. The height of the active zone of the column filled with discs and doughnuts is 1.2 m, and its diameter is 50 mm. The distance between a disk and a doughnut is 25 mm. The agitation is induced by means of a lateral pulsator located at the column bottom.

The continuous heavy phase flow (Qc) is oxalic acid in water, and the dispersed phase flow (the solvent) is tributylphosphate (TBP) which is furthermore hardly soluble in water (0.039 mass %). Since tributylphosphate has a relatively high viscosity (3.56×10^{-3} Pa.s) and specific gravity close to unity (0.98), it is necessary to mix it with a diluent (dodecane) in order to facilitate good phase separation. Faizal et al. (1991) selected a mixture of 60 vol.% tributylphosphate + 40 vol.% dodecane saturated with water (4.67 mass.%) as final solvent. Dodecane was chosen as inert diluent because of its low viscosity (1.15×10^{-3} Pa.s), its low specific gravity (0.75) and its insolubility in water.

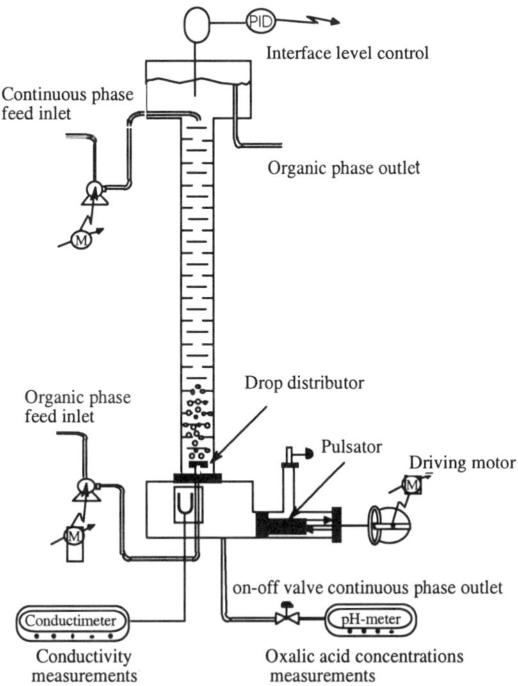

Figure 8. Schematic diagram of the pulsed column.

The light phase (TBP + dodecane) is fed into the column below the active part of the column and predispersed by means of a distributor. The dispersed phase flow (Qd) rises through the column, coalesces at the interface in the upper settling zone and leaves the extractor at the top. The continuous heavy phase (water + oxalic acid) is fed into the column below the upper settling zone and flows through the column counter-currently to the dispersed phase. The flow-rates are measured and controlled respectively by flow-meters and pumps. The pulse frequency (Fr) is controlled by a d.c. motor, the pulse amplitude (A_P) has been kept constant during this study.

To prevent flooding at the top of the column, the interface level in the settling zone is detected by a capacity probe and a PID controller acts on a valve controlling the continuous phase discharge (raffinate). An industrial pH-meter is used for on-line measurement of the composition of final raffinate. The initial concentration (Xi) of oxalic acid in the continuous phase inlet is less than 2 % in mass. Finally, a

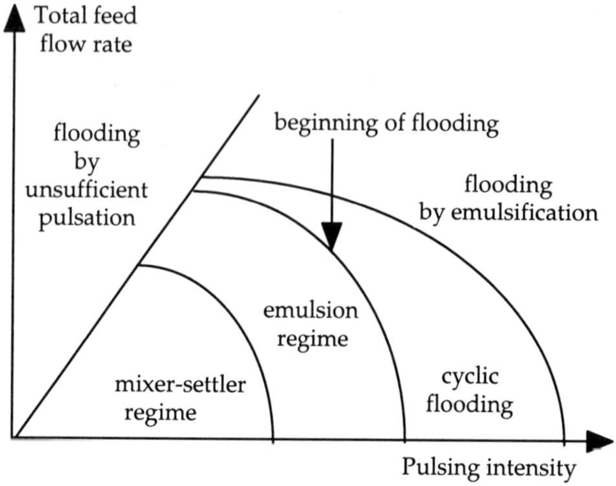

Figure 9. Regimes of pulsed column

Macintosh Series IIX computer is attached to the equipment through a National Instruments NB-MIO 16 interface. A Supervisory Control and Data Acquisition (SCADA) system has previously been programmed for the column.

4.3. Analysis Of The Column Behaviour

Previous studies in our laboratory [Casamatta, 1981] had defined an optimal-behaviour zone corresponding to specific hydrodynamic conditions. It has been proved that, whatever the liquid-liquid system, the operation of the column under optimal conditions implies maintaining it near the flooding point (Fig. 9). As indicated in Fig. 9, five types of phase-dispersion behaviour have been observed in pulse columns as a function of feed flowrates (Qc and Qd) and the pulsating conditions (A_pFr) [Sege and Woodfield, 1954]. The optimal behaviour is defined in terms of column efficiency by the minimal amount of oxalic acid remaining in the raffinate (continuous-phase outlet) which corresponds indeed to the beginning of flooding.

This phenomenon is characterised by the appearance of a "fluidised like" swarm of dispersed phase drops just below the distributor and it is located between two operating regimes (see Fig. 9): the emulsion regime and the cyclic flooding regime. In this study, the objective is to minimise the concentration (Xr) of oxalic acid in the raffinate. It has been established that the flooding appearance could be detected by measuring the conductivity of the liquid medium at a place located just below the distributor. The conductivity fluctuates between two limits: the upper value is the aqueous-phase conductivity and the lower value is the dispersed-phase conductivity. The control purpose is to maintain the column in its optimal behaviour zone in spite of flowrates and physical properties of solvent and solute fluctuations.

4.4. Design Of The Neural Controller

The multivariable control of the column consists in the computation of the pulsation frequency and the solvent flowrate in order to maintain the column in the vicinity of flooding and to obtain a specific product quality.

A control scheme has been designed which objective is to maintain the column in its optimal-behaviour zone. The measured variables (controlled variables) are then the conductivity below the distributor and the final raffinate pH which represents the concentration of oxalic acid in the continuous phase outlet. The control actions are the pulsation frequency (Fr) and the solvent flowrate (Qd).

Owing to the interactions between hydrodynamic and mass transfer phenomena, the neural controller implements two interconnected networks (Fig. 14), based on the inverse modelling of the liquid-liquid extraction column. The first network (Fig. 12) computes the pulsation frequency to be applied to the pulsator in order to maintain the conductivity close to the desired set-point. The second network (Fig. 13) computes the solvent flowrate to be applied to the dispersed phase pump in order to obtain a given product specification and a desired conductivity at the bottom of the column.

The ranges of the input values for the neural networks corresponding to different operating conditions and step responses are presented in Table 1. A number of open-loop experiments have been performed (25 hours), involving essentially variations of the solvent flowrate and of the pulsation frequency in order to form the data base for the neural nets learning phase. Some of these input variations are shown in Fig. 10. The step responses of the pH (representing the concentration of oxalic acid in the raffinate), and the conductivity to pulsation frequency and solvent flowrate variations are presented in Fig. 11.

Table 1. Variation range for operating conditions and response of the process

Parameters	Min value	Max value
Qc (l/h)	0.0	40
Qd (l/h)	0.0	32
Fr (Hz)	0.5	2
Xi (mass.%)	0.5	2
pH	1.0	3
Cond (S)	0.0	1.9

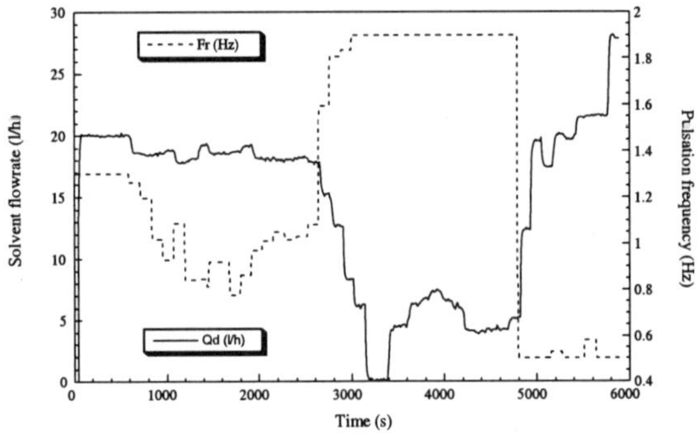

Figure 10. Dynamic steps of pulsation intensity and solvent flowrate.

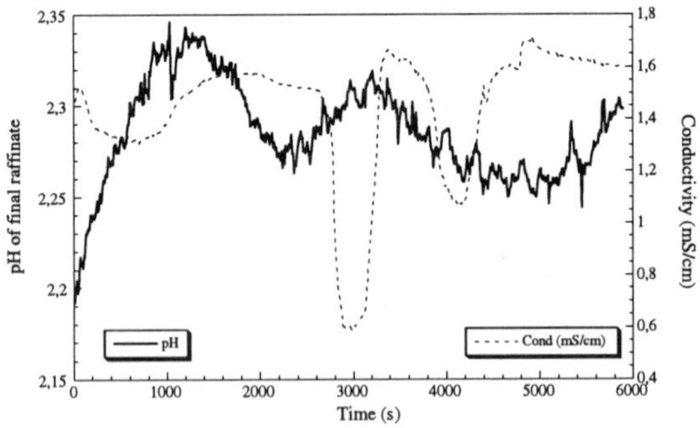

Figure 11. pH and conductivity step responses with Qc = 20 l/h and Xi = 0.5 wt%.

Analysis of the dynamic behaviour of the plant led us to consider two different sampling periods according to the phenomenon concerned: 10 s for the first network and 40 s for the second one.

The developed multivariable controller consists of two interconnected neural networks. These neural nets have been trained off-line. The first one (Fig. 12) is devoted to the computation of frequency (Fr(t)), the input layer includes 8 nodes (Qc(t-1), Qd(t-1), Fr(t-1), Cond(t-1), Qc(t), Qd(t), Cond(t), Cond(t+1)) and the hidden layer 10 nodes. The second network (Fig. 13), allows the determination of the solvent flowrate (Qd(t)), there are 12 nodes (Qc(t-1), Qd(t-1), Fr(t-1), pH(t-1),Cond(t-1), Qc(t), Fr(t), Xi, pH(t), Cond(t), pH(t+1), Cond(t+1)) in the input layer and 9 nodes in the hidden layer. The design methodology of these neural controllers is based on the process inverse dynamics modelling presented section 2.4: the learning data base is generated in an open-loop structure and learning of the neural network is carried out by considering the future process outputs as the references. Therefore, during the learning phase, pH(t+1) and Cond(t+1) represent the measured values of pH of the final raffinate and the conductivity at (t+1). During the process control, these values are replaced by the corresponding desired set-points.

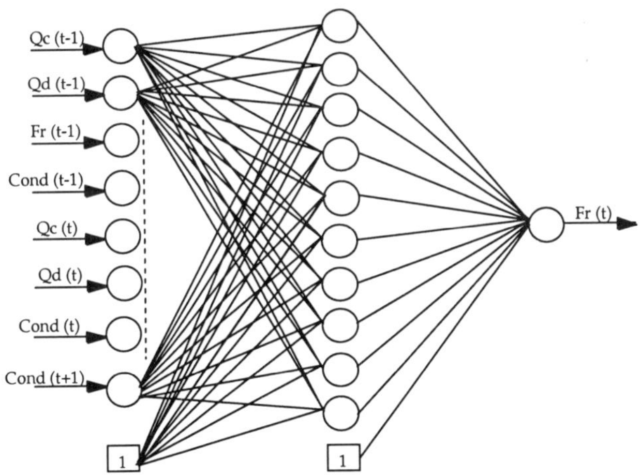

Figure 12. Neural network for the prediction of pulsation frequency.

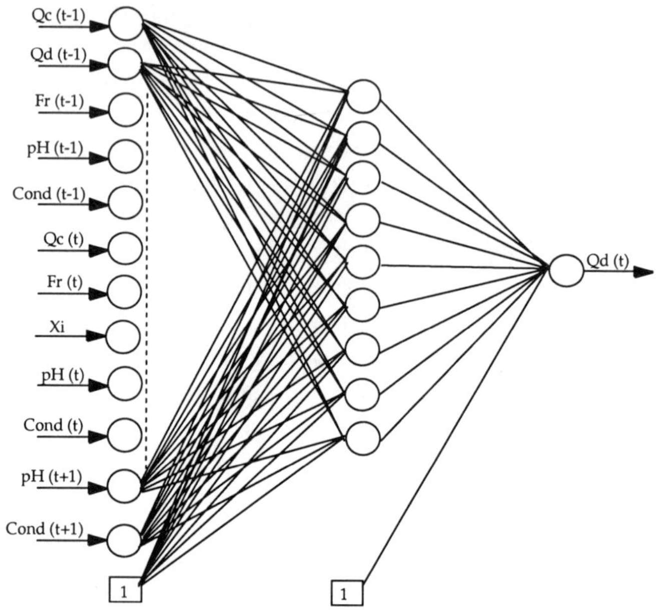

Figure 13. Neural network for the prediction of solvent flowrate.

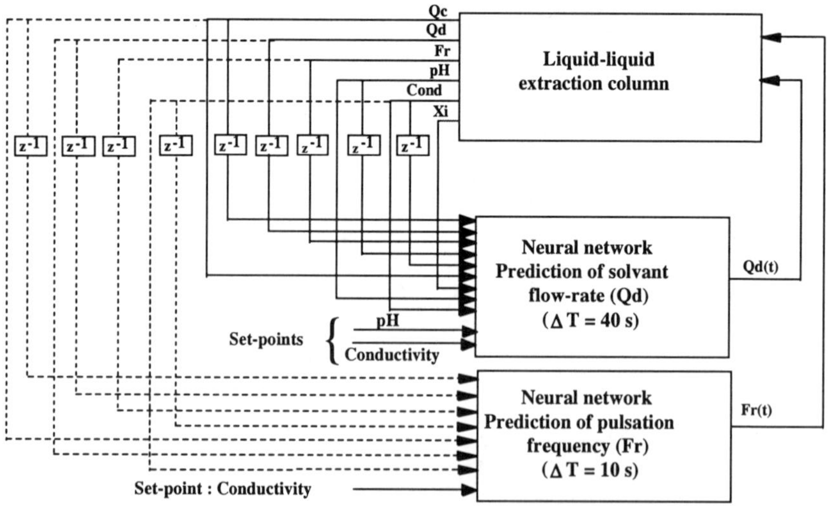

Figure 14. Block control diagram of the liquid-liquid extraction column.

4.5. Close-Loop Experiments

Initially, the column is manually brought near its optimal operating point, then the column is switched over the microcomputer. The control scheme, presented in Fig. 14, represents the neural controller constituted by the two interconnected neural networks. At every sampling period $\Delta T = 10$ s, the pulsation frequency is computed and applied to the column in order to maintain the conductivity close to the desired set-point (a specific hydrodynamic state). In another time, the solvent feed flowrate is computed at every sampling period $\Delta T = 40$ s and also applied to the process to get a given product specification (a weak oxalic acid concentration in the raffinate). The output variables are measured at time t and the control variables are calculated and applied to the column at $t + \tau$, where τ is the computation time related to the neural networks (less than 1 s).

Several real-time control experiments were performed on the column. To illustrate the performance of such an approach and to show the robustness of the neural controller, two control experiments are presented. The continuous phase feed is constituted by an aqueous solution of 0.5 wt% of oxalic acid. The dispersed phase is a mixture of T.B.P. and dodecane. The conductivity set-point was chosen equal to 1.4 mS/cm (value corresponding to the limit between the emulsion regime and cyclic flooding). The final concentration set-point was chosen equal to 0.05 wt% (corresponding to pH = 2.5).

For the first experiment, Figs. 15, 16, 17 and 18 represent respectively the time variations of the pulse frequency (action), the conductivity (controlled variable), the continuous phase flow and the dispersed phase flow (action), and finally pH of the final raffinate (controlled variable). It can be noticed that the neural controller performs well. The conductivity remains close to the desired value (Fig. 16) in every case. The difference between the measured output (conductivity) and the desired value (1.4 mS/cm) is less than 0.2 mS/cm (with the exception of the beginning of the experiment), in spite of a continuous phase flowrate change (13% - see Fig. 17). It can be noted that the control of the pH in the outlet stream has also been correctly performed by the elaborated control system (Fig. 18). The pH increases slowly, until it reaches the desired set-point, in spite of the decrease of the continuous phase feed (Fig. 18).

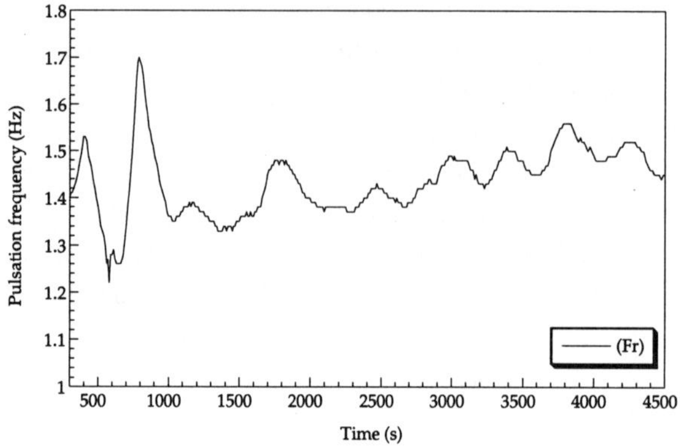

Figure 15. Time evolution of the pulse frequency.

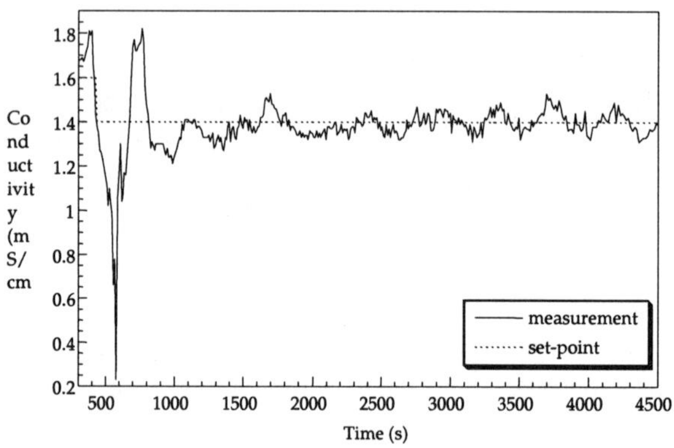

Figure 16. Time evolution of the conductivity.

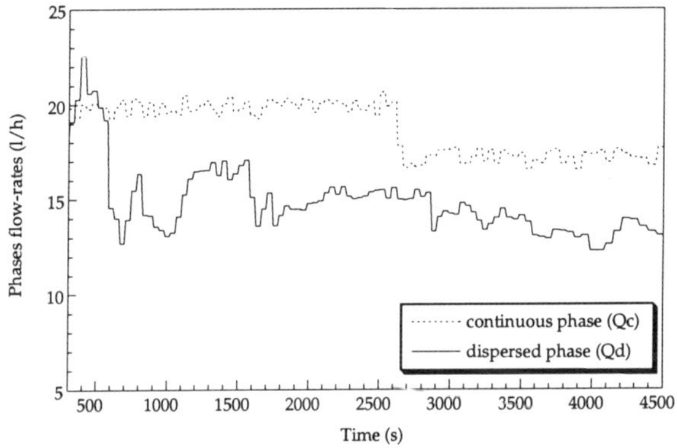

Figure 17. Time evolution of continuous and dispersed phase flowrates.

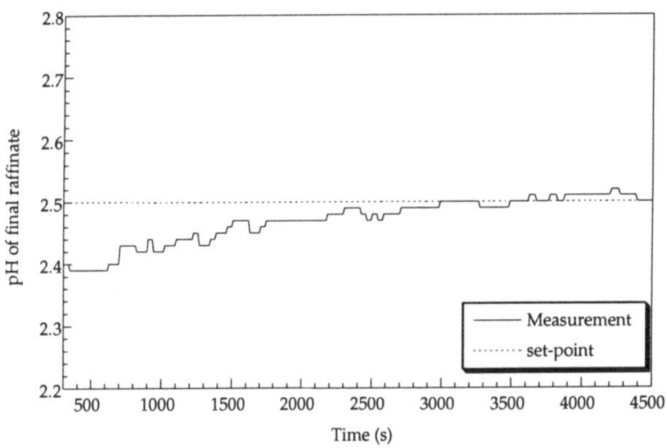

Figure 18. Time evolution of the final raffinate pH.

For the second experiment step variations are operated on the desired value of pH in the raffinate. The set-point started from 2.2 to reach 2.8 at the end of the experiment. These set-point modifications allow demonstrating the tracking performances of the controller. Figures 19 to 22 represent the time variations of the pulse frequency (action), the conductivity (controlled variable), the continuous phase flow and the dispersed phase flow (action), and the pH of the final raffinate (controlled variable) respectively. The conductivity (Fig. 20) is maintained close to the desired value in spite of fluctuations. The change in the continuous phase flowrate (18%) is quite important (Fig. 21). Between 4200 and 4600 s, a decrease of the conductivity has been registered (Fig. 20) following an increase of the solvent flowrate which was computed by the controller to compensate the change in the continuous phase flowrate (Fig. 21).

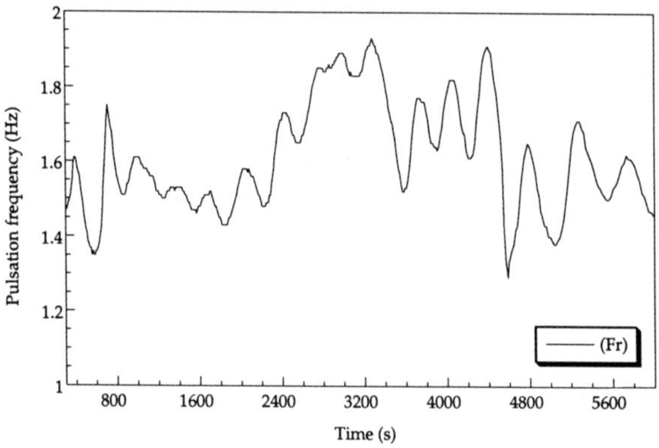

Figure 19. Time evolution of the pulse frequency.

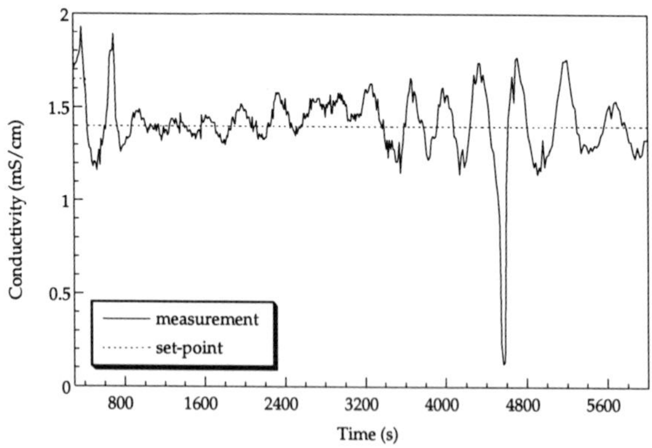

Figure 20. Time evolution of the conductivity.

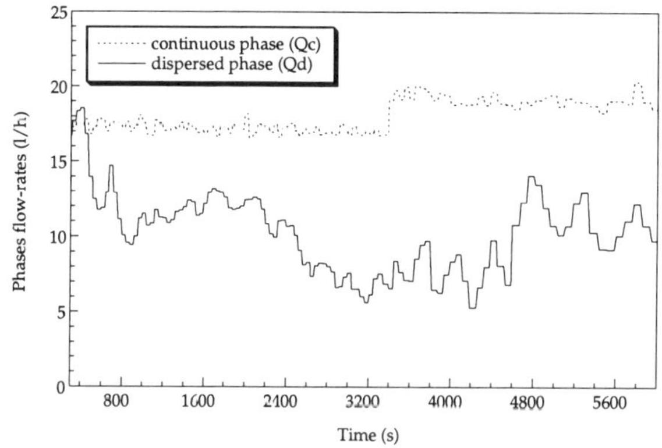

Figure 21. Time evolution of continuous and dispersed phase flowrates.

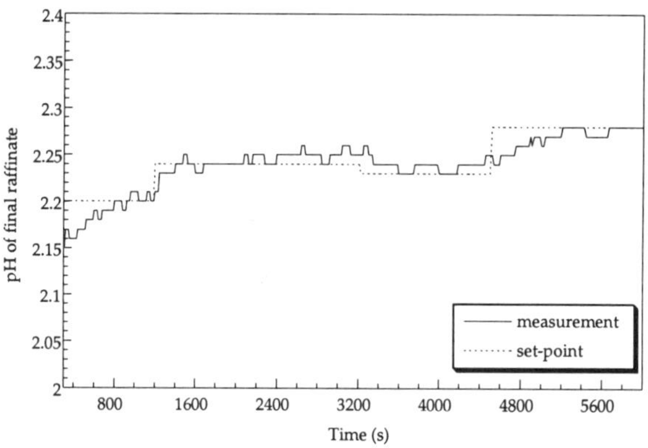

Figure 22. Time evolution of the final raffinate pH

4.6. Conclusions

This section presents the development of a multivariable neural controller, based on two interconnected neural networks designed by inverse modelling. This controller has been successfully applied to a liquid-liquid extraction column, which presents a highly non-linear behaviour and time-varying dynamics. The results illustrate the efficiency of such a control methodology. It is important to notice that the control scheme has allowed considering two different sampling periods for the models, which is a key-point for this type of multivariable controller.

5. Design of A Global Strategy Based on Neural Networks For The Control of LPCVD Reactors

5.1. Introduction

A wide variety of thin films for microelectronic use can be deposited by low-pressure chemical vapour deposition (LPCVD). This technique is particularly used, in its most straightforward form, for the deposition of intrinsic polycrystalline silicon films from silane. The most popular equipment to implement this process is by far, the horizontal, tubular hot-wall reactor. It consists of a quartz tube, lying horizontally in a three-zone furnace. Both reactor ends are cooled, often by water cooling, in order to facilitate door tightness. The substrates are circular wafers, polished on one side, concentrically stacked inside the reactor hot zone and normal to the flow of gases. Gases, diluted or undiluted in an inert gas, enter the reactor and flow through the tube, up to the pumping system and are then exhausted. LPCVD processes are typically carried out at pressures less than 150 Pa and at temperatures of about 600 °C (580 °C to 630 °C) for polysilicon deposition.

The main concern of industrials is to obtain films of uniform thickness along the whole line of wafers. Up to now, the design and the selection of the operating conditions of LPCVD reactor are still performed by semi-empirical methods and trial-and-error procedures. In this paper, we present a new approach of LPCVD reactor modelling and thermal control based on the use of NN [Fakhr-Eddine, 1998].

5.2. LPCVD Reactor

A LPCVD reactor can be divided into three parts: the entrance and exit zones, where the doors are kept cold by water circulation and the central heated zone. A mathematical model describing the complete behaviour of low-pressure chemical vapour deposition reactors (CVD1) has been developed in our laboratory [Azzaro et al., 1992]. The reactor is assumed to be at steady state.

The overall stoichiometry of the reaction of polysilicon deposition from silane can be expressed as follows :

$$SiH_4 \text{ -----> } Si + 2\, H_2 \qquad (5)$$

To establish the model, it has been assumed that a LPCVD reactor can be considered as a series of continuously stirred tank reactors in which the gases are perfectly mixed, each reactor being constituted by an interwafers space, the corresponding internal wall of the tube and by the corresponding part of the internal elements (wafer supports, etc....). The main heat transfer mechanism is radiation between solid surfaces (wafers and walls). In the case of pure polysilicon deposition, it is assumed that there are no radial variations. With such assumptions, deposition naturally leads to uniform layers across each wafer.

In each cell, the consumption of silane by reaction (5) results in silicon deposition on three places: on the tube wall, on the wafer carrier boat or other internal elements, and on the wafers surfaces. In the isothermal part of the load, only silane consumption is observed and the growth rate decreases along the reactor. The parameters governing the deposition rate and therefore the deposit thickness are the reactor temperature, the reactor pressure and the gases flow rates. At the reactor entrance, only SiH_4 is present, but along the reactor H_2 is produced and its flow rate increases.

The geometrical parameters of the pilot reactor (Fig. 23) used to obtain the results presented in this paper are the following:

- Tube length: 2 m
- Tube diameter: 153 mm diameter
- 100 wafers of 100 mm diameter in the isothermal part
- Interwafers distance: 10 mm

The behaviour of LPCVD reactor for silicon deposition from silane is very well predicted by CVD1 model [Azzaro et al., 1992] for various operating conditions. Hence, a comparison between neural networks and CVD1 performances is sufficient in a first attempt to evaluate the feasibility of such an approach.

5.3. Modelling of the LPCVD Reactor by Neural Networks

A first objective is to provide on-line sensors of film thickness. The LPCVD reactor has been lumped in a succession of basic elements or cells including 10 wafers.

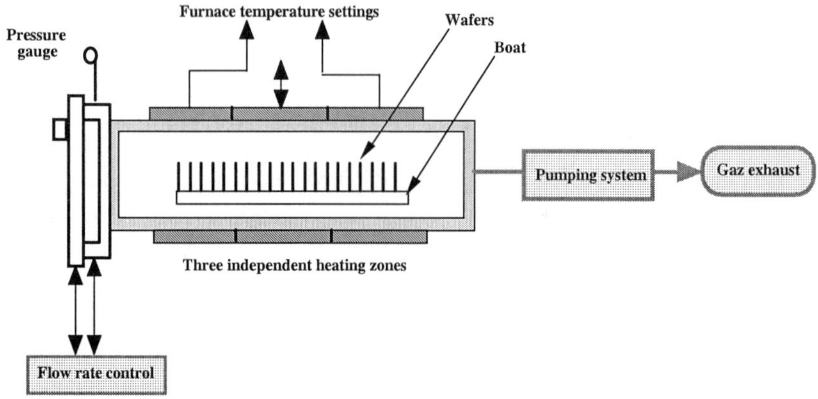

Figure 23. Low-pressure chemical vapour deposition equipment.

According to a previous physical analysis of the reactor, it is clear that the behaviour is the same along the reactor. Therefore, it is possible to model all the different cells by an unique NN model. This NN is composed of three layers. The inputs consist of a set of scaled values corresponding to the operating conditions: temperature, Pressure, SiH_4 flowrate and H_2 flowrate. The gases flow rates are expressed in sccm and correspond to $cm^3.min^{-1}$ for normal temperature and pressure conditions. The output layer corresponds to the deposition rate on two wafers of the basic element (number 3 and 7).

The learning data base has been generated by using the simulation code CVD1 [Azzaro et al., 1992]. Several isothermal runs have been carried out according to different operating conditions: temperature has been varied from 550 to 650 (°C), the pressure from 0.07 to 2 (Torr) and the input flow rate of SiH_4 from 150 to 600 (sccm). A data base of 125 examples has been elaborated which lead to 1250 examples corresponding to a single cell of 10 wafers. Then, 1000 examples have been used to form the learning data base and 250 to form the test data base. The best learning results have been obtained for a number of neurones in the hidden layer equal to 15 [Fakhr-Eddine et al., 1996].

The NN modelling accurately a cell of 10 wafers, it is possible to simulate the whole reactor by a succession of cells. Nevertheless, to go from one element to the next one, the values of the gases flow rates entering the next element have to be computed. This computation is carried out according to the values of the gases flow rates entering the cell and the film deposition rates computed by the NN.

An algebraic network has then been established according to mass balances equations deduced from the consumption of SiH4 and the production of H2 in the cell (equation 1).

Let us consider a cell including a given number (n_W) of wafers. The wafer surface is given by:

$$s_W = 2 \pi r_W^2 \quad (6)$$

where r_W is the wafer radius. The interwafer zone surface is:

$$s_i = 2 \pi d_{WW} r_t \quad (7)$$

where d_{WW} is the interwafer distance and r_t the reactor radius. Therefore, the total silicon deposit surface in a cell including n_W wafers is:

$$s_t = n_W (s_W + s_i) \quad (8)$$

The neural network computes the deposition rates on the wafers 3 and 7, the average growth rate in the cell can be approximate by:

$$V_{dSi} = (V_3 + V_7) / 2 \quad (9)$$

Consequently, the number of moles of Si deposited per second in a cell is given by:

$$F_{Si} = V_{dSi} * vm_{Si} * s_t \quad (10)$$

where vm_{Si} is the molar volume of Si.

The SiH_4 and H_2 flow rates are given for the normal conditions of temperature and pressure ($T_0 = 273.15$ K and $P_0 = 101325$ Pa). Therefore, according to the well-know relationship:

$$P_0 V = nRT_0 \text{ or } P_0 Q = FRT_0 \quad (11)$$

with R = 8.314 $J.K^{-1}.mole^{-1}$. The volumic flow rate of SiH_4 (in sccm) which has been consumed in the cell is computed by:

$$Q_{SiH4} = F_{Si} (RT_0 / P_0) * 6\ 10^7 \quad (12)$$

The SiH_4 flow rate entering the next cell is then given by :

$$D_{SiH4}(n+1) = D_{SiH4}(n) - Q_{SiH4}(n) \tag{13}$$

According to equation (1), the reaction produces two moles of H2 for one mole of SiH4 consumed.

$$D_{H2}(n+1) = D_{H2}(n) + 2 * Q_{SiH4}(n) \tag{14}$$

The LPCVD reactor is represented by a hybrid structure associating the NN and the algebraic networks. Globally, it is represented by a succession of 10 NNs and 9 algebraic networks according to Fig. 24.

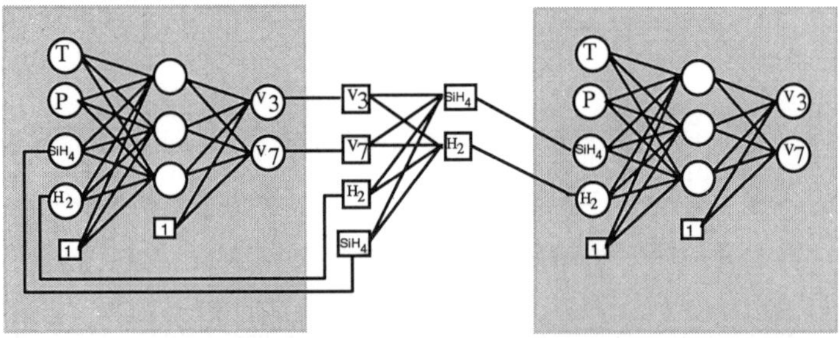

Figure 24. Architecture of the network model

5.4. Modelling Results

To exemplify the validity of the developed methodology, longitudinal evolutions of the growth rate along the wafers line have been computed by CVD1 and the network model for different values of temperature. When a uniform temperature is applied to the reactor, the growth rate profile is decreasing along the reactor. It is then interesting to impose a temperature ramp down the length of the reactor to offset reactant depletion. To simulate a non-isothermal functioning of the LPCVD reactor, the temperature profile has been discretised assuming a constant temperature in each cell of 10 wafers. This temperature is then changed for each cell. A very good agreement is obtained between the network model results and the CVD1 computations, as shown by Fig. 25. Let us recall that the weights of the NN which simulates the cell of 10 wafers have been determined using a learning data base established with isothermal examples.

5.5. Optimisation

As said before, the main concern of microelectronic industrials is to obtain films of uniform thickness along the whole line of wafers. To solve this problem, an optimisation procedure has been developed. It consists in computing on-line the basic elements temperature profile inside the reactor. In order to obtain the same deposition rate for the overall wafers load, this optimisation is carried out in order to minimise the function F_{obj} given by:

$$F_{obj} = \frac{1}{2} \sum_{j=2}^{N} \left[\left(V3Si_{ref} - V3Si(j) \right)^2 + \left(V7Si_{ref} - V7Si(j) \right)^2 \right] \quad (15)$$

where N represents the number of basic elements and the number of variables to optimise (i.e. the N basic element temperatures). $V3Si_{ref}$ and $V7Si_{ref}$ are respectively the deposition rate on the wafers 3 and 7 of the central basic element of the load, used as references.

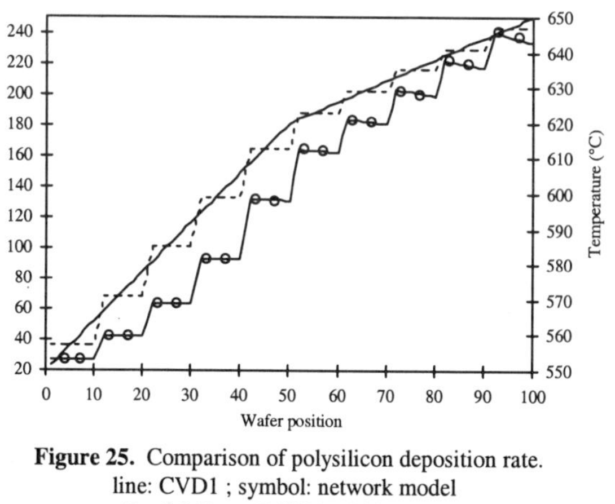

Figure 25. Comparison of polysilicon deposition rate.
line: CVD1 ; symbol: network model
P = 0.3 Torr - DSiH4 = 300 sccm.

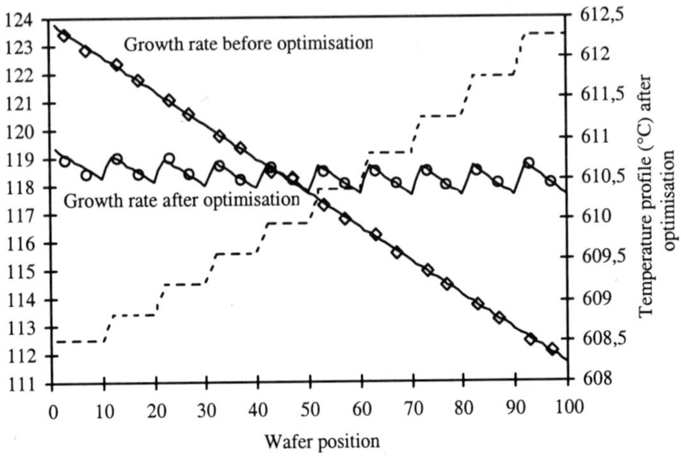

Figure 26. Optimisation of the temperature profile (P = 0.3 Torr - Tini = 610 °C - DSiH4 = 300 sccm).

In Fig. 26, the temperature profile obtained after optimisation is presented (dotted line). The continuous lines (with symbols) represent the evolution of the deposition rates for two cases: with an optimised temperature profile and for a constant temperature. When uniform temperature is applied inside the reactor, the growth rate profile is decreasing. The determination of a temperature profile inside the reactor by optimisation, allows a uniform polysilicon deposition rate.

However, set-up of a temperature profile inside the reactor represents a delicate control problem. The second section of this paper describes the treatment of the thermal control of the LPCVD reactor based on the use of NN controllers designed by inverse modelling.

5.6. Thermal Control of the LPCVD Reactor by Neural Networks

Among the more sensitive parameters of a chemical vapour deposition (CVD) operation are the wafers surface temperatures. However, temperature control of such a unit creates problems respectively in regulation to obtain the specified space profile and in pursuit to follow the desired evolution.

The experimental equipment is presented in Fig. 27. It consists of a horizontal quartz tube, heated by an electrical resistance organised in three zones, regulated independently by three PID's. The wafers load is centred in the heated zone of the reactor and is considered as a succession of three compartments corresponding to three heating-zone. The first and the last compartments include 30 wafers, whereas the central one contains 40 wafers. As a whole 100 wafers are treated at each run. Three thermocouples of K type (Chromel-Alumel) set in the middle of each compartment respectively at 82.5, 100 and 117.5 cm, allow the measurement of the compartment temperatures T1, T2 and T3. A computer equipped with A/D and D/A converters provides real-time data acquisition and control.

The three PID controllers are not directly connected to the temperatures measured inside the reactor. Up to now, these controllers are controlling the temperature of the electrical resistance. Therefore, only empirical knowledge of the thermal behaviour of the reactor allows the operator to obtain a temperature profile inside the reactor after a trial and error procedure. The objective of this work was to develop a controller, which can directly control the space temperature profile inside the reactor. To do this, three thermocouples have been set in the reactor, and a neural network has been developed in order to make a link between the measured temperatures and the set-points given to these PID's controllers acting on the electrical resistances.

Open-loop experiments (Fig. 28) have clearly shown that the three zones of the furnace do not behave independently but are strongly interacting. Therefore, to control the temperatures of the three zones simultaneously, it is necessary to implement a multivariable controller able to modify, at the same time, the control actions of the three heating zones. In practice, the three PID's independently control the electrical resistances, and the control actions which must be computed by the multivariable controller are the set-points given to these PID's.

5.7. Design of the Neural Controller

The neural network controller has been designed using the inverse dynamics modelling methodology (see section 2.4). The learning data base is obtained by applying input values to the plant in an open-loop structure. After a successful learning, the neural network is integrated in a feedback control loop. At this step, the neural controller must be able to compute the inputs (the manipulative variables) to apply to the process from the knowledge of its current state and the desired future state. The input units, which coded the future state of the plant during the learning phase, are then replaced by the future desired set-points.

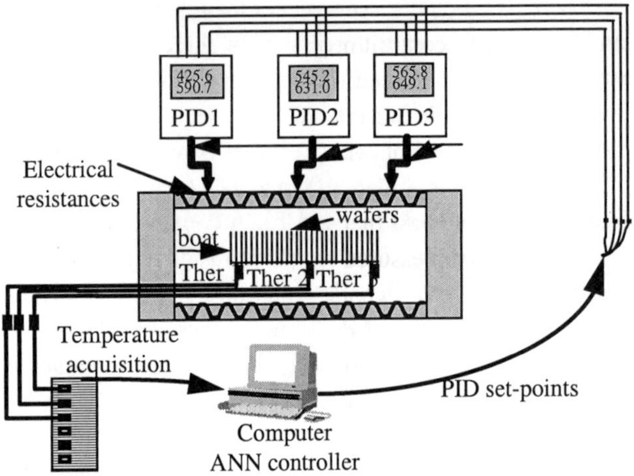

Figure 27. The experimental pilot-plant LPCVD reactor

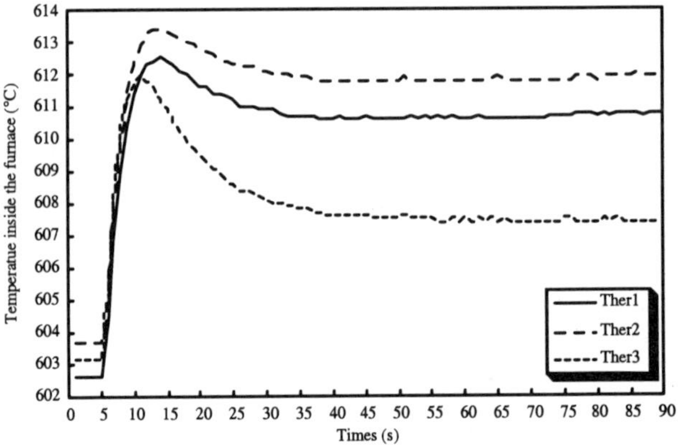

Figure 28. Dynamic response of the furnace to a step input of 10 °C applied to the PID of the central heating zone (zone 2) - step input applied at the 5th iteration

For the neural controller (NC), the notion of set-point must be introduced, i.e. the network must be able to compute the three PID's set-points in order to track a given temperature profile inside the reactor furnace. The output layer comprises three neurones to compute these PID's set-points ($PID_1(t)$, $PID_2(t)$, $PID_3(t)$) to be given to the local PID's controlling the electrical resistances of the furnace.

Concerning the input layer, a set of values are used to take into account the present thermal state of the reactor characterised by the three measured temperatures and the three PID's set-points (i.e. the values previously computed by the neural controller). Moreover, in order to model the thermal behaviour of the reactor, informations concerning past thermal state of the reactor were also included in the input neurones (i.e. past measured temperatures ($T_1(t-1)$, $T_2(t-1)$, $T_3(t-1)$)).

The learning data base is composed of a set of experimental temperatures data. Several experiments, of a total duration of roughly 13 hours, involving temporal step variations of the PID's set-points [PID1(t), PID2(t), PID3(t)] of variable lengths and amplitudes, have been carried out in order to obtain temperatures evolutions inside the reactor in the normal operating conditions (550°C to 650°C).

To improve the information quality, the PID's actions were disturbed of ± 1% of their nominal values. With a sampling period of 60 s, a data base of 780 examples has been elaborated, from which 520 examples have been selected to form the learning data base and 260 to form the test data base. The best learning results have been obtained for 11 neurones in the hidden layer (see Fig. 29).

The interest of the NN temperature control has easily been demonstrated by comparing the desired and the real temperature profiles obtained in the furnace during several 1.5 hours experiments. Result of a temperature control corresponding to a load of 100 wafers positioned in the centre of the hot zone inside the reactor, is presented Fig. 30.

A very good agreement between the desired values (solid lines) and the measured values (dotted lines) is observed for the three zones inside the reactor. In more details, light oscillations can be observed at the beginning of the control procedure, which then disappear.

5.8. Global Strategy

To achieve film thickness control in an experimental LPCVD reactor pilot plant in order to get a defined an uniform deposition thickness on the wafers all along the reactor, a global software has been elaborated, constituted by the hybrid network model which is used as a software sensor of the deposition rate, an optimisation

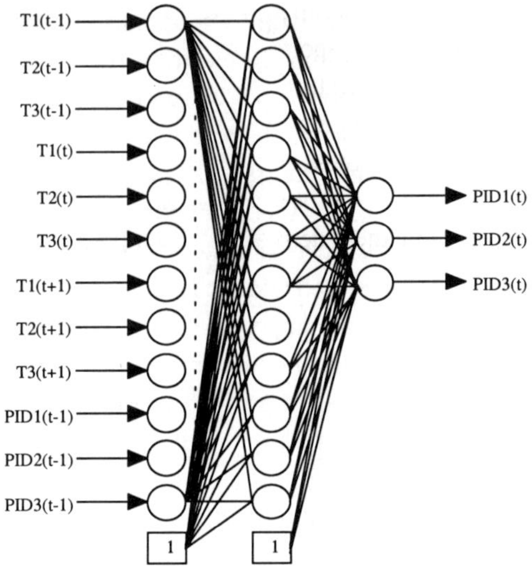

Figure 29. Architecture of the neural controller

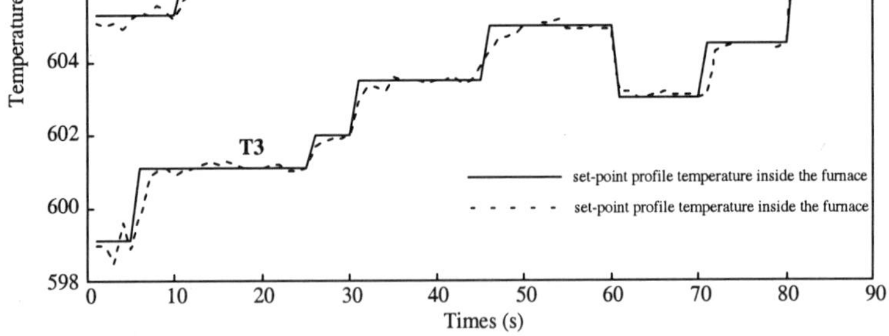

Figure 30. LPCVD reactor temperature control by the neural controller

algorithm which allows to determine on-line the required temperature profile inside the reactor (used as set point-point for the neural controller) and the NC which ensures tracking of this temperature profile.

The average thickness computed by the networks model is evaluated at every iteration. If the desired thickness is to be obtained at the next step, the process is stopped. In the other case, the procedures are repeated until the desired thickness are reached. The average thickness deposition is computed by:

$$Ep_{\text{arrêt}} = \sum_{j=1}^{N} \frac{V_3(j) + V_7(j)}{20} \qquad (16)$$

where N represents the number of basic elements of 10 wafers (N = 10), and V_3, V_7 are respectively the polysilicon deposition rates computed by the hybrid networks model for the wafers number 3 and 7.

To illustrate the efficiency of the global software, time longitudinal evolutions of the thickness deposition on the wafers are presented on Fig. 31. The different curves correspond to the longitudinal evolution of the silicon thickness deposition computed on-line by the hybrid networks model at different iterations. We can observe that, rapidly, the silicon thickness deposition is uniform over all the wafers of the load due to the optimisation procedure and the action of the neural controller.

Validation of all the computations is given by the measures carried out at the final step. Effectively, for the last iteration, the values computed by the hybrid networks model are compared with 6 experimental thickness deposition measures. A very good agreement between experimental and computed thickness deposition values is observed. The desired final thickness was 1620 Å. and the values plotted on Fig. 31, show that, at the end of the run, a uniform thickness deposition is obtained very close to this desired value.

5.9. Conclusions

In this paper, LPCVD model and controller have been developed based on the use of NNs. Firstly, a NN model has been determined to compute the deposition rate on two wafers in a cell including 10 wafers. By associating this NN to an algebraic network, a hybrid networks model has been realised which allows computing the deposition rate profile along the reactor. A new approach for LPCVD reactor temperature control has been developed based on the use of a neural controller (NC) designed by inverse modelling. The NC has been designed to compute the set-points that must be given to the three PID's controlling the furnace zones to obtain a convenient space-time temperature profile inside the reactor. Good results have been obtained for the control of space-time temperature profiles inside a pilot LPCVD reactor. Finally, a global software has been elaborated to achieve film thickness control in an experimental LPCVD reactor pilot plant. The aim of the experiments was to get a defined a uniform deposition thickness on the wafers all along the reactor. The software is constituted by the hybrid networks model which is used as a software sensor of the deposition rate, an optimisation algorithm which allows to determine on-line the required temperature profile inside the reactor (used as set points for the NC) and the NC which ensures tracking of this temperature profile. Experimental results are presented which confirm the efficiency of the whole control strategy.

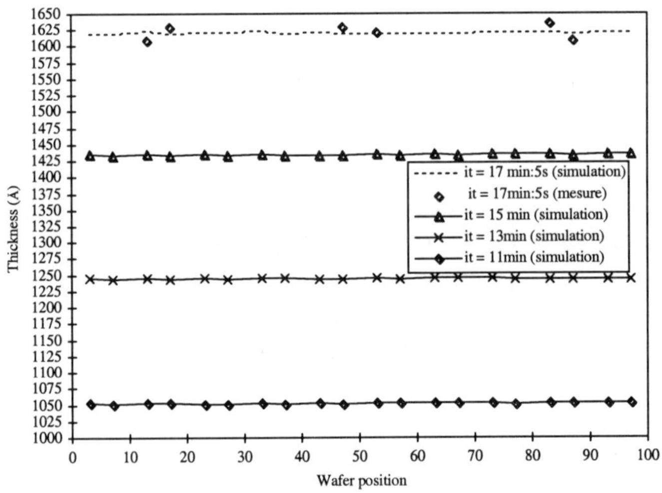

Figure 31. Evolution in time of the silicon thickness deposition on the wafers ($P=0{,}3$ Torr - $DSiH_4=300$ sccm - $T_{compl}=600$ °C).

6. Conclusions

The above results show that artificial neural networks provide an exciting opportunity to rapidly develop controllers of complex processes. Dynamic modelling capabilities of artificial neural networks have been exploited to build direct process controllers through the so-called inverse modelling methodology. The use of this methodology allows to rapidly elaborate controllers of complex and non-linear processes.

The chosen examples go from a single input – single output process to a multivariable one with strong interactions between the different variables and time dynamics. It is important to notice that a preliminary analysis of the process behaviour is fundamental to correctly choose the inputs of the neural network. For example, the influence of time-delay has been clearly demonstrated. On the other hand, the design of special architectures can be necessary, for example by considering interconnected neural networks, to properly take into account the different dynamics of phenomena.

Finally, the different applications presented in this chapter demonstrate that a good understanding of the process behaviour plays a key-role in the success of the development of neural networks as complex chemical processes controllers.

References

Azzaro, C., et al., *Chem. Engng Sci.* **47** (1992), 3827–3838.

Baum, J. and Haussler, D., *Neural Comput.* (1989), 151-160.

Bulsari, A. B., (Ed.), *Neural Networks for Chemical Engineers* (Elsevier, Amsterdam, 1995).

Casamatta, G., *Comportement de la population des gouttes dans une colonne d'extraction : transport, rupture, coalescence, transfert de matières*, Ph.D. thesis (I.N.P. Toulouse, 1981).

Casamatta, G. and A. Vogelpohl, *Ger.Chem.Engng.* **8** (1985), 96-103

Chouai, A., *Application des réseaux de neurones à la modélistion et à la commande multivariable des colonnes d'extractionliquide-liquide*, Ph.D. thesis, (I.N.P. Toulouse, 1999).

Delgrange, N., et al., *Desalination.* **118** (1998), 213-227.

Dennis, J.E., Jr., and Schnabel, R.B., *Numerical methods for unconstrained optimization and nonlinear equations* (Prentice-Hall, Englewood Cliffs, New Jersey, 1983).

Dirion J.L., *Contribution à la mise en œuvre de réseaux de neurones pour la modélisation et la conduite thermique de réacteurs batch*, Ph.D. thesis, (I.N.P. Toulouse, 1993).

Dirion, J.L., et al., *Comput. Chem. Engng.* **19** (1995), s797-s802.

Dirion J. L., et al., *Chem. Eng. Proc.* **35** (1996), 225-234.

Faizal M., et al., *in Proc. of Fourth World Congress of Chemical Engineering*, Kalsruhe (1991)

Fakr-Eddine, K., et al., *Comput. Chem. Engng.* **20** (1996), S521-S526.

Fakr-Eddine, K., *Elaboration d'un capteur logiciel à base de réseaux de neurones pour la régulation thermique et la conduite des réacteurs de LPCVD*, Ph.D. thesis, (I.N.P. Toulouse, 1998).

Grondin-Perez, B., et al., *Entropie.* **201** (1996), 49-56.

Hamachi, M., *Mesure dynamique de l'épaisseur du depôt à l'aide d'un capteur optique et modélisation par réseau de neurones de la microfiltration tangentielle de suspensions*, Ph.D. thesis, (I.N.P. Toulouse, 1997).

Hamachi, M., et al., *Chem. Eng. Proc.* **38** (1999), 203-210.

Le Lann, M.V., et al., *Adaptive Model Predictive Control, Methods of Model Based Process Control*, ed. Berber, R., (Kluwer Academic Publishers, Dordrecht, 1995), 426-458.

Morris, A.J., et al., *Trans IChem.* **72** (1994), 3 - 19.

Nahas E.P., et al., *Comput. Chem. Engng.* **16** (1992), 1039-1057.

Psichiogos, D.C. and Ungar, L.H., *Ind. Eng. Chem. Res.* **30** (1991), 2564-2573.

Scheflan, L. and Jacobs, M.B., *The Handbook of Solvents.* (2nd edn, Krieger Publishing Company, New York, 1973).

Sege, G. and Woodfield, F.W., *Chem.Eng.Progr.* **50** (1954) 396.

Thibault, J.and Grandjean, B.P.A., *IFAC International Symposium Advanced Control of Chemical Process*, Toulouse, France, 1991, 295-304.

Watrous, R. L., in *Proc. of IEEE First Int. Conf. Neural Networks* (1987) 619-627.

Zaldivar, J.M. and Hernandez, H., *Chem. Eng. Proc.* **31** (1992), 173-180.

15. INTELLIGENT MODELING AND OPTIMIZATION OF PROCESS OPERATIONS USING NEURAL NETWORKS AND GENETIC ALGORITHMS: RECENT ADVANCES AND INDUSTRIAL VALIDATION

L. PUIGJANER
Chemical Engineering Department, Universitat Politècnica de Catalunya
ETSEIB, Diagonal 647, 08028 Barcelona, Spain

Artificial Neural Networks (ANN) have been used as black-box models for many systems during the past years. Specifically, neural networks have been used advantageously in the Chemical Processing Industries (CPI) in a number of ways. Successful applications reported range from enhanced productivity by kinetic modeling, to improved product quality, and the development of models for market forecasting. Typically, a main objective in ANN modeling is to accurately predict steady-state or dynamic process behavior to monitor and improve process performance. Furthermore, they also can help in process fault diagnosis. The black-box character of neural net models can be enriched by available mathematical knowledge. This approach has been extended to consider nonlinear time-variant processes. The potential of neural network technology faces rewarding challenges in two key areas: evolutionary modeling and process optimization including qualitative analysis and reasoning. Recent work indicates that evolutionary optimization of non-linear time-dependent processes can be satisfactorily achieved by combining neural network models with genetic algorithms. Industrial validation studies indicate that present solutions point to the right direction, but additional effort is required to consolidate and generalize the results obtained.

1. Introduction

Just ten years ago, the only widely reported commercial application of ANN technology outside the financial industry was the airport baggage explosive detection system [1]. Since that time, scores of industrial and commercial applications have come into use, although the details of most of these systems are considered corporate secrets and are kept in secrecy. This hastening trend is due in part to the availability of an increasingly wide array of dedicated neural network hardware [2].

The first successful applications of adaptive neural networks were developed by Widrow and Hoff almost forty years ago. They employed single-neuron linear networks trained by the LMS algorithm [3]. These linear networks are easy to train and have found widespread commercial application over the past three decades.

Significant applications include: telecommunications (modems in the high-speed transmission of digital data through telephone chamois), control of sound and vibration (used in air-conditioning and automotive systems, and in industrial applications), and particle accelerator control (Standford Linear Accelerator Center). Unlike their linear counterparts, nonlinear neural networks have found commercial applications only recently. This is largely because the most useful neural network algorithm (backpropagation) did not become widely known until the beginning of the last decade [4]. The potential use of nonlinear networks is much broader than their linear counterparts, since they are best suited for applications involving complex nonlinear relationships for which acceptable classical solutions are unavailable. Such is the case of chemical process industries (CPI).

In the chemical process industries nonlinear models are typically required for process control, process optimization and prediction of process behavior. When theoretical modeling is difficult, data-driven modeling offers a unique opportunity [5, 6, 7, 8]. Successful industrial applications reported range from enhanced productivity by kinetic modeling [9], to improved product quality [10, 11, 12], and to the development of a realistic projection for a product's market [13]. Further use of neural network technology is in the inversion of very complex simulation models to know what range of plant operating conditions would result for a desired range of product properties [14].

Special attention has been given to neural network applications in process control such as nonlinear process identification and control [15, 16], adaptive process control [17, 18], process scheduling in real time [19] and using hybrid models to control chemical processes models [20, 21, 22]. Using neural network technology with data from chemical plant monitoring offers the prospect of better quality control. As the network is updated continuously with new data to increase its knowledge of the process, its output then can be used by the plant's process control system to set operating conditions for the new performance [5, 7, 23, 24].

Neural networks also can help in process fault diagnosis. The gradual degrading of process equipment performance through its life time can lead to deviations in the process variables and eventual breakdown. The causes of such deviations and/or equipment malfunction can be investigated via neural networks [25, 26, 27, 28].

The black-box character of neural net models can be enriched by available mathematical knowledge [29]. In this way real-time simulation can be effectively achieved. This approach has been extended to consider nonlinear time variant processes. In this case, it is necessary to continuously update the parameters in the network. Continuously updating and on-line adaptation raise a number of issues including the general approach for updating, the numerical method for recursive updating and the speed of updating. It has been demonstrated that the use of neural networks in conjunction with recursive least squares can be used effectively in industrial cases of some complexity [30, 31].

The potential of neural network technology faces rewarding challenges in two key areas: evolutionary modeling and process optimization. This is specially true for multiproduct and multipurpose flexible facilities where the production resources are confronted with a rapidly varying scenario. Very recent work [32,33] indicates that evolutionary optimization of nonlinear time-dependent processes can be satisfactorily achieved by combining neural network models with genetic algorithms. Industrial validation studies indicates that present solutions point to the right direction, but additional effort is required to consolidate and generalize the results obtained.

This work focuses in recent advances reported in dynamic process modeling. Specifically, a hybrid system which combines the potential of neural networks to recognise partners in the process variables together with the advantages of genetic algorithmic techniques for accurate process variables prediction purposes, is described in detail. In this way, a continuously updated process modeling can be obtained, which can be further used for product recipe improvement and in on-line production scheduling situations and real-time optimisation. Examples of industrial applications of substantial complexity are presented which demonstrate the feasibility of the proposed process modeling scheme and its potential for future developments.

2. A Hybrid Approach to Process Modeling

There is an increasing interest in developing modeling methods that successfully address the process dynamics and control. In this sense the analysis of time series has become an important subject in present industrial process modeling approaches, since it is able to provide accurate future, predicted values.

The ARIMA model allows to predict the value of *yt* in a time series by combining an autorregresive filter (AR), which uses the previous values of the series to produce the estimated forecast, and a moving average filter (MA) to produce the forecast from the previous series prediction errors (Fig. 1).

In the Box and Jenkins methodology (1976), the following iterative approach to model building for forecasting is proposed:

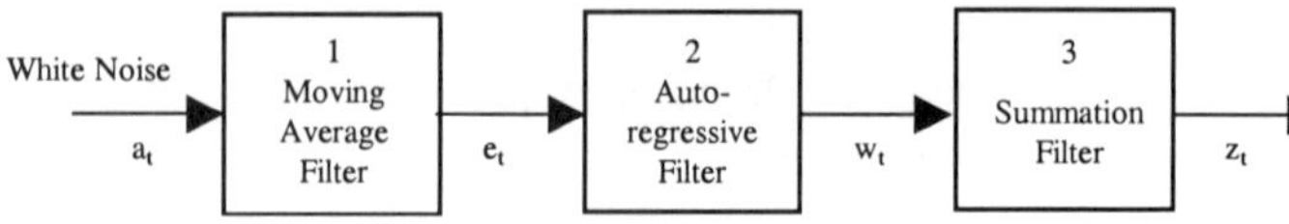

Figure 1. The ARIMA model block diagram.

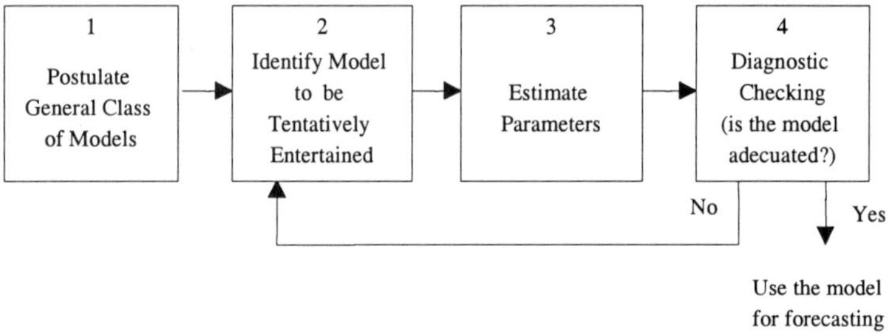

Figure 2. Iterative approach to model building by forecasting.

1. Fix a useful class of models from the interactions between theory and practice.
2. Identify subclasses of these models to be tentatively considered.
3. This tentatively considered model is fitted to data and its parameters estimated.
4. Diagnostic checking to know if this is an adequate model.

If any inadequacy is found, the iterative cycle of identification, estimation and diagnostic checking is repeated until a representation is found (Fig. 2).

In the classical approach, a discrete linear transfer function is considered to obtain the dynamic system response *yt* from an input *xt* in the presence of noise *Nt* (Fig. 3). This methodology suffers from both the expertise needed to follow the alternative steps to obtain the model, and the absence of automatic tools to estimate the model parameters. Furthermore, if the system analyzed were nonlinear, complex classical nonlinear methodologies would be required which demand increased experience.

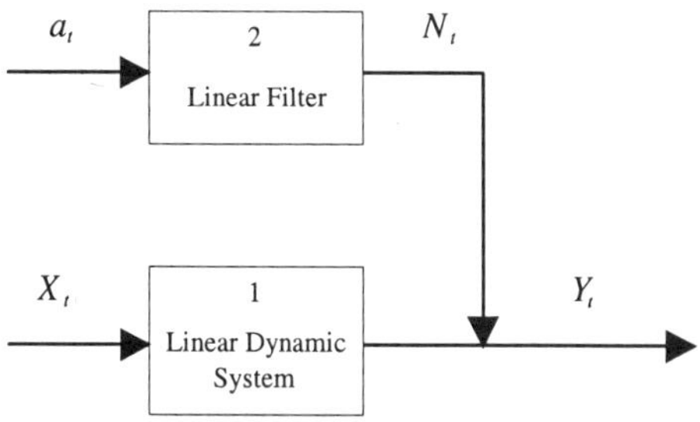

Figure 3. Dynamic System Response *yt* from an input *xt* in the presence of Noise *Nt*.

In principle, artificial neural networks should be very useful because of their ability to model complex nonlinear processes, even when process understanding is very limited [34]. However the ability of neural networks to learn non-parametric approximations to arbitrary functions is their strength, but it is also a weakness. A typical neural network involves hundreds of internal parameters, which can lead to overfitting and poor generalization. Moreover, interpretation of such models is difficult [35]. Present approaches try to combine "a-priori" knowledge with neural networks. These approaches exploit the knowledge available prior to receiving process data and attempt to reduce the dependence on noisy, sparse data. Alternative approaches have been summarized by Thompson and Kramer [36], and are given in Table 1.

The use of prior knowledge about the process is used to structure the neural network model. In the modular design approaches, neural network models are interconnected following the topological and functional structure of the process. Such is the hierarchical network proposed by Mavrovouniotis and Chang [35]. The resulting modular architecture has fewer parameters, easier training, reduction of infeasible input/output interactions and easier interpretation of model behavior. Semiparametric approaches combine a parametric model in series or parallel with the neural network. First principles models, existing empirical correlations or known mathematical transformations are the basis for parametric models.

In the serial approach (Fig. 4), the neural network estimates the process parameters which are used in the parametric model [37]. In this way, the internal structure of a hybrid neural network model clearly identifies the contribution of each part of the model to its predictions. As a result, the number of potential error sources can be drastically reduced and the adaptation improved [16].

Table 1. Approaches Combining Prior Knowledge with Neural Networks [36].

	Model Structure			Training	
		Semiparametric			
Approach	**Modular**	**Serial**	**Parallel**	**Inequality Constraints**	**Objective Function**
Advantages	May Improve Interpretability Easier to train	Guaranteed Output Behavior	Network compensates for discrepancies between data and inexact parametric model	Consistent output	Preferred functional behavior. Improved generalization
Disadvantages	Output behavior not guaranteed Unstructured subnetworks	Unstructured networks	Output behavior not guaranteed	More difficult to train	Difficult to determine appropriate form

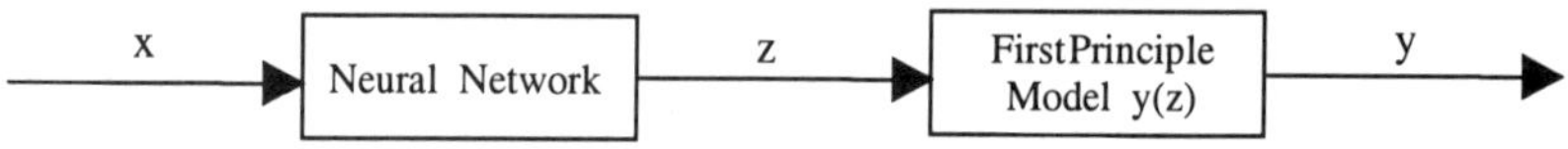

Figure 4. Serial semiparametric model.

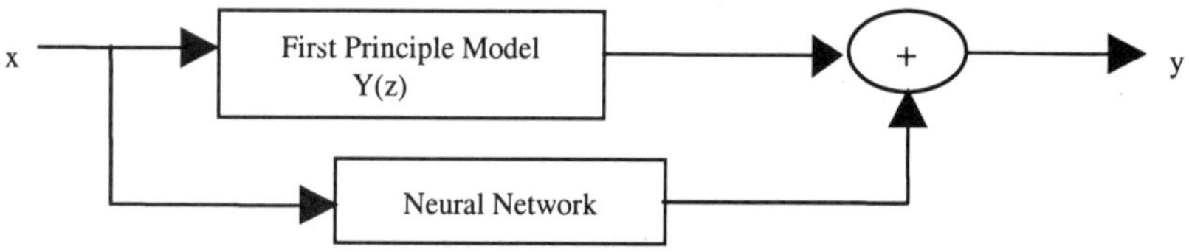

Figure 5. Parallel semiparametric approach.

The parallel semiparametric arrangement uses the combined output of the neural network and the First Principles model to determine the total model output (Fig. 5). The neural network is trained on the residual between the data and the parametric model to compensate for any uncertainties that arise from the inherent process complexity [38].

Additionally, model training approaches use prior knowledge to set inequality constraints on the model, which may involve the inputs and outputs of the network, as well as the model parameters. In this sense, prior knowledge dictates the form of the parameter estimation problem. This reduces the feasibility region of the parameter space, and the account of data required for their optimal estimates.

In an attempt to create a general methodology that combines many forms of prior knowledge with neural networks for modeling chemical processes, a hybrid model using nonparametric radial basis function network (RBFN) has been proposed [36]. The model structure is shown in Fig. 6. In this structure, a parametric "default" model in parallel with a RBFN are combined in series with a parametric output model. The default model accounts for parametric model behavior that holds in the absence of data. The neural network captures unknown functional relationships between the inputs and outputs. The output model enforces the explicit functional relationship between the inputs and outputs.

The authors successfully apply this modeling scheme to synthesize structure of a fed-batch penicillin fermentation. The process state at time t is defined by three state variables (penicillin concentration, biomass concentration and substrate concentration) and other three inputs are exogenous variables (substrate concentration in the feed, dilution rate and time increment) The three output variables are state variables at time $t + \Delta t$ (Fig. 7).

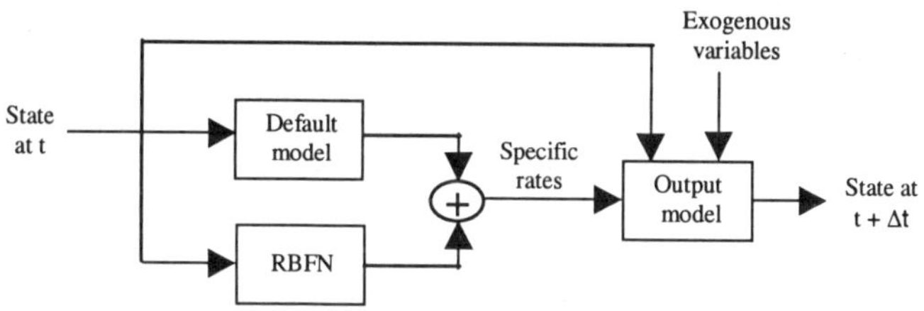

Figure 6. Hybrid model structure.

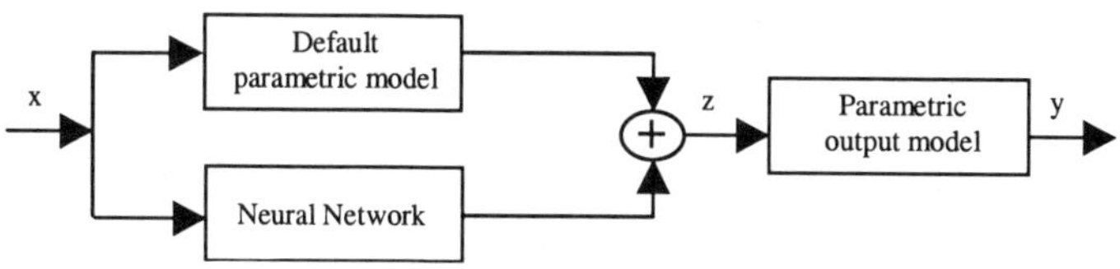

Figure 7. Hybrid model for penicillin fermentation study [36].

3. Dynamic Modeling and Control Hybrid Approach

The hybrid modeling methodology has been extended to consider real-time situations. Shubert et al. [39] combined the serial model approach with a fuzzy expert system to model a real-time fed-batch baker's yeast production. Although this model offered better interpolation and range extrapolation properties than pure black-box neural network models; however, the dimensional extrapolation properties were not studied. Therefore, it is not possible to relate a-priori the application domain of the model to the required domain for the identification data.

A serial semiparametric modeling arrangement has been proposed that combines the neural network model with the general structure of first principles dynamic models, based on macroscopic balances for application in biochemical processes [40]. This approach results in accurate models with reliable extrapolation properties using only a limited data set for identification. Furthermore, the proposed model is tested for its ability to function well in a model-based predictive controller (MPC). The strategy is demonstrated on the modeling and control of a pressure vessel, for which real time results are presented.

The candidate model is compared with pure neural network models and with a serial semiparametric model containing a polynomial, with respect to its interpolation and extrapolation properties (Fig. 8).

In order to clarify the origin of the improved dimensional extrapolation of the obtained serial model, it is also compared with a parallel semiparametric model (Fig. 9).

In all cases, the future pressure $y(k+1)$ is predicted on the basis of the actual pressure $y(k)$ and two inputs (valve position $u_1(k)$ and gas flow rate $u_2(k)$). Possible inaccuracies in the model predictions are caused by the parameter K, which is associated with the friction in the outlet. The inaccurate known terms of a macroscopic balance, like conversion kinetics and friction factors, can be modeled by a neural network, and the identification data covered only the input-output space of the inaccurately known terms.

The model predictive control (MPC) requires a dynamic model which can predict with reasonable accuracy over a horizon. Standard feedforward network architecture generally perform poorly over a trajectory because errors are amplified when inaccurate network outputs are recycled to the input layer.

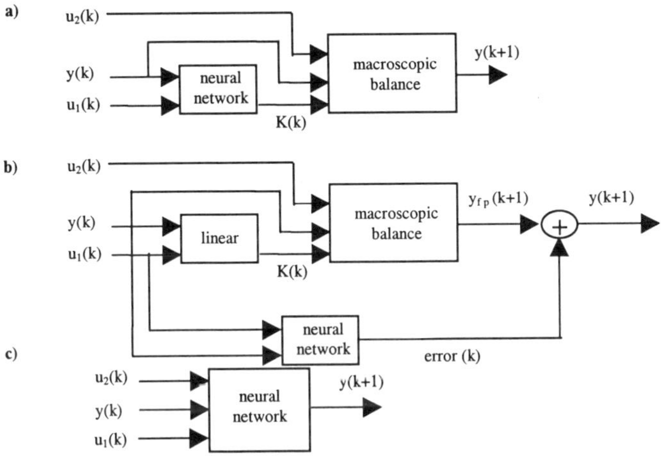

Figure 8. Different model configurations.
a) principles model, containing a linear correlation for *K(k)*;
b) black box neural network reference model for single-input single-output mode;
c) serial gray box model with a polynomial for *K(k)* [40].

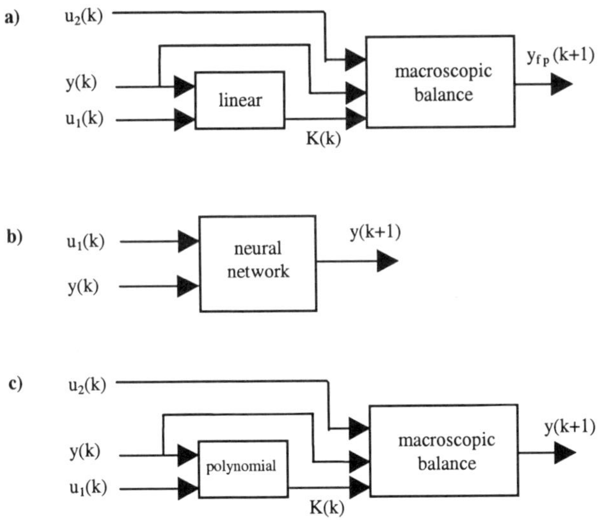

Figure 9. Alternative model configurations.
a) Serial hybrid model with neural network for *K(k)*;
b) parallel hybrid model with first principles model and a neural network model;
c) black box neural network reference model [40].

To improve prediction over a horizon, time-lag recurrent networks have been proposed [18]. A network that is trained in this mode has the ability to predict process behavior with a consistent degree of accuracy (Fig. 10). This kind of network has been successfully used in MPC [18, 24].

The general philosophy of neural model predictive control is the same as that of any MPC. The control consists in the optimisation of an objective function where the prediction model considers a dynamic neural network function. A general scheme of the controller is shown in Fig. 11. At every sampling step, the past and current measurements of the controller and manipulated variables are fed into the dynamic neural network model. Using the last vector of recommended manipulated variables, the model calculates the trajectory of the process outputs over the horizon. The prediction are input into the optimizer where the objective function is evaluated. The optimizer computes a new set of manipulated variables and passes them back to the neural network model. The iteration continues until the calculation converges. This model predictive control has been applied to an industrial packed bed reactor, where the neural network model-based controller can achieve tighter temperature control for disturbance rejection.

A different dynamic neural network architecture is used by [8]. Inspired from biological control systems, intrinsically dynamic neurons are the processing elements in the networks architecture. This results in a network which incorporates dynamic elements with continuous feedback.

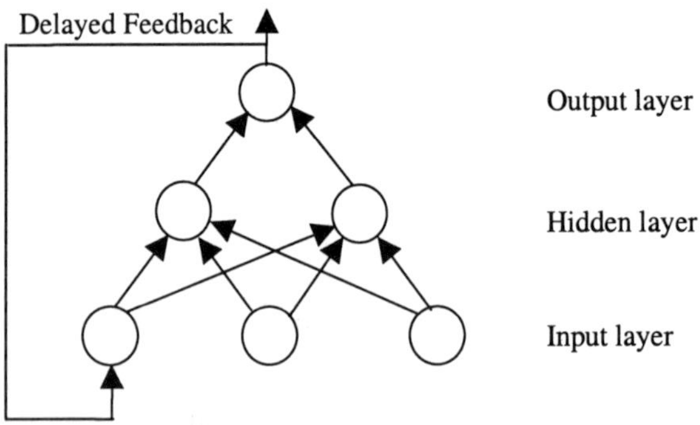

Figure 10. Architecture of a recurrent neural network.

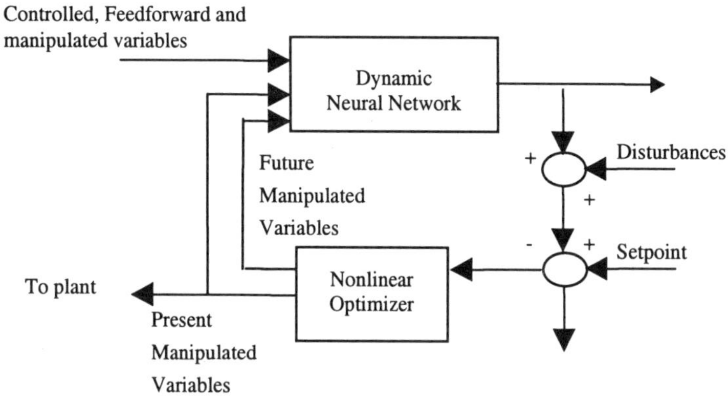

Figure 11. Controller structure.

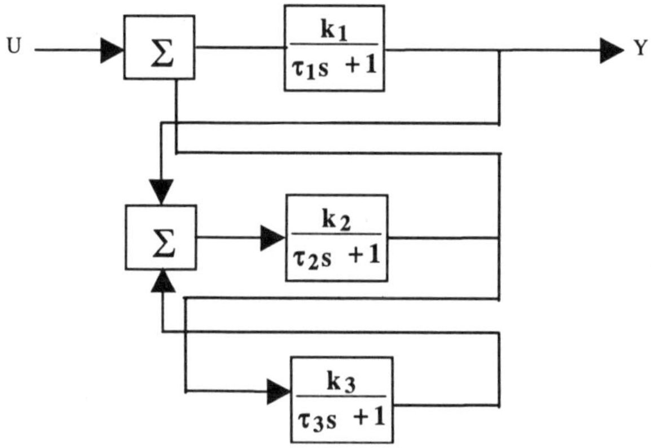

Figure 12. Generalized three neuron structure for dynamic neural network [8].

The dynamic neural network architecture belongs to the Hopfield network type [41], which is enriched with an independently nonlinear gain and time constant in a single neuron, giving rise to rich behavior with relatively few neurons. The generalized three neuron structure is shown in Fig. 12. Although several architectures are possible, in this case, each neuron receives the external input, but only one (the neuron whose output is the network output) receives outputs from the other two.

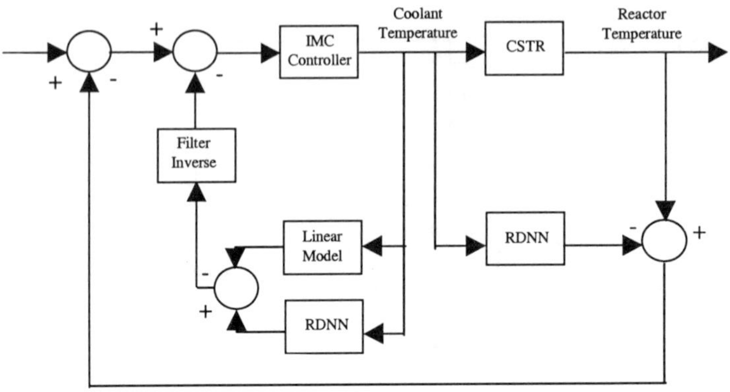

Figure 13. Closed-cool control structure for the case study [42].

The so-called biologically motivated dynamic network (RDNN) module can be implemented in a model-based control scheme, such as Internal Model Control (IMC) or Model Predictive Control (MPC). The control structure for a catalyzed reaction carried out in a well mixed stirred tank reactor [42] is shown in Fig. 13, and is composed of two parts:

1. the dynamic model (RDNN) which contributes to a feedback signal representing the difference between the true process and the modeled output; and
2. a model inverse loop which contains the RDNN model, a linear approximation to the RDNN model and a linear IMC controller.

4. Evolutionary Modeling

The development of a modeling technology for the optimization of process operations, taking into account energy, productivity, environment and economic issues, requires an integrated view of different problems affecting the competitiveness of process industries, leading to the study and development of new optimization methods that will integrate the synthesis, control and operation objectives and will treat steady state and dynamic models.

The use of Neural Networks as modeling methodology, implies the acquisition and management of large sets of plant measurements, leading to a final result without any formal relation with applicable physical laws. However, the simple structure of a neural network model should potentially permit its use in a wide range of situations, from evolutionary modeling to global plant multiobjective optimization, where other more comprehensive, but mathematically complex approaches have shown limited success. Even more, the generality and process independence of their structure favors the versatility of the computational tools based on this models.

Specifically, flexible manufacturing containing continuous, batch and semicontinuous processes offer a formidable challenge to the development of new methodologies and tools leading to improved process performance [46]. The effort is well justified, given the significant position of time-dependent processes in today's overall industrial texture. The inherent versatility of such processes makes them very attractive, since they allow the production of special chemicals with excellent yields and permit a rapid change from one process to another with minor modifications. However, this flexible processing network creates very complex situations at various levels of interrelated decision-making structures [43].

Production with batch and/or semicontinuous processes involves sequences of operations, defined by product recipes, which require precise synchronization and planning to meet the demand specified for each product, and to maintain the production facilities with high productivity levels at all times.

Present trends in batch process operations planning point out the need for off-normal conditions re-scheduling provisions in present scheduling algorithms. Unexpected events and/or off-nominal product specifications must be taken into account to update production planning, and provide for alternate routes when machine failure or other bottlenecking problems may occur. A hierarchical decision-making structure for the production planning in single-site production plants has been recently proposed [47]. This system assures a continuous flow of information between three closely interrelated production levels:

- the plant management level, which involves decisions on allocating the available resources among the various products under demand, with eventual retrofit considerations and re-scheduling activities;
- the recipe level, which decides recipe initialisation, modification and any necessary correction;
- the process level, which implements decisions on standard regulation actions and sequence control, and provides real time information for decision-making at upper levels.

The solution approach [48] considers an adaptative re-scheduling knowledge-based strategy which results in successive recipe improvements, reduced lead times, and improved and more consistent product quality. The overall platform includes (Fig. 14):

- an expert process supervisory system which uses fuzzy logic for diagnosis in abnormal situations, and suggests batch changes during normal operation and eventual re-scheduling;
- a relational database management system (RDBMS) which is updated and enriched with knowledge and information provided at several levels and from different sources;
- a plant modeling system which is successfully improved and adapted with better knowledge of current process situations;
- a recipe catalogue updating system built on external information (legislation, patents, etc.) or internal information (recipe improvements, expert knowledge acquisitions, etc.); and
- a scheduling system supported by the multi-level expert decision-making framework.

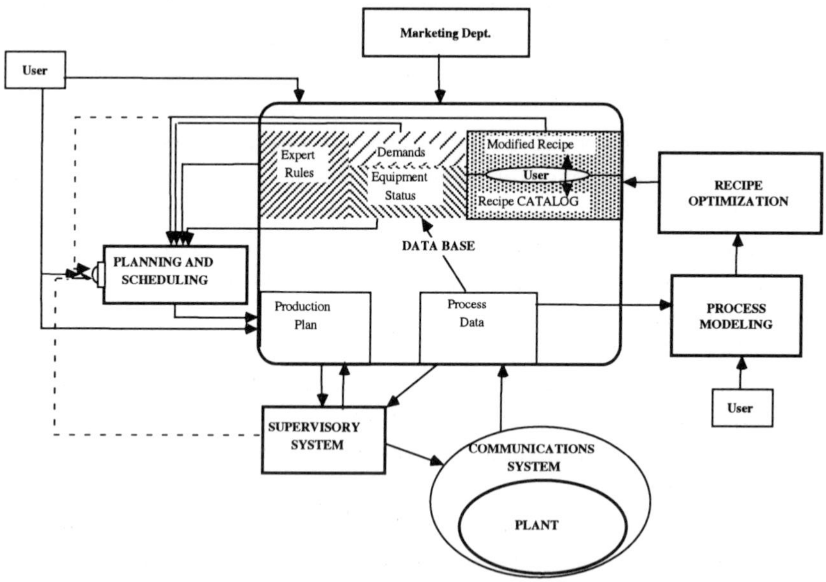

Figure 14. Configuration of the proposed schedule optimization and recipe adaptation platform.

A key element in the above strategy is plant modeling updating. Very recently, knowledge-based modeling is emerging as a realistic and promising support technique to solve routine/predictable problems at industrial scale. The potential of neural networks to recognize patterns in the process variables through a training procedure is also becoming a practical promise. A hybrid expert system/neural network has been proposed which exploits the advantages of each [49]. Towards this end, a new kind of neural network system has been developed which overcomes present limitations by integrating genetic algorithmic techniques, so that it can be used for accurate process variables prediction purposes. In this way a continuously updated process modeling can be further used for product recipe improvement, in an on-line scheduling scheme or for any of the other decision-making scenarios outlined above.

5. A Hybrid Approach to Evolutionary Modeling

A hybrid approach has been proposed to model the process automatically from historical and present data, including the building of the neural network structure itself and the parameters estimation using the genetic algorithm (GA) paradigm [50].

The feedforward structure of recurrent Neural Networks has been modified by using a new Non-linear Back-Propagation algorithm (NLBP). By using a nonlinear expression in the learning algorithm, the derivative involved in the procedure of updating the weights is avoided. An adaptative method was created to accelerate the backpropagation convergence [33]. The neural network model proposed is shown in Fig. 15. Using Fig. 3 as an illustrative base, the first component, the linear Dynamic System is substituted by a Nonlinear Dynamic System (Module 1) where the regressive relationship between the inputs xt and where the output yt is found; Module 2 contains a set of p neurons connected to the linear output to obtain their autoregressive relationship; a set of q neurons are connected to the linear output to find the relationship with the time residuals Yt (Module 3). All the modules are connected to the linear output (Module 4) (Fig. 15).

Model generalization is secured by splitting the learning pattern set in two, a *learning set*, and a *testing set*. The first is used for direct pattern estimation, and the second set is referred as internal validation test and is used to determine the stopping point of the training process. The cost function E_{ap} continues to be used for the learning set and E_{test} to evaluate the second set.

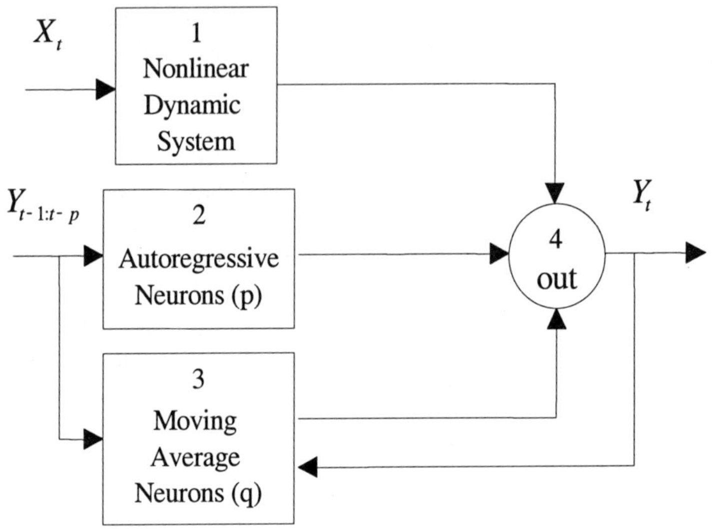

Figure 15. Neural Network Model Proposed.

When the learning process begins, both functions, E_{ap} and E_{test} have a monotone decrease and, usually after some epochs, the second function begins to grow, which indicates a decline on the generalization competence. Since local minimum will eventually appear, a sound heuristic solution has been developed consisting in saving automatically the set of parameters which give the least value of the expression $E_{ap}+E_{test}$ (Fig. 16). The testing set is chosen in the range of 15-30% of the total patterns to obtain a good generalization capability.

In order to avoid *over-parametrization*, the Akaike Information Criterion (AIC) and the Minimum Descriptor Length (MDL) have been employed. Therefore, to evaluate the neural model the expression $E_{ap}+E_{test}$ is used, but taking into account the parsimony principle by adding a penalty term. In this way, a good enough model is obtained containing the least number of parameters. The MDL criterion has been found to obtain the best results. The value of the expression $E_{ap}+E_{test}$ and taking into account the residual variance and the number of parameters is returned to the hybrid system controller, the genetic algorithm.

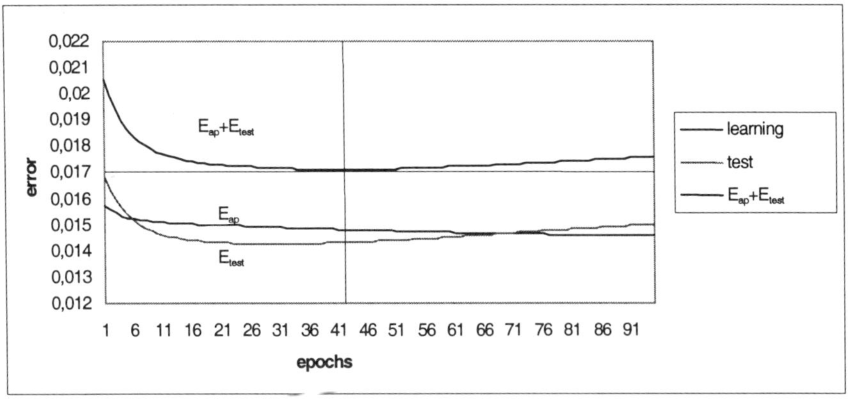

Figure 16. Evolution with time of the stopping criteria.

6. Genetic Algorithm

From an initial population of genomes generated randomly, where each one represents the genetic characteristic of a neural network successive populations are generated in the reproduction process and according to their fitness function, thus improving the results over time [44]. Employing these techniques the task consists of encoding a neural network on a string and then manipulating a population of these strings using the corresponding operator: reproduction, crossover and mutation. The final structure of the hybrid model proposed is shown in Fig. 17.

The general procedure is schematically given in Fig. 18. The first module corresponds to the Data Analysis component. In this part data are analysed to have a previous idea of the general structure, and four decisions can be taken:

- depending on the complexity, the maximal number of hidden units could be fixed. Using the Flexible Genetic Algorithm (AGF) could help to have an idea of the number of linear hidden units [53],
- the possible direct linear connections linking the inputs with the output by a multivariant linear regression or a step regression,
- the possible data transformation to obtain the best model,
- the analysis of the input data is a guide to determine:
  - if there are redundant variables and some of them need to be eliminated,
  - the input variables order according with the linear relationship with the output to take only a subset to create the direct linear connections,

- if the sampling interval is too narrow existing too much information to reduce the number of patterns.

This module can determine the length of the string used in the Genetic Algorithm (GA) and its composition using a flexible structure. Using these heuristics, the time of processing can be reduced and the search will be guide in a wide spectrum containing the optimal model to be found.

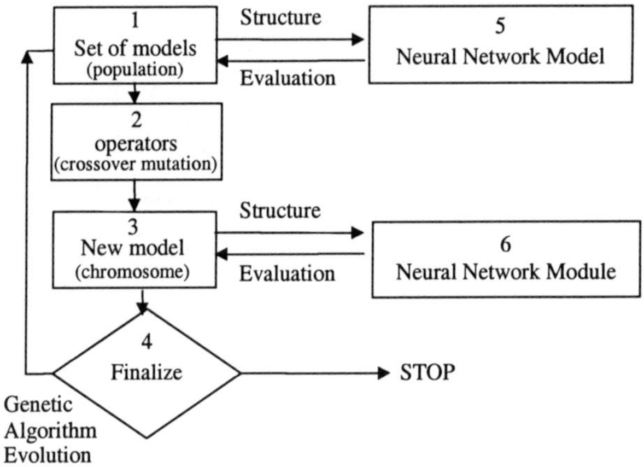

Figure 17. The hybrid System: GA and NN.

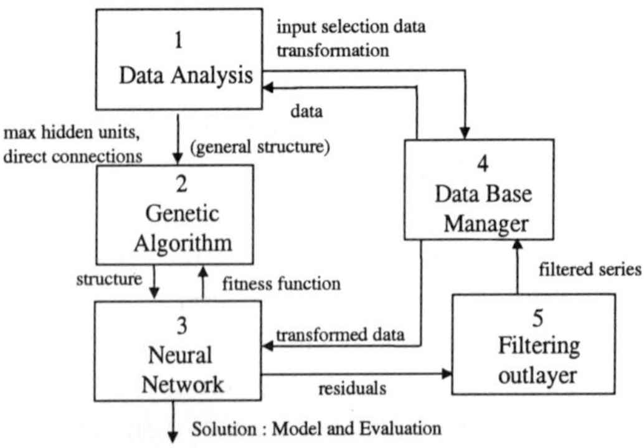

Figure 18. Overall solution approach.

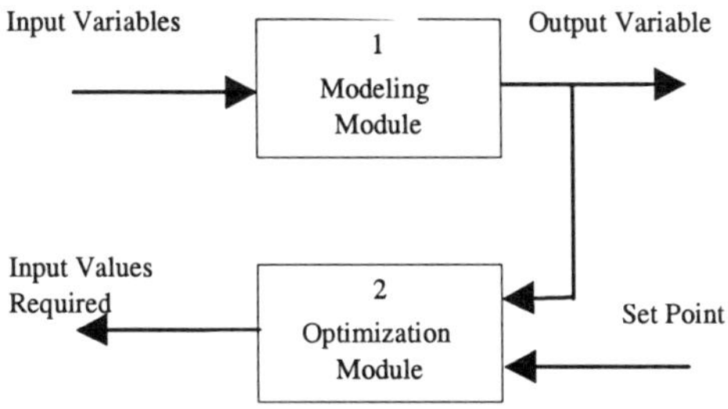

Figure 19. Optimization Procedure.

The second module is a Genetic Algorithm. This module provides to the NN module the structure to be evaluated and using an iterative process the best structure built is returned as the system solution.

We select this searching method because it guarantees a good enough searching in the space of states.

The module 3, the most important, estimates the parameters and returns to the GA module the fitness function taking into account the parsimony principle. The used stopping criterion allows us good generalisation of the NN structure found.

In the module 4 all the information is processed depending on the Data Analysis result. The NN module uses the training set transformed to the best form.

Finally, module 5 is the outlayer detection module that identifies outlayers and tests their influence on the average error and selects the data to be considered.

This scheme offers a first approximation to link some heuristics to determine a goodNN model, and to obtain it automatically.

Furthermore, the Hybrid Modeling Module (Hybrid System), can be combined with an optimization module to calculate the required status to produce the desired output in online operation (Fig. 19). A software prototype (ENESIMO) has been built on the above concepts and is successfully being tested in a variety of industrial scenarios, as indicated in the next section.

6. Industrial Case Studies

The evolutionary modeling methodology presented in this work has been tested and implemented in a variety of industrial scenarios. Furthermore, it can be used online to achieve real plant optimization by integrating the dynamics of the process and its scenario in the actual decision-making of the plant operation. Selected industrial applications are summarized in the following case studies:

6.1. Case Study 1: Malt Manufacturing

In this example, the process considered consists of barley malting and is based on one of the largest malt manufacturing industries in Spain.

The barley malting process employs usually a batch-wise procedure. In this specific factory, the processing stages can be grouped into five sectors. The most time and energy consuming step corresponds to the germination process, which must be conducted under rigorous temperature and humidity controls. The quality of the final product (beer) depends largely on a correct germination process and on the proper procedure to stop this germination by drying.

The germination process and the drying process have been chosen here as samples of the methodology employed and expected results. The germination process (germination1) has been modelled and the neural network simulation produces the chamber outlet dry air temperature as a function of five relevant process variables: 1) time (h); 2) outside temperature; 3) outside relative humidity; 4) the inlet air temperature; 5) the humid air temperature.

Using the hybrid system (Neural Network -Genetic Algorithm) it was found that the best identified genome for germination1 has 5-4-1 structure (inputs: 5, hidden: 4, outputs: 1) with a residual sum of squares of 0.02665 (Table 2, row 1). The hidden units have a sigmoidal activation function and the output has a linear activation function. Not only are the results better, but also one should note that a good neural model was found without knowledge-of or formulation of a mathematical model.

Table 2. Neural Modeling results

data process	net structure	learning error	test error	square sum	network param	learning patterns
germination1	5-4-1	0.02851	0.02651	0.02665	29	1749
germination2	6-4-1	0.01758	0.00951	0.00710	30	1748
drying1	9-3-1	0.02677	0.03272	0.02923	33	1178
drying2	10-4-1	0.01758	0.02354	0.01884	37	1177

(a)

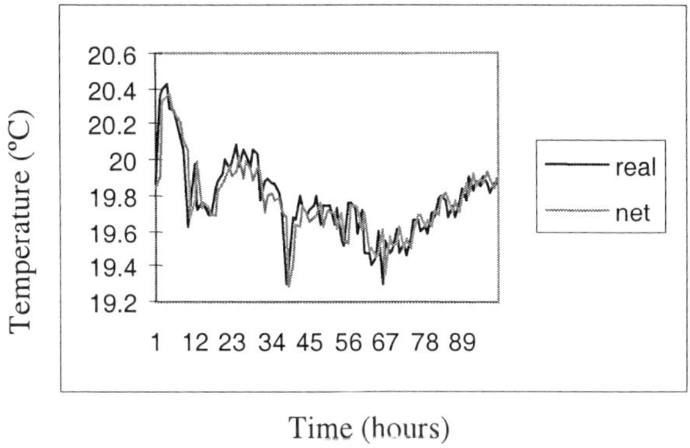

(b)

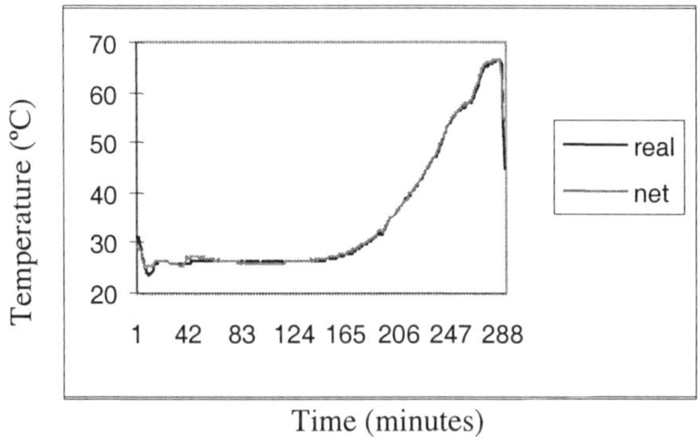

Figure 20. Real values and Neural Network results for the germination (a) and drying process (b). Time (a.hours, b.minutes) vs. Outlet air Temperature (°C).

The hybrid system has the possibility to test which autorregresive input or output improves the result. In this case (germination 2), it finds that adding an input, y_{t-1}, the result is improved (Table 2, row 2). Figure 20a shows the performance of the neural model (net) versus the real values (real).

The drying process (drying1) has been modelled and the neural network simulation produces the chamber outlet dry air temperature as a function of nine significant process variables: (1) the offset time from process time start; (2) the outside temperature; (3) the outside relative humidity; (4) the inlet air temperature; (5) the outlet air temperature; (6) the heat exchanger air temperature; (7) the outlet wet-bulb air temperature; (8) the inlet air pressure; (9) the outlet air pressure. The model is used for predicting and controlling the behaviour of any of the 5 drying chambers in the malting process. In this case the genetic algorithm to stop the learning process finds the least value of the expression MDL (function of $E_{ap} + E_{test}$). The last identified genome has 9-3-1 structure for drying 1 (inputs: 9, hidden: 3, outputs:1) with a residual sum of squares of 0.02923 (Table 2, row 3). The hidden units have a sigmoidal activation function and the output has a sigmoidal activation function. Figure 20b shows a good agreement between model values and experimental data.

In the last case, drying2, the system finds automatically that adding an input, y_{t-1}, the result is also improved (Table 2, row 4). The input data in all cases have been standardised. This becomes useful, since the pattern values are in different ranges and after this linearization procedure, all input neurons will have a mean value near zero and similar standard deviation. Thus the initial values of the network parameters are random values near zero.

6.2. Case Study 2: Power House - Cold Utility System

In this case, the cold facility of a Power House servicing a polymer manufacturing plant located in the vicinity of Barcelona, is considered. A double objective is intended: first to obtain a reliable model using the Neural Network Hybrid System (ENESIMO) and use it to simulate real time scenarios, and second to optimise the recipe for the best management of the cold utility generation (minimum cost).

The power plant (cold utility) consists of three compression units (U42-0, U42-1, U42-2) and two absorption machines (U42-3, U42-4) that keep the process cooling agent (brine) at the required temperature of –6ºC approximately (Fig. 21). The cold produced is consumed up to 85-90% in the fiber manufacturing, and the rest in the polymerization section. A variable demand causes variations in the brine temperature at the outlet of the plant. Cold utility generation was presently adjusted manually and proportionally to the temperature changes observed. In both cases (absortion and compression units) the neural network based simulation produces the cold generated by the corresponding unit (Mfrig/h) as a function of five main variables. In the case of compression units the following main variables were considered: water temperature and flow, brine flow and temperature, and the gas (freon) flow. Standard operating conditions for the compression system are given in Fig. 22.

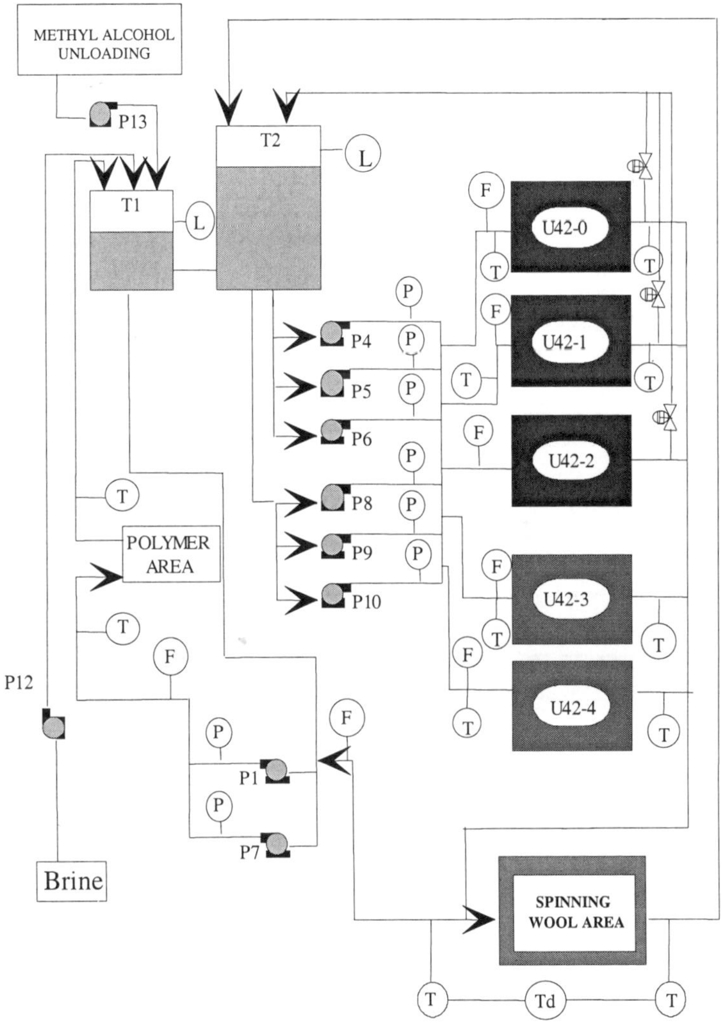

Figure 21. Cold utility plant.

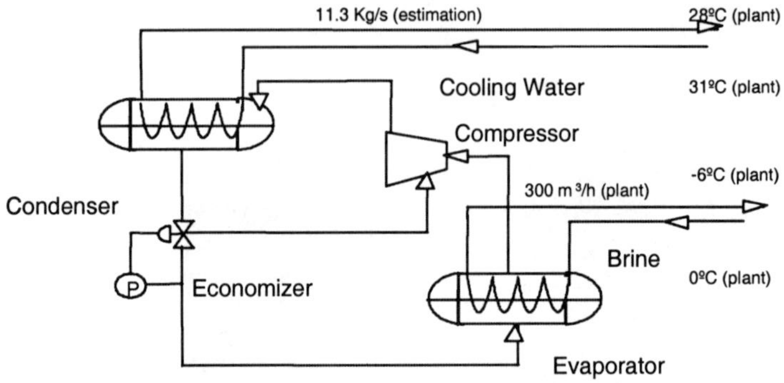

Figure 22. Compression Unit.

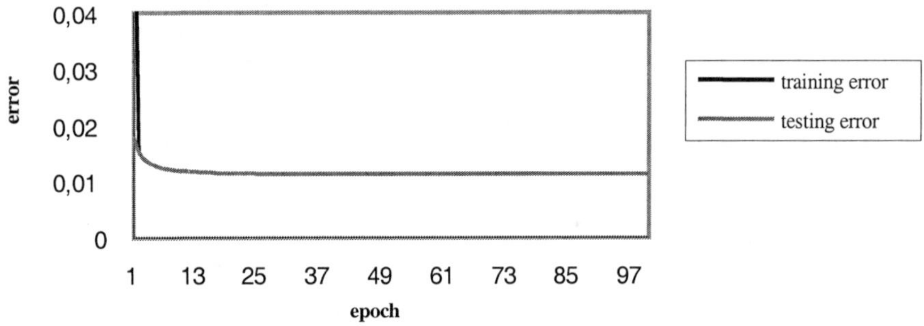

Figure 23. Learning and testing error for the 5-3-1 structure.

Training and testing results are shown in Fig. 23. The best Neural Network Structure found after training is 5-3-1 (inputs: 5, hidden: 3, outputs: 1) with a learning and test error of less than 0.011.

The hidden units have a sigmoid activation function and the output is linear. The hybrid system finds the best configuration automatically after 26 generations of the GA, using as fitness function the Minimum Descriptor Length (MDL) criteria. The model obtained predicts cold production with precise accuracy compared to the real plant operation.

The same study has been conducted for the absorption system (Fig. 24). The plant has two absorption units. In the modeling procedure, over 50,000 patterns were used from a variety of cold production conditions. The process variables were three flow rates (brine, cooling water, vapour) two temperatures (brine and water) and concentration of the ammonia in the condenser/evaporator zone.

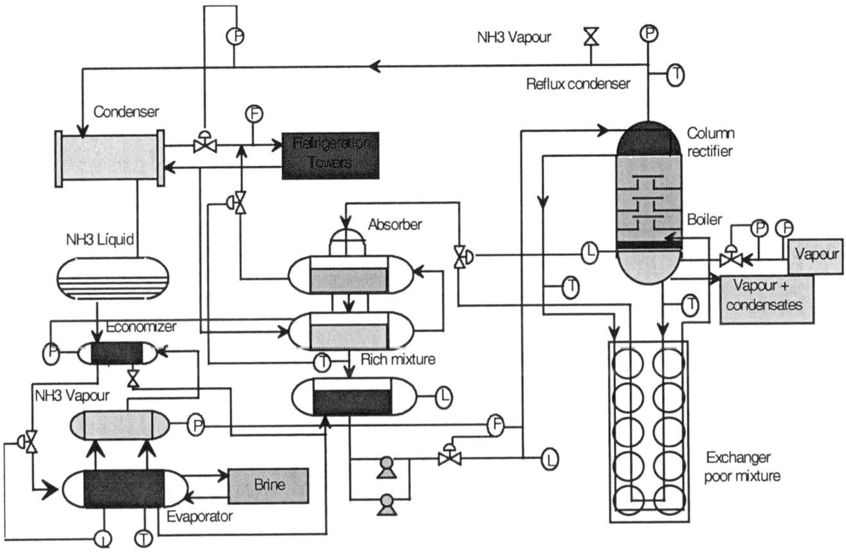

Figure 24. Absortion System.

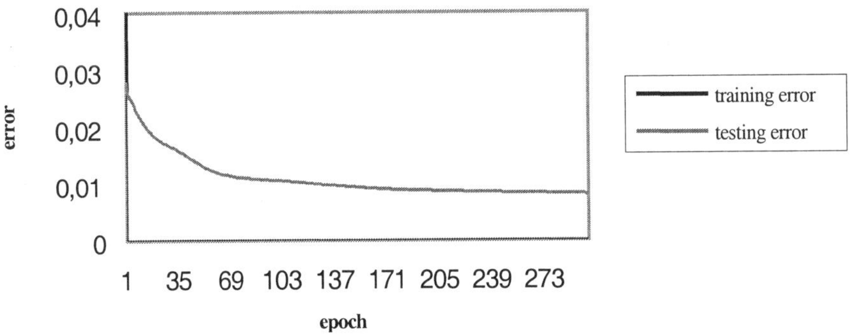

Figure 25. Simulation results.

The network structure found in this case is 6-3-1. The squared sum error (training and testing is now less than 0.008). Simulation results are shown in Fig. 25.

Optimization studies were also carried out to determine the optimum operation management of the cold utility system. Table 3 shows the results obtained under variable (cold) demand conditions (from 3 to 11 Mfrig/hr), giving the best plant operation scenario (minimum cost) in each case.

It can be observed that when the cold demand is at 3 Mfrig/h or less, the solution found is unique and the minimum cost equipment is used. The linearly increasing cold demand is closely met at optimal cost in every other case as shown at the right, while at left the same linear trends is observed for an increasing cost as that of the cold production (Fig. 26).

The simulator/optimizer ENESIMO was also used to set optimum operation conditions on-real-time operation. A Sample of the results obtained are given in Fig. 27.

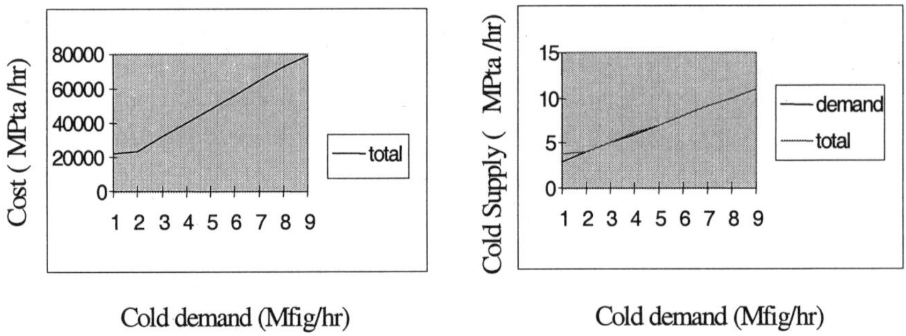

Figure 26. Optimization results.

6.3. Case Study 3 : Real Time Optimization – Gasification Plant

An integrated platform has been created that incorporates optimisation and production planning techniques in conjunction with real time plant measurements and control aiming at product quality enhancement and waste reduction [45, 43].

The system architecture has three layers. The first is a supervisory control level which includes techniques for diagnosis that consider an artificial neural network based supplement of a fuzzy system in a block oriented configuration (Fig. 28). The second is the co-ordination level, which provides real time information for decision making at upper levels. The third level involves decisions on allocating the available resources to the various products under demand. (Fig. 29).

Table 3. Cold utility optimum management under varying demand (Mfrig/h) and associated costs (Mptas/hr)

Cold	Abs1		Abs2		Comp1	
demand	Cold	Cost	Cold	Cost	Cold	Cost
3	1.88	11068	1.88	11068	0	0
4	1.88	11068	2.15	12771	0	0
5	1.88	11068	2.15	12771	0	0
6	1.88	11068	2.15	12771	0	0
7	1.88	11068	2.15	12771	0	0
8	1.88	11068	2.15	12771	0	0
9	1.88	11068	1.88	11068	1.26	10184
10	1.88	11068	2.15	12771	2	16080
11	2.5	15325	2.5	15325	2	16080

Cold	Comp2		Comp3		Cold	Cost
demand	Cold	Cost	Cold	Cost	Total	Total
3	0	0	0	0	3.76	22136
4	0	0	0	0	4.03	23839
5	0	0	1.06	8576	5.09	32415
6	0	0	2	16080	6.03	39919
7	1.06	8576	1.93	15544	7.02	47959
8	2	16080	2	16080	8.03	55999
9	2	16080	2	16080	9.02	64480
10	2	16080	2	16080	10.03	72079
11	2	16080	2	16080	11	78890

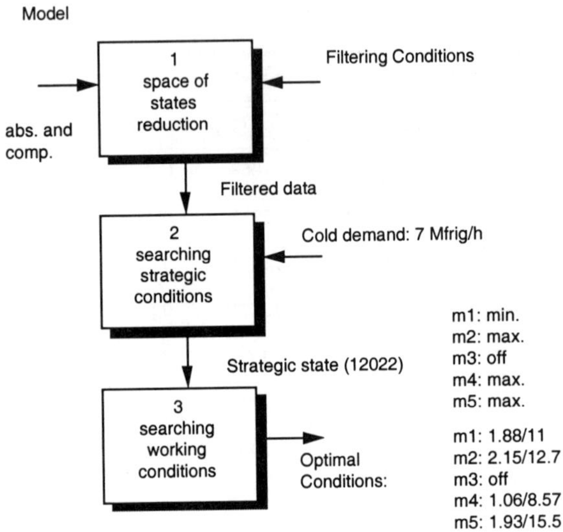

Figure 27. Setting process optimum operation conditions.

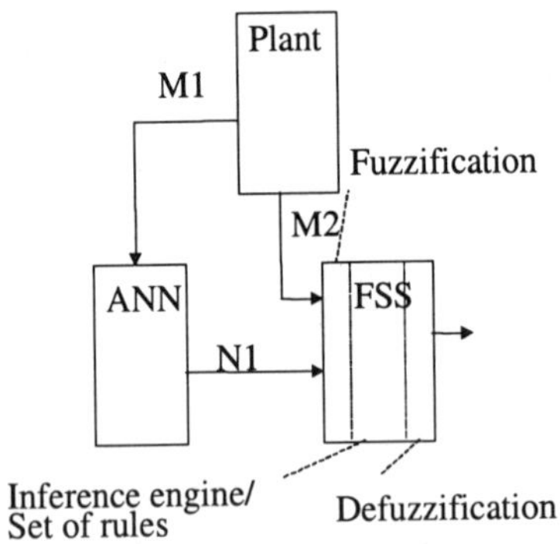

Figure 28. ANN-based modeling in a fuzzy system

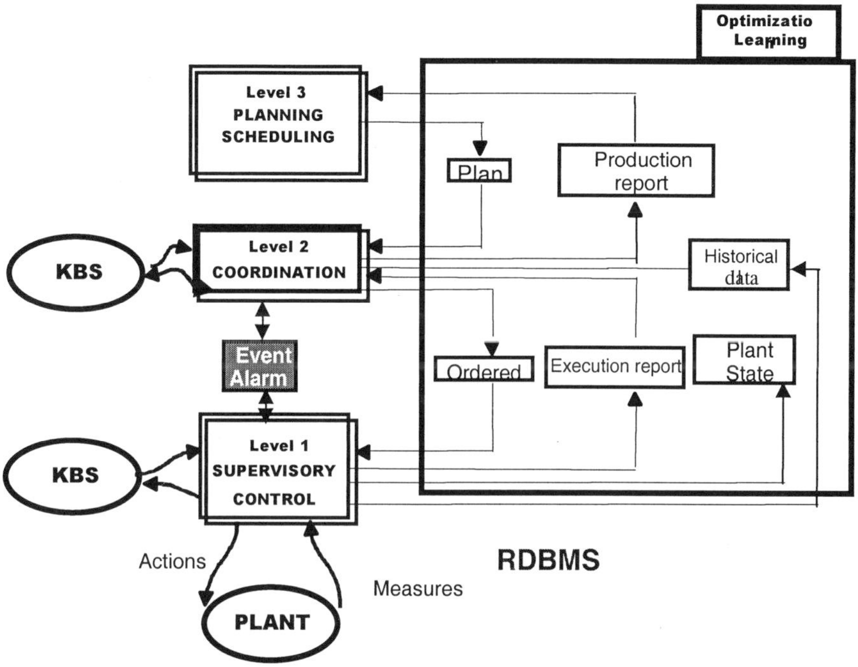

Figure 29. Real-time optimization system

The whole system exchanges information in two ways, by the communications network system and by the database management system (RDBMS). The communications network system incorporates a local control network supported by distributed control system vendors (DCS), a control network consisting in a real-time client interface and advanced control system, and the information network providing real-time data from long term operation, on-line plant data, planning and scheduling information [45].

The architecture described has been implemented in a fluidised bed gasifier plant. The plant performance is optimised online in terms of energy and gas quality. The plant layout appears in Fig. 30. The solid feed is introduced at the bottom of the reactor over the gas distributor. The gasifying agent (air + stream) is fed at the reactor bottom side at 650ºC, allowing the solid fluidisation. An online gas analyser is connected to the outlet gas stream for continuous monitoring of the gas composition.

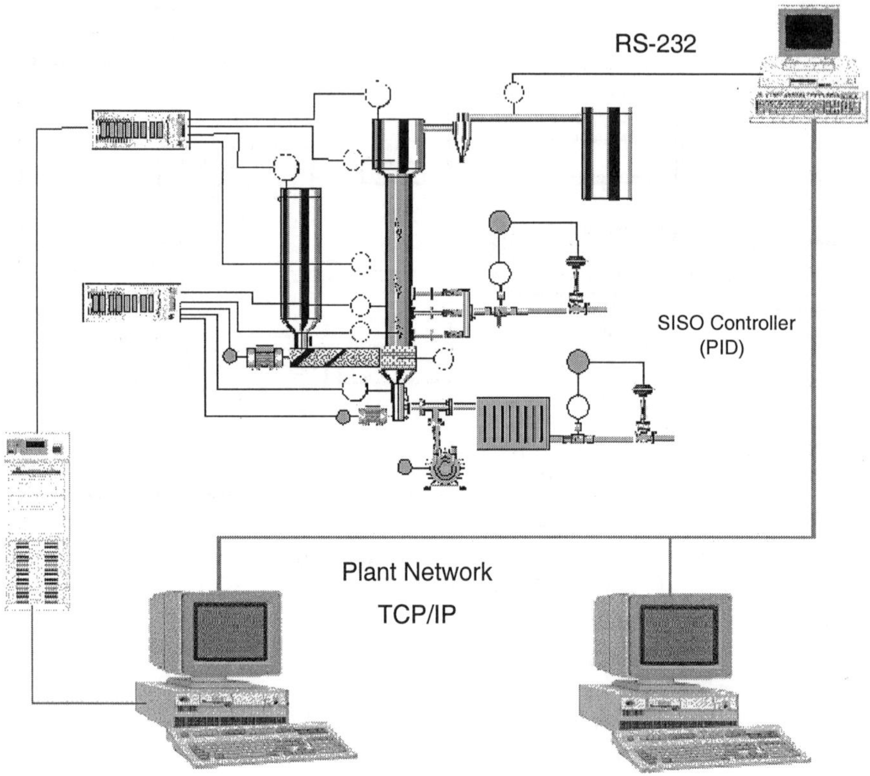

Figure 30. Plant layout and system integration

The system has four inputs (coal feed, airflow, heating power and water flow) and three outputs (gas composition, reactor temperature and differential pressure drop across the bed).

The advanced control system uses a model based control (MBC) strategy that incorporates the hybrid modelling system described before (ENESIMO). The identification of the plant dynamic response was carried out by performing a set of gasification runs in open loop to generate the data needed to build the dynamic neural network model of the reactor. Process dynamics to changes of input variables were analysed and data conditioning and filtering improved substantially the dynamic response. The best ANN model when optimised by the GA corresponds to a network structure with 5 neurones in the hidden layer. Fig. 31 shows one sample

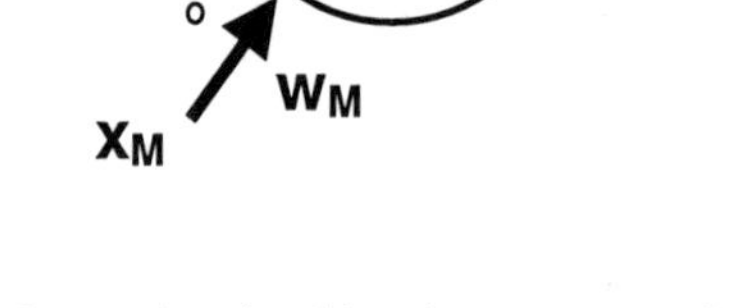

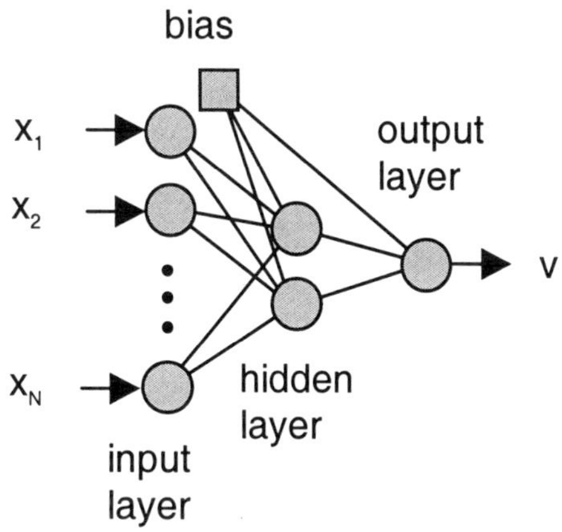

$$\xi = \tfrac{1}{2} \Sigma_j (T_{o,j} - v_{o,j})^2 \tag{5}$$

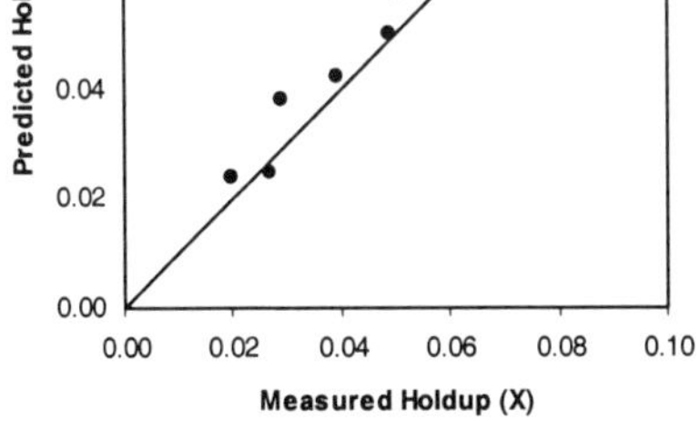

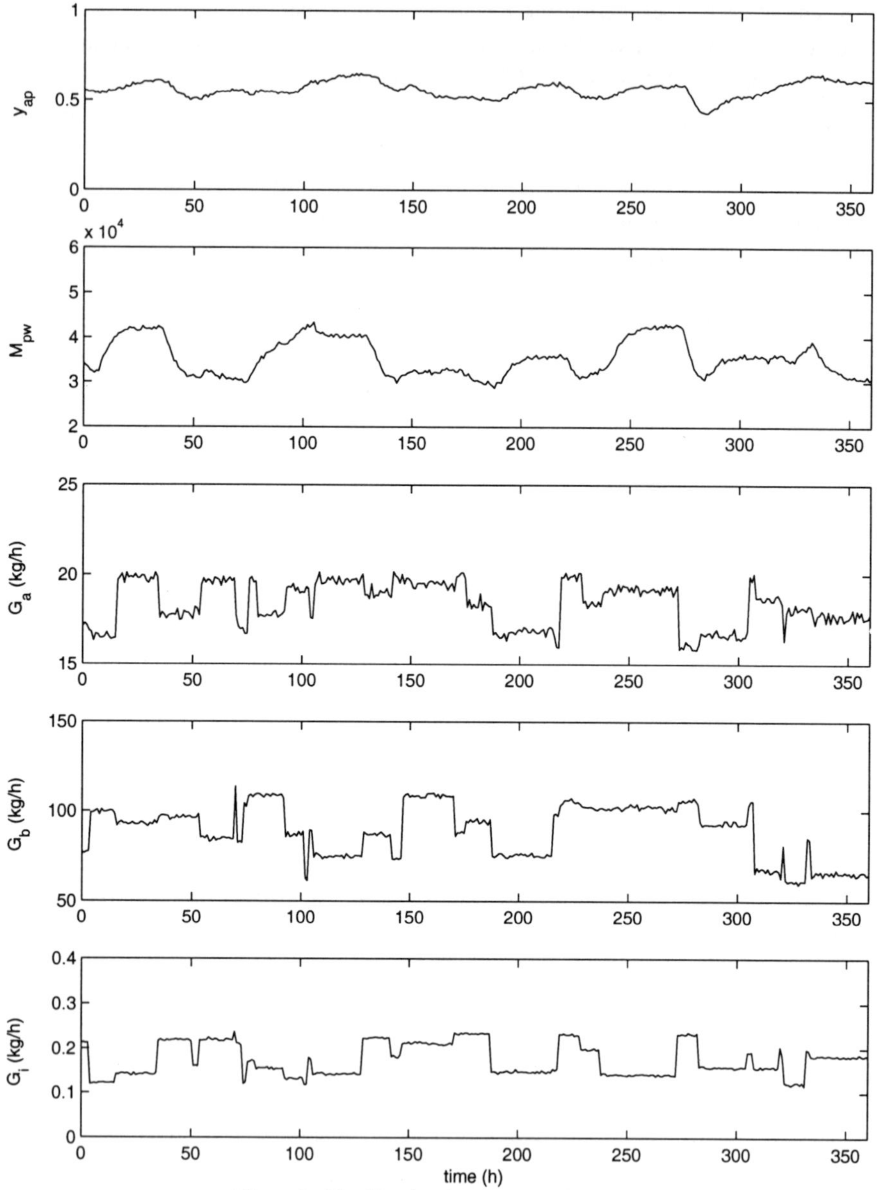

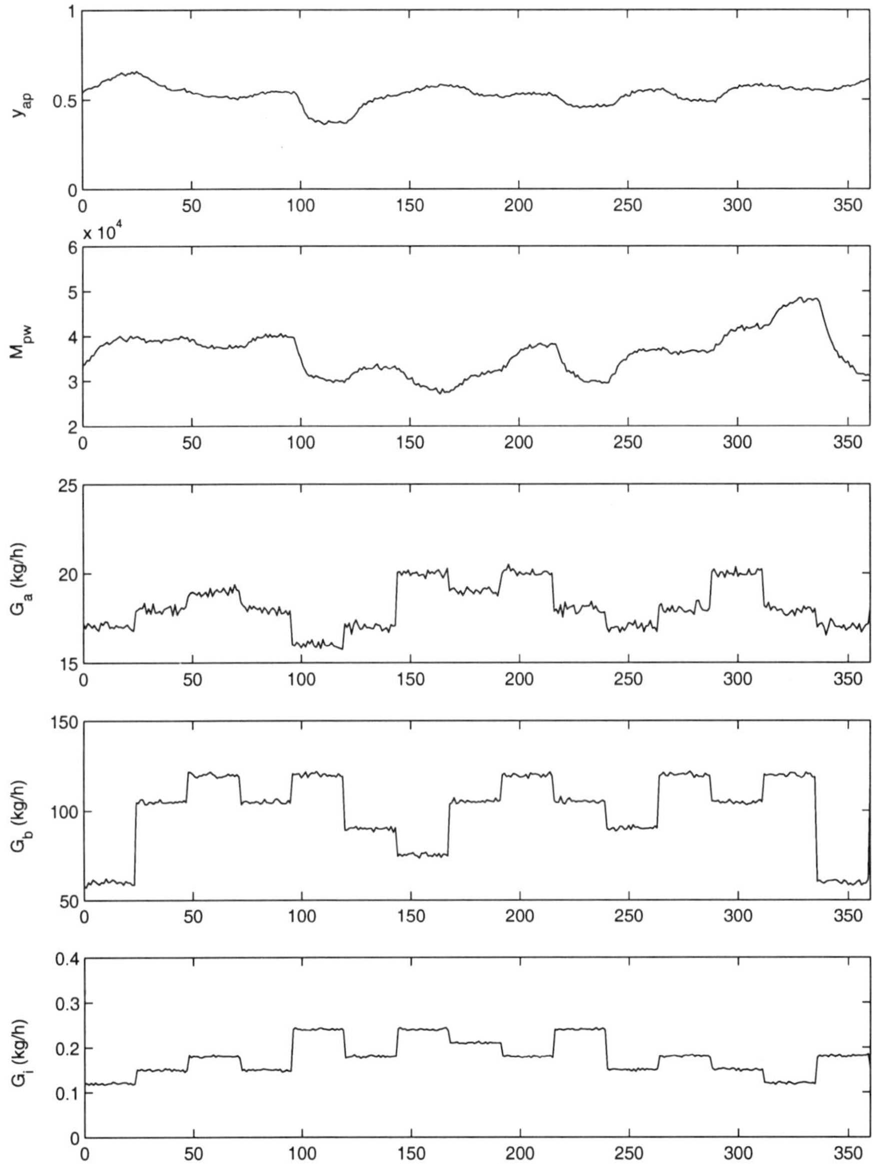

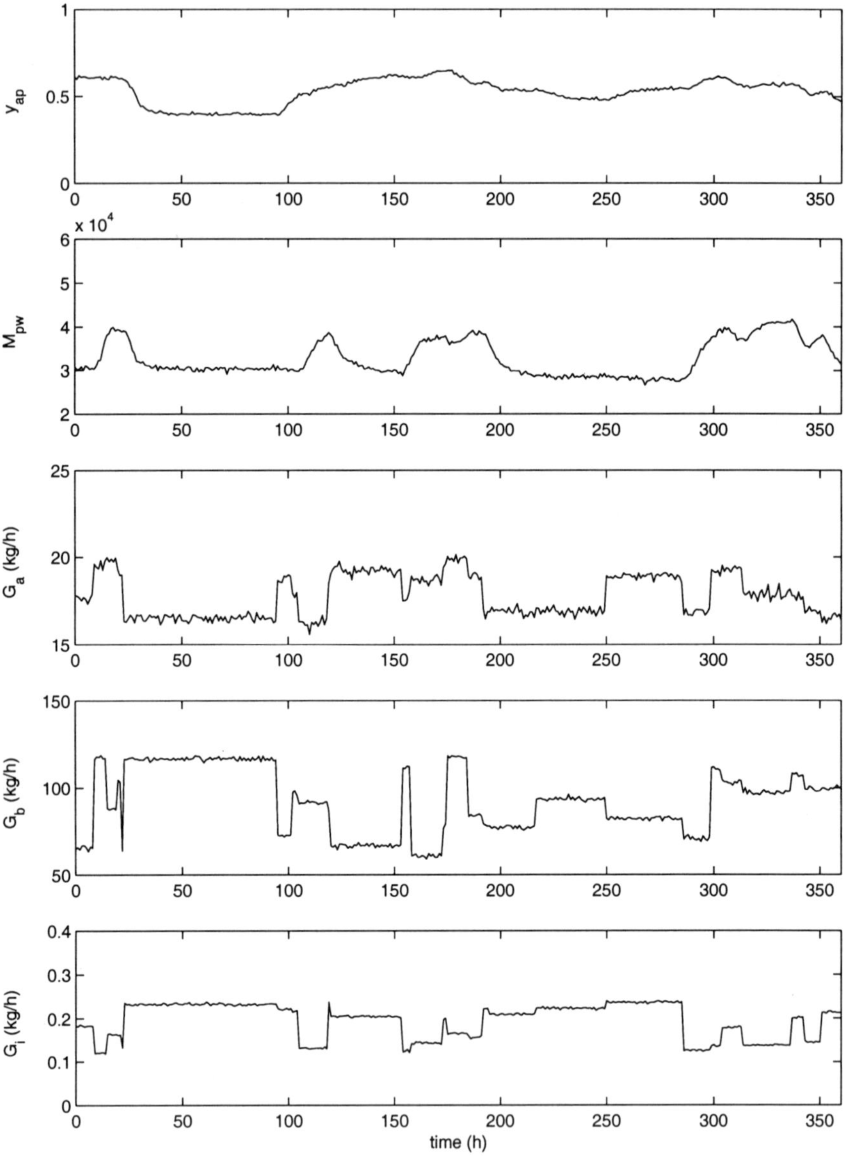

$$\begin{aligned}\mathbf{x}(i)=\big[\hat{y}_1(i-1)\ \cdots\ \hat{y}_1(i-m_1)\ \cdots\ \hat{y}_{n_y}(i-1)\ \cdots\ \hat{y}_{n_y}\left(i-m_{n_y}\right)\ u_1(i-\theta_1-1)\\ \cdots\ u_1(i-\theta_1-n_1)\ \cdots\ u_{n_u}\left(i-\theta_{n_u}-1\right)\ \cdots\ u_{n_u}\left(i-\theta_{n_u}-n_{n_u}\right)\big]^T\end{aligned} \tag{7}$$

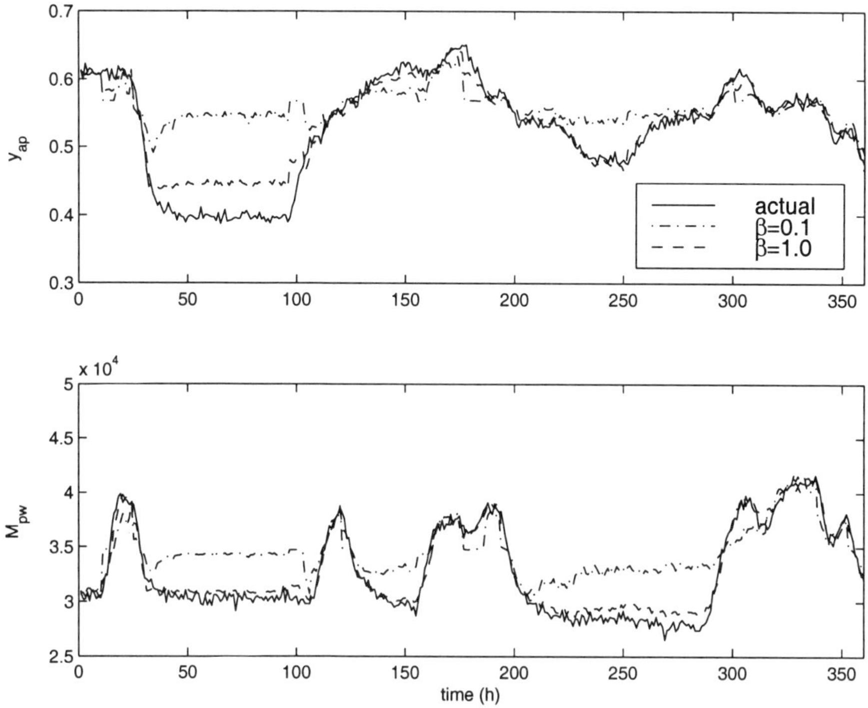

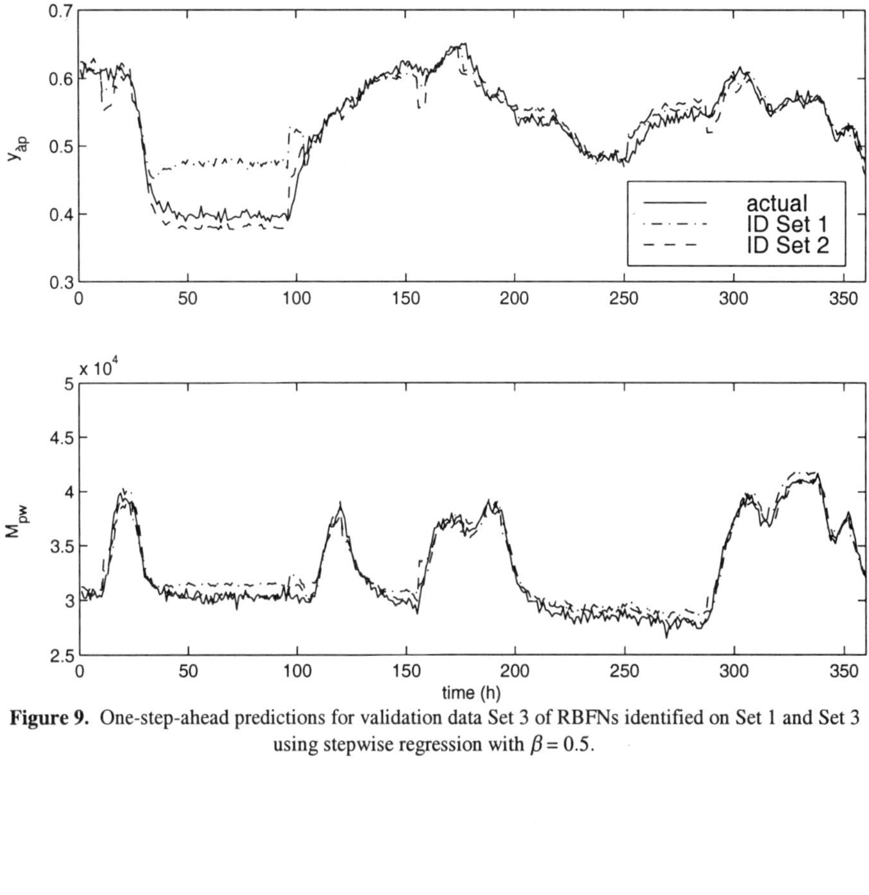

$$p(D \mid \alpha, \beta, H) = \int_{\text{all } \mathbf{w}} p(D \mid \mathbf{w}, \beta, H) p(\mathbf{w} \mid \alpha, H) \, \mathrm{d}\mathbf{w} \tag{6}$$

At this point a simplifying approximation is required to solve the above integrand analytically—that the posterior parameters distribution is Gaussian distributed. This is equivalent to assuming that the function $S(\mathbf{w}) = \beta E_D + \alpha E_W$ is quadratic. $S(\mathbf{w})$ can therefore be written as a second order Taylor expansion about the most probable parameters $\mathbf{w}^{mp}$, thus: $S(\mathbf{w}) = S(\mathbf{w}^{mp}) + 0.5\Delta\mathbf{w}^{\mathrm{T}}\mathbf{A}\Delta\mathbf{w}$. $\mathbf{A}$ is the Hessian $\mathbf{A} = \nabla\nabla^{\mathrm{T}}S(\mathbf{w}^{mp}) = \beta\nabla\nabla^{\mathrm{T}}E_D{}^{mp} + \alpha\nabla\nabla^{\mathrm{T}}E_W{}^{mp}$; and $\Delta\mathbf{w} = \mathbf{w}—\mathbf{w}^{mp}$. Eq. 7 can therefore be expressed in the analytical form

$$p(D \mid \alpha, \beta, H) = \frac{\exp(-S(\mathbf{w}^{mp}))}{Z_D(\beta) Z_W(\alpha)} \int_{\text{all } \mathbf{w}} \exp(-0.5\Delta\mathbf{w}^{\mathrm{T}} \mathbf{A} \Delta\mathbf{w}) \, \mathrm{d}\mathbf{w} \tag{8}$$

$$= \frac{\exp(-S(\mathbf{w}^{mp}))}{Z_D(\beta) Z_W(\alpha)} (2\pi)^{K+M/2} \mid \mathbf{A} \mid^{-1/2} \tag{9}$$

$$X_i = \left[\chi_{i_1}(x_i)\ \ \chi_{i_2}(x_i)\ \ \cdots\ \ \chi_{i_{c_i}}(x_i)\right]^T \tag{3}$$

Figure 1. Fuzzy sets of a fuzzy model with two inputs: $\chi_{1_i}(x_1), i = 1,2,3$ (above); $\chi_{2_i}(x_2), i = 1,2$ (below).

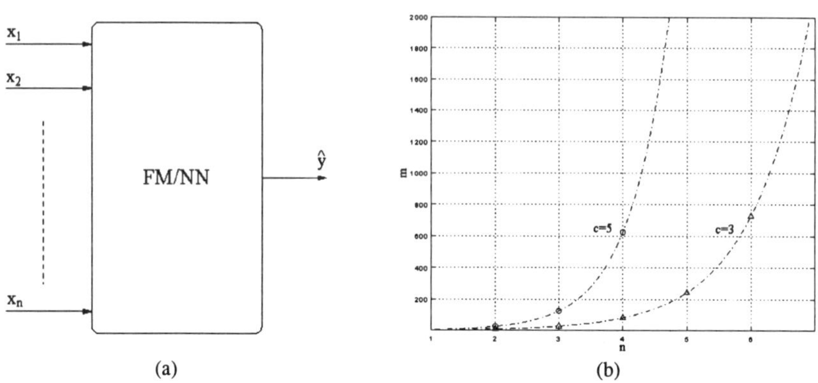

$$X_j = \begin{bmatrix} \mathcal{X}_{j1}(x_j) & \mathcal{X}_{j2}(x_j) & \cdots & \mathcal{X}_{jc}(x_j) \end{bmatrix}^T, \quad j = 1, \cdots, n \tag{6}$$

$$\hat{Y}_h = \begin{bmatrix} \mathcal{Y}_{h1}(\hat{y}_h) & \mathcal{Y}_{h2}(\hat{y}_h) & \cdots & \mathcal{Y}_{h_c}(\hat{y}_h) \end{bmatrix}^T, \quad h = 1, \cdots, n-2 \tag{7}$$

T_r (k-2)	T_r (k-1)	T_r (k)
□ Input	T_j (k-1)	T_j (k)
■ Output	Qr (k-1)	**Q_r (k)**

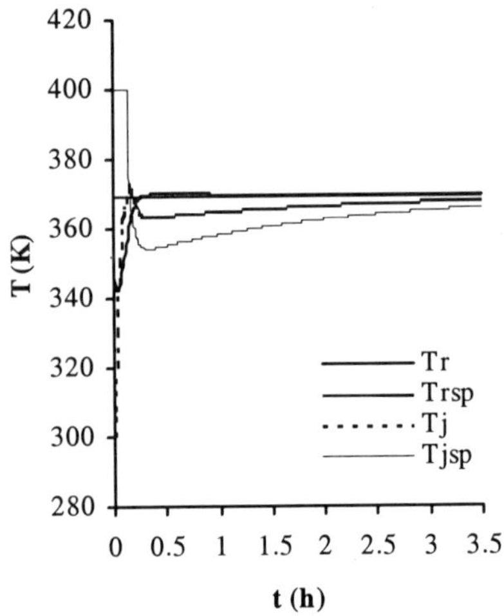

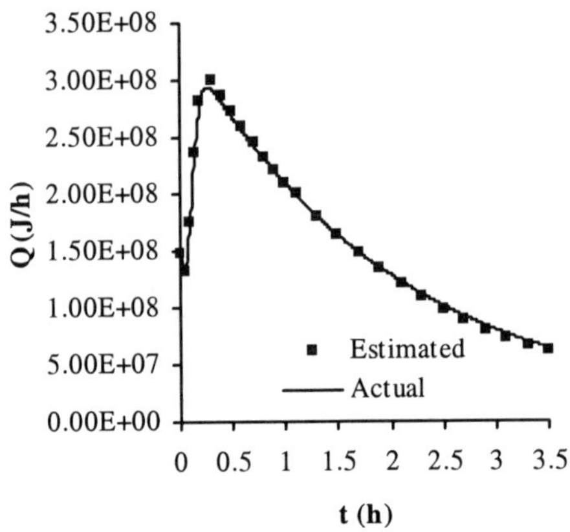

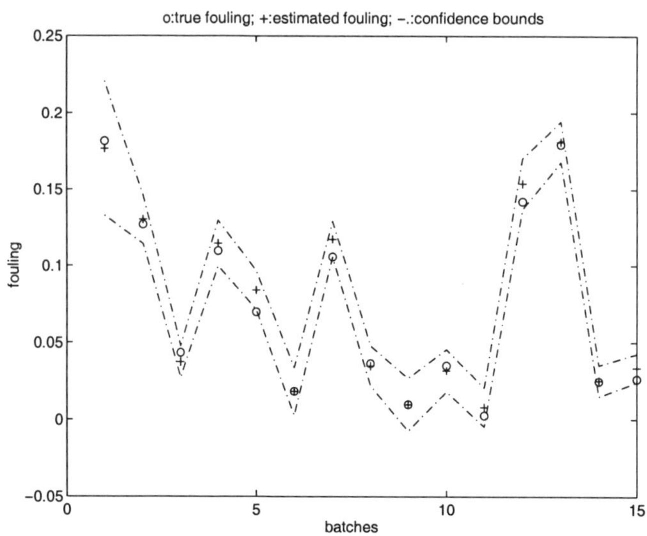

$$\text{The magnitude of noise} = \frac{\text{The magnitude of noise from the noise generator}}{C_{noise}} \tag{5}$$

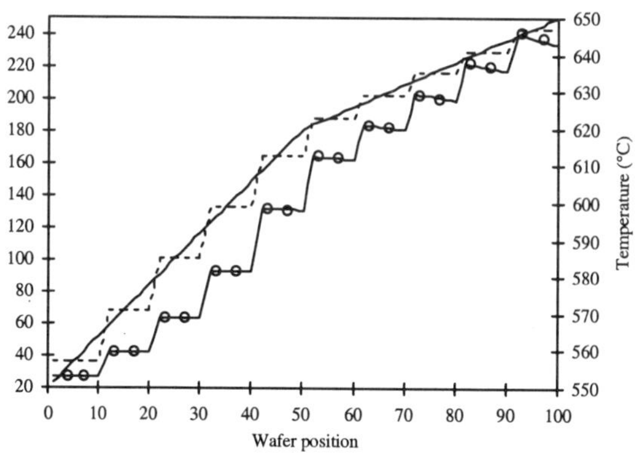

of the good agreement between model (solid line) and experimental data (dotted line).

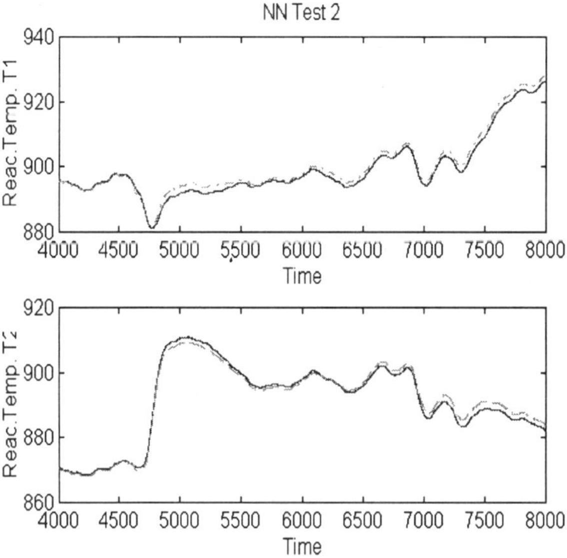

Figure 31. Reactor temperature profile at two sampling points. Comparison between experimental data (dotted line) and the ANN model (continuous line).

7. Future Directions

Developments of Neural Network applications in chemical engineering and processing reveal an exponential growth in recent times, and the industrial interest in present achievements corroborate an optimistic forecast. However, such developments have been largely confined to the solution of selected process components with specific solutions. Future developments should include:

- Use of Principal Component Analysis and heuristic approaches to further automatise the data analysis and selection as fully integrated in the neural network model building process (Fig. 18).
- Further development of an evolutionary modeling framework of process operations based on neural network structures specifically designed for multi input/output modeling applications and recurrent nonlinear backpropagation

connections for control applications, leading to real-time models that will address operational problems and support decisions for maximum efficiency and robustness of process operations.

- Research efforts towards inductive solutions in engineering problems (Fig. 32). Inductive programming improves the economics of software production by decreasing software engineering time. Inductive programming can help to ease the search for the representative *training set*.
- Qualitative analysis and reasoning should be further explored and exploited using neural network knowledge representation to better understand the systems behavior. Using qualitative information at some stage may then resort to more detailed quantitative reasoning only when it is necessary to do so to solve ambiguities in the production.

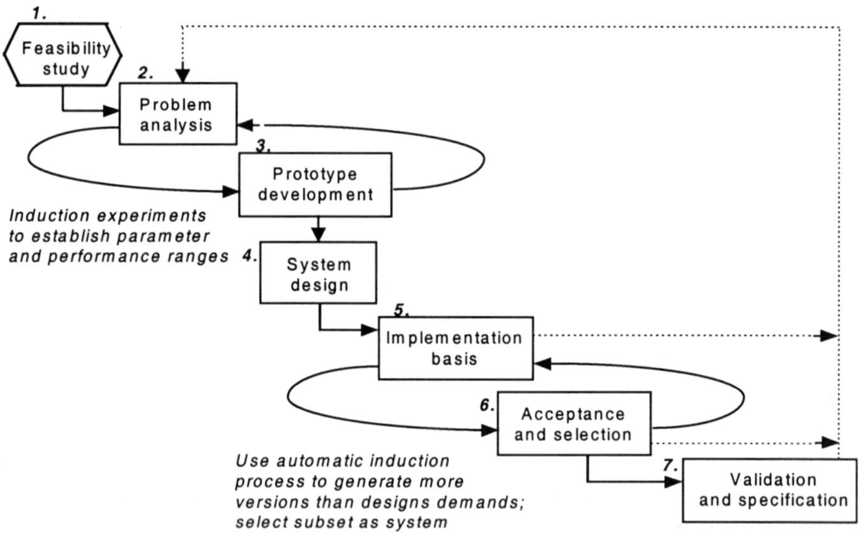

Figure 32. Inductive software engineering with neural networks.

References

1. Shea, P. M. and Lin, V., in *Proceedings of the International Joint Conference on Neural Networks*, Washington D. C. **II** (1989), 31.
2. Widrow, B., et al., *Communications of the ACM*. **37** (1994), 93-105.
3. Widrow, B and Lehr, M. A., in *Proceedings of IEEE 78,* **9** (1990), 1415-1442.
4. Rumelhart, D. E. et al., *Parallel Distributed Processing* (The MIT press, 1986) 1, Chap. 8.
5. Bhat, N. and McAvoy, T. J., *Comput. Chem. Eng.* **14** (1990), 573-582.
6. Klemes, J. and Ponton, J. W. in *Proc. 4th International Symposium on Process System Engineering-PSE'91*, Montebello, Quebec, Canada, **IV** (1991), IV.3.1-IV.3.12.
7. Pollard, J. F. et al., *Comput. Chem. Eng.* **16** (1992) 253-270.
8. Shaw, A. M. et al., *Comput. Chem. Eng.* **21** (1997), 371-386.
9. Galván, I. M. et al., *Comput. Chem. Eng.* **20** (1996), 1451-1466.
10. Pulley, R. A. et al., in *ESCAPE-4: 4th European Symposium in Computer Aided Process Engineering, eds.* Perris, T. and Perkins, J. (IChemE., Rugby, U.K., 1994), 399-403.
11. Braunbilla, A. and Trivella, F., *Hydrocarbon Processing*. **92** (1996), 61-66.
12. Guglielmi, N. et al., *IEEE Trans. on Neural Networks*. **7** (1996), 206-213.
13. Chitra, S. P., *Chem. Eng. Prog.* **89**, (1993), 44-52.
14. Nerrand, O. et al., *IEEE Trans. on Neural Networks*. **5** (1994), 178-184.
15. Huang, Y. W. et al., *Biotechnol. Prog.*, **9** (1993), 401-415.
16. Psichogios, D.C. and Ungar, L. H. *AIChE J.*, **38** (1992), 1499-1511.
17. Cooper, J. D. et al., *AIChE J.*, **38** (1992) 42-54.
18. Temeng, K. O. et al., *J. Proc. Control*, **5** (1995), 19-27.
19. Cavalieri, S. and Mirabella, O., *IEEE Trans. on Neural Networks*, **7** (1995), 1272-1285.
20. Lee, M. and Park, S., *AIChE J.*, **38**, 193-200 (1992).
21. Tani, T. et al., *IEEE Trans.on Fuzzy Systems*. **4** (1966), 360-368.
22. Chen C. and S. Peng, *J. of Proc. Cont.*, **9** (1999), 493-503.
23. Ydstie, B. E., *Comput. Chem. Eng.* **14** (1990), 583-599.
24. Palau, A., A. et al., *Comput. Chem. Eng.* **20S** (1996), 297-302.
25. Venkatasubramanian, V. and Chan, K., *AIChE J.* **35** (1989), 1993-2002.
26. Quantrille, T. and Lin, Y., *Artificial Intelligence in Chemical Engineering* (Academic Press, San Diego, CA, 1991), 466-481.
27. Zhao, J. et al., *Comput. Chem. Eng.* **23** (1989), 83-92.
28. Marcu, T., *IEEE Control Systems*. **19** (1999), 72-79.

29. Ploix, J. L. and Dreytus, G., in *ICANN'95*, Paris, October (1995).
30. Nikravesh, M. et al., *Comput. Chem. Eng*. **20** (1996), 1277-1290.
31. Puigjaner, L. et al., in *ICANN'95*, Paris, October (1995).
32. Puigjaner, L. and Espuña, A., in *I-CIMPRO'96*, Eindhoven (Holland), June 3-4 (1996).
33. Delgado, A. et al., in *Fifth World Congress of Chemical Engineering*, San Diego, CA (USA), July 14-18 (1996).
34. Mah, R. S. H. and Chakravarty, V., *Comput. Chem. Eng*., **16** (1992), 371-378.
35. Mavrovouniotis, M. L. and Chang, S., *Comput. Chem. Eng*., **16** (1992), 347-370.
36. Thompson, M. L. and Kramer, M. A., *AIChE J.*, **40** (1994), 1328-1340.
37. Jordan, M. I. and Rumelhart, D. E., *Cognitive Sci.*, **16** (1992), 307.
38. Su, H.-T. et al., in *IFAC Symp. on Dynamics and Control of Chemical Reactors*, 327 (1992).
39. Schubert, J. et al., *J. Biotechnol.*, **35** (1994), 51.
40. Van Caan, H.J.L. et al., *AIChE J.* **42** (1996), 3403-3418.
41. Hopfield J. J. and Tank, D., *Science*. **233** (1986), 625.
42. Engell, S. and Klatt, K. U. in *Proceedings of American Control Conference*, San Francisco, 294 (1993).
43. Puigjaner, L., Espuña, A., *Comput. Chem. Eng.* **22** (1998), 87-107.
44. Goldberg, D. E. *Genetic Algorithm in Search Optimization and Machine Learning* (Addison-Wesley , 1989).
45. Nougues, J.M. et al., in *Workshop on Chemical Engineering Mathematics*, 10, Bad Honeff, Germany (1998).
46. Puigjaner, L. and Espuña, A., in *Trends in Chemical Engineering*, Council of Scientific Research Integration, Trivandrum, **1** (1994), 77-91.
47. Puigjaner, L. et al., *J. of Proc. Cont.* **4** (1994), 281-290.
48. Puigjaner, L., *Comput. Chem. Eng.* **23** (1999), S929-S943.
49. Espuña, A. et al., *Computers in Industry*. **36** (1998), 271-278.
50. Goldberg, D. E., *Genetic Algorithm in Search, Optimization and Machine Learning* (Addison-Wesley, N.Y., 1989).
51. Akaike, H., *Information Theory and Extension of the Maximum Likelihood Principle* (Akademia Kiado, Budapest , 1973), 267-810.
52. Rissanen, J., *Automatica*. **14** (1978), 464-471.
53. Delgado, A., *Neural Networks. Contribution to the theory and practical applications*, PhD Thesis, (UPC, Barcelona, 1998).

Acknowledgements

The author wishes to acknowledge support of this research work from the European Community (Imagine, Contract No. 7220-ED-081) and the CICYT-MEC (project REALISSTICO, Contract No. QUI99-1091).